# LONDON MATHEMATICAL SOCIETY LECTURE NOTE SERIES

Managing Editor: Professor N.J. Hitchin, Mathematical Institute,
University of Oxford, 24–29 St Giles, Oxford OX1 3LB, United Kingdom

The titles below are available from booksellers, or, in case of difficulty, from Cambridge University Press.

| | |
|---|---|
| 46 | *p*-adic Analysis: a short course on recent work, N. KOBLITZ |
| 59 | Applicable differential geometry, M. CRAMPIN & F.A.E. PIRANI |
| 66 | Several complex variables and complex manifolds II, M.J. FIELD |
| 86 | Topological topics, I.M. JAMES (ed) |
| 88 | FPF ring theory, C. FAITH & S. PAGE |
| 90 | Polytopes and symmetry, S.A. ROBERTSON |
| 96 | Diophantine equations over function fields, R.C. MASON |
| 97 | Varieties of constructive mathematics, D.S. BRIDGES & F. RICHMAN |
| 99 | Methods of differential geometry in algebraic topology, M. KAROUBI & C. LERUSTE |
| 100 | Stopping time techniques for analysts and probabilists, L. EGGHE |
| 104 | Elliptic structures on 3-manifolds, C.B. THOMAS |
| 105 | A local spectral theory for closed operators, I. ERDELYI & WANG SHENGWANG |
| 107 | Compactification of Siegel moduli schemes, C.-L. CHAI |
| 109 | Diophantine analysis, J. LOXTON & A. VAN DER POORTEN (eds) |
| 113 | Lectures on the asymptotic theory of ideals, D. REES |
| 116 | Representations of algebras, P.J. WEBB (ed) |
| 119 | Triangulated categories in the representation theory of finite-dimensional algebras, D. HAPPEL |
| 121 | Proceedings of *Groups - St Andrews 1985*, E. ROBERTSON & C. CAMPBELL (eds) |
| 128 | Descriptive set theory and the structure of sets of uniqueness, A.S. KECHRIS & A. LOUVEAU |
| 130 | Model theory and modules, M. PREST |
| 131 | Algebraic, extremal & metric combinatorics, M.-M. DEZA, P. FRANKL & I.G. ROSENBERG (eds) |
| 138 | Analysis at Urbana, II, E. BERKSON, T. PECK, & J. UHL (eds) |
| 139 | Advances in homotopy theory, S. SALAMON, B. STEER & W. SUTHERLAND (eds) |
| 140 | Geometric aspects of Banach spaces, E.M. PEINADOR & A. RODES (eds) |
| 141 | Surveys in combinatorics 1989, J. SIEMONS (ed) |
| 144 | Introduction to uniform spaces, I.M. JAMES |
| 146 | Cohen-Macaulay modules over Cohen-Macaulay rings, Y. YOSHINO |
| 148 | Helices and vector bundles, A.N. RUDAKOV *et al* |
| 149 | Solitons, nonlinear evolution equations and inverse scattering, M. ABLOWITZ & P. CLARKSON |
| 150 | Geometry of low-dimensional manifolds 1, S. DONALDSON & C.B. THOMAS (eds) |
| 151 | Geometry of low-dimensional manifolds 2, S. DONALDSON & C.B. THOMAS (eds) |
| 152 | Oligomorphic permutation groups, P. CAMERON |
| 153 | L-functions and arithmetic, J. COATES & M.J. TAYLOR (eds) |
| 155 | Classification theories of polarized varieties, TAKAO FUJITA |
| 158 | Geometry of Banach spaces, P.F.X. MÜLLER & W. SCHACHERMAYER (eds) |
| 159 | Groups St Andrews 1989 volume 1, C.M. CAMPBELL & E.F. ROBERTSON (eds) |
| 160 | Groups St Andrews 1989 volume 2, C.M. CAMPBELL & E.F. ROBERTSON (eds) |
| 161 | Lectures on block theory, BURKHARD KÜLSHAMMER |
| 163 | Topics in varieties of group representations, S.M. VOVSI |
| 164 | Quasi-symmetric designs, M.S. SHRIKANDE & S.S. SANE |
| 166 | Surveys in combinatorics, 1991, A.D. KEEDWELL (ed) |
| 168 | Representations of algebras, H. TACHIKAWA & S. BRENNER (eds) |
| 169 | Boolean function complexity, M.S. PATERSON (ed) |
| 170 | Manifolds with singularities and the Adams-Novikov spectral sequence, B. BOTVINNIK |
| 171 | Squares, A.R. RAJWADE |
| 172 | Algebraic varieties, GEORGE R. KEMPF |
| 173 | Discrete groups and geometry, W.J. HARVEY & C. MACLACHLAN (eds) |
| 174 | Lectures on mechanics, J.E. MARSDEN |
| 175 | Adams memorial symposium on algebraic topology 1, N. RAY & G. WALKER (eds) |
| 176 | Adams memorial symposium on algebraic topology 2, N. RAY & G. WALKER (eds) |
| 177 | Applications of categories in computer science, M. FOURMAN, P. JOHNSTONE & A. PITTS (eds) |
| 178 | Lower K- and L-theory, A. RANICKI |
| 179 | Complex projective geometry, G. ELLINGSRUD *et al* |
| 180 | Lectures on ergodic theory and Pesin theory on compact manifolds, M. POLLICOTT |
| 181 | Geometric group theory I, G.A. NIBLO & M.A. ROLLER (eds) |
| 182 | Geometric group theory II, G.A. NIBLO & M.A. ROLLER (eds) |
| 183 | Shintani zeta functions, A. YUKIE |
| 184 | Arithmetical functions, W. SCHWARZ & J. SPILKER |
| 185 | Representations of solvable groups, O. MANZ & T.R. WOLF |
| 186 | Complexity: knots, colourings and counting, D.J.A. WELSH |
| 187 | Surveys in combinatorics, 1993, K. WALKER (ed) |
| 188 | Local analysis for the odd order theorem, H. BENDER & G. GLAUBERMAN |
| 189 | Locally presentable and accessible categories, J. ADAMEK & J. ROSICKY |
| 190 | Polynomial invariants of finite groups, D.J. BENSON |
| 191 | Finite geometry and combinatorics, F. DE CLERCK *et al* |
| 192 | Symplectic geometry, D. SALAMON (ed) |
| 194 | Independent random variables and rearrangement invariant spaces, M. BRAVERMAN |
| 195 | Arithmetic of blowup algebras, WOLMER VASCONCELOS |
| 196 | Microlocal analysis for differential operators, A. GRIGIS & J. SJÖSTRAND |
| 197 | Two-dimensional homotopy and combinatorial group theory, C. HOG-ANGELONI *et al* |

198   The algebraic characterization of geometric 4-manifolds, J.A. HILLMAN
199   Invariant potential theory in the unit ball of $\mathbb{C}^n$, MANFRED STOLL
200   The Grothendieck theory of dessins d'enfant, L. SCHNEPS (ed)
201   Singularities, JEAN-PAUL BRASSELET (ed)
202   The technique of pseudodifferential operators, H.O. CORDES
203   Hochschild cohomology of von Neumann algebras, A. SINCLAIR & R. SMITH
204   Combinatorial and geometric group theory, A.J. DUNCAN, N.D. GILBERT & J. HOWIE (eds)
205   Ergodic theory and its connections with harmonic analysis, K. PETERSEN & I. SALAMA (eds)
207   Groups of Lie type and their geometries, W.M. KANTOR & L. DI MARTINO (eds)
208   Vector bundles in algebraic geometry, N.J. HITCHIN, P. NEWSTEAD & W.M. OXBURY (eds)
209   Arithmetic of diagonal hypersurfaces over finite fields, F.Q. GOUVÊA & N. YUI
210   Hilbert C*-modules, E.C. LANCE
211   Groups 93 Galway / St Andrews I, C.M. CAMPBELL et al (eds)
212   Groups 93 Galway / St Andrews II, C.M. CAMPBELL et al (eds)
214   Generalised Euler-Jacobi inversion formula and asymptotics beyond all orders, V. KOWALENKO et al
215   Number theory 1992–93, S. DAVID (ed)
216   Stochastic partial differential equations, A. ETHERIDGE (ed)
217   Quadratic forms with applications to algebraic geometry and topology, A. PFISTER
218   Surveys in combinatorics, 1995, PETER ROWLINSON (ed)
220   Algebraic set theory, A. JOYAL & I. MOERDIJK
221   Harmonic approximation, S.J. GARDINER
222   Advances in linear logic, J.-Y. GIRARD, Y. LAFONT & L. REGNIER (eds)
223   Analytic semigroups and semilinear initial boundary value problems, KAZUAKI TAIRA
224   Computability, enumerability, unsolvability, S.B. COOPER, T.A. SLAMAN & S.S. WAINER (eds)
225   A mathematical introduction to string theory, S. ALBEVERIO, J. JOST, S. PAYCHA, S. SCARLATTI
226   Novikov conjectures, index theorems and rigidity I, S. FERRY, A. RANICKI & J. ROSENBERG (eds)
227   Novikov conjectures, index theorems and rigidity II, S. FERRY, A. RANICKI & J. ROSENBERG (eds)
228   Ergodic theory of $\mathbb{Z}^d$ actions, M. POLLICOTT & K. SCHMIDT (eds)
229   Ergodicity for infinite dimensional systems, G. DA PRATO & J. ZABCZYK
230   Prolegomena to a middlebrow arithmetic of curves of genus 2, J.W.S. CASSELS & E.V. FLYNN
231   Semigroup theory and its applications, K.H. HOFMANN & M.W. MISLOVE (eds)
232   The descriptive set theory of Polish group actions, H. BECKER & A.S. KECHRIS
233   Finite fields and applications, S. COHEN & H. NIEDERREITER (eds)
234   Introduction to subfactors, V. JONES & V.S. SUNDER
235   Number theory 1993–94, S. DAVID (ed)
236   The James forest, H. FETTER & B. GAMBOA DE BUEN
237   Sieve methods, exponential sums, and their applications in number theory, G.R.H. GREAVES et al
238   Representation theory and algebraic geometry, A. MARTSINKOVSKY & G. TODOROV (eds)
239   Clifford algebras and spinors, P. LOUNESTO
240   Stable groups, FRANK O. WAGNER
241   Surveys in combinatorics, 1997, R.A. BAILEY (ed)
242   Geometric Galois actions I, L. SCHNEPS & P. LOCHAK (eds)
243   Geometric Galois actions II, L. SCHNEPS & P. LOCHAK (eds)
244   Model theory of groups and automorphism groups, D. EVANS (ed)
245   Geometry, combinatorial designs and related structures, J.W.P. HIRSCHFELD et al
246   p-Automorphisms of finite p-groups, E.I. KHUKHRO
247   Analytic number theory, Y. MOTOHASHI (ed)
248   Tame topology and o-minimal structures, LOU VAN DEN DRIES
249   The atlas of finite groups: ten years on, ROBERT CURTIS & ROBERT WILSON (eds)
250   Characters and blocks of finite groups, G. NAVARRO
251   Gröbner bases and applications, B. BUCHBERGER & F. WINKLER (eds)
252   Geometry and cohomology in group theory, P. KROPHOLLER, G. NIBLO, R. STÖHR (eds)
253   The q-Schur algebra, S. DONKIN
254   Galois representations in arithmetic algebraic geometry, A.J. SCHOLL & R.L. TAYLOR (eds)
255   Symmetries and integrability of difference equations, P.A. CLARKSON & F.W. NIJHOFF (eds)
256   Aspects of Galois theory, HELMUT VÖLKLEIN et al
257   An introduction to noncommutative differential geometry and its physical applications 2ed, J. MADORE
258   Sets and proofs, S.B. COOPER & J. TRUSS (eds)
259   Models and computability, S.B. COOPER & J. TRUSS (eds)
260   Groups St Andrews 1997 in Bath, I, C.M. CAMPBELL et al
261   Groups St Andrews 1997 in Bath, II, C.M. CAMPBELL et al
263   Singularity theory, BILL BRUCE & DAVID MOND (eds)
264   New trends in algebraic geometry, K. HULEK, F. CATANESE, C. PETERS & M. REID (eds)
265   Elliptic curves in cryptography, I. BLAKE, G. SEROUSSI & N. SMART
267   Surveys in combinatorics, 1999, J.D. LAMB & D.A. PREECE (eds)
268   Spectral asymptotics in the semi-classical limit, M. DIMASSI & J. SJÖSTRAND
269   Ergodic theory and topological dynamics, M.B. BEKKA & M. MAYER
270   Analysis on Lie Groups, N.T. VAROPOULOS & S. MUSTAPHA
271   Singular perturbations of differential operators, S. ALBEVERIO & P. KURASOV
272   Character theory for the odd order function, T. PETERFALVI
273   Spectral theory and geometry, E.B. DAVIES & Y. SAFAROV (eds)
274   The Mandlebrot set, theme and variations, TAN LEI (ed)
276   Singularities of plane curves, E. CASAS-ALVERO
279   Topics in symbolic dynamics and applications, F. BLANCHARD, A. MAASS & A. NOGUEIRA (eds)
281   Explicit birational geometry of 3-folds, ALESSIO CORTI & MILES REID (eds)

London Mathematical Society Lecture Note Series. 281

# Explicit Birational Geometry of 3-Folds

Edited by

Alessio Corti
*University of Cambridge*

Miles Reid
*University of Warwick*

CAMBRIDGE
UNIVERSITY PRESS

CAMBRIDGE UNIVERSITY PRESS
Cambridge, New York, Melbourne, Madrid, Cape Town,
Singapore, São Paulo, Delhi, Mexico City

Cambridge University Press
The Edinburgh Building, Cambridge CB2 8RU, UK

Published in the United States of America by Cambridge University Press, New York

www.cambridge.org
Information on this title: www.cambridge.org/9780521636414

© Cambridge University Press 2000

First published 2000

*A catalogue record for this publication is available from the British Library*

ISBN 978-0-521-63641-4 Paperback

# Contents

Alessio Corti and Miles Reid: Foreword     1

Klaus Altmann: One parameter families containing three
    dimensional toric Gorenstein singularities     21

János Kollár: Nonrational covers of $\mathbb{CP}^m \times \mathbb{CP}^n$     51

Aleksandr V. Pukhlikov: Essentials of the method of
    maximal singularities     73

A. R. Iano-Fletcher: Working with weighted complete
    intersections     101

Alessio Corti, Aleksandr Pukhlikov and Miles Reid: Fano
    3-fold hypersurfaces     175

Alessio Corti: Singularities of linear systems and 3-fold
    birational geometry     259

Miles Reid: Twenty five years of 3-folds – an old person's
    view     313

Index     345

# Foreword

Alessio Corti          Miles Reid

## 1   Introduction

This volume is an integrated collection of papers working out several new directions of research on 3-folds under the unifying theme of *explicit birational geometry*. Section 3 summarises briefly the contents of the individual papers.

Mori theory is a conceptual framework for studying minimal models and the classification of varieties, and has been one of the main areas of progress in algebraic geometry since the 1980s. It offers new points of view and methods of attacking classical problems, both in classification and in birational geometry, and it raises many new problem areas. While birational geometry has inspired the work of many classical and modern mathematicians, such as L. Cremona, G. Fano, Hilda Hudson, Yu. I. Manin, V. A. Iskovskikh and many others, and while their results undoubtedly give us much fascinating experimental material as food for thought, we believe that it is only within Mori theory that this body of knowledge begins to acquire a coherent shape.

At the same time as providing adequate tools for the study of 3-folds, Mori theory enriches the classical world many times over with new examples and constructions. We can now, for example, work and play with hundreds of families of Fano 3-folds. From where we stand, we can see clearly that the classical geometers were only scratching at the surface, with little inkling of the gold mine awaiting discovery.

The theory of minimal models of surfaces works with nonsingular surfaces, and the elementary step it uses is Castelnuovo's criterion, which allows us to contract $-1$-curves (exceptional curves of the first kind). A chain of such contractions leads us to a minimal surface $S$, either $\mathbb{P}^2$ or a scroll over a curve, or a surface with $K_S$ numerically nonnegative (now called *nef*, see 2.2 below). These ideas were well understood by Castelnuovo and Enriques a century ago, and are so familiar that most people take them for granted. However, their higher dimensional generalisation was a complete mystery until the late 1970s, and may still be hard to grasp for newcomers to the field. It involves a suitable category of mildly singular projective varieties, and the crucial new ingredient of *extremal ray* introduced by Mori around 1980. As

1

we discuss later in this foreword, extremal rays provide the elementary steps of the minimal model program (the divisorial contractions and flips of the Mori category that generalise Castelnuovo's criterion) and also the definition of Mori fibre space (that generalise $\mathbb{P}^2$ and the scrolls), our primary object of interest.

Higher dimensional geometry, like most other areas of mathematics, is marked by creative tensions between abstract and concrete on the one hand, general and special on the other. Contracting a $-1$-curve on a surface is a concrete construction, whereas a Mori extremal ray and its contraction is abstract (compare Remark 2.4.1). The "general" tendency in the classification of varieties, exemplified by the work of Iitaka, Mori, Kollár, Kawamata and Shokurov, includes things like Iitaka–Kodaira dimension, cohomological methods, and the minimal model program in substantial generality. The "special" tendency, exemplified by Hudson, Fano, Iskovkikh, Manin, Pukhlikov, Mori and ourselves, includes the study of special cases, for their own sake, and sometimes without hope of ever achieving general status.

By *explicit*, we understand a study that does not rest after obtaining abstract existence results, but that goes on to look for a more concrete study of varieties, say in terms of equations, that can be used to bring out their geometric properties as clearly as possible. For example, the list of Du Val singularities by equations and Dynkin diagrams is much more than just an abstract definition or existence result, and can be used for all kinds of purposes. This book initiates a general program of explicit birational geometry of 3-folds (compare Section 5). On the whole, our activities do not concern themselves with 3-folds in full generality, but work under particular assumptions, for example, with 3-folds that are hypersurfaces, or have only terminal quotient singularities (see below). The advantage is that we can get a long way into current thinking on 3-folds while presupposing little in the way of technical background in Mori theory.

Treating 3-folds and contractions between them in complete generality would lead us of necessity into a number of curious and technically difficult backwaters; these include many research issues of great interest to us, but we leave them to more appropriate future publications (see however Section 5 below). Making the abstract machinery work in dimension $\geq 4$ is another important area of current research, but the geometry of 4-folds is presumably intractable in the explicit terms that are our main interest here.

# 2    The Mori program

This section is a gentle introduction to some of the ingredients of the 3-fold minimal model program, with emphasis on the aspects most relevant to our current discussion. Surveys by Reid and Kollár [R1], [Kol1], [Kol2] also offer introductory discussions and different points of view on Mori theory. At a technically more advanced level, we also recommend a number of excellent (if somewhat less gentle) surveys: Clemens, Kollár and Mori [CKM], Kawamata, Matsuda and Matsuki [KMM], Kollár and Mori [KM], Mori [Mo] and Wilson [W].

## 2.1   Terminal singularities

It was understood from the outset that minimal models of 3-folds necessarily involve singular varieties (one reason why is explained in 2.5). The *Mori category* consists of projective varieties with *terminal singularities*; the most typical example is the cyclic quotient singularity $\frac{1}{r}(a, r - a, 1)$. Here $a$ is coprime to $r$, and the notation means the quotient $\mathbb{C}^3/(\mathbb{Z}/r)$, where the cyclic group $\mathbb{Z}/r$ acts by

$$(x, y, z) \mapsto (\varepsilon^a x, \varepsilon^{r-a} y, \varepsilon z),$$

and $\varepsilon$ is a primitive $r$th root of 1. The most common instance is $\frac{1}{2}(1, 1, 1)$, the cone on the Veronese surface. The effect of saying that this point is terminal is that if we first resolve it by blowing up, then run a minimal model program on the resolution, we will eventually need to contract down everything we've blown up, taking us back to the same singularity.

There are a few other classes of terminal singularities, including isolated hypersurface singularities such as $xy = f(z, t) \subset \mathbb{C}^4$, where $f(z, t) = 0$ is an isolated plane curve singularity, and a combination of hypersurface and quotient singularity, for example, the hyperquotient singularity obtained by dividing the hypersurface singularity $xy = f(z^r, t)$ by the cyclic group $\mathbb{Z}/r$ acting by $\frac{1}{r}(a, r - a, 1, 0)$. At some time you may wish to look through some sections of Reid [YPG] (especially Theorem 4.5) for a more formal treatment. But for most purposes, the cyclic quotient singularity $\frac{1}{r}(a, r - a, 1)$ is the main case for understanding 3-fold geometry, and if you bear this in mind, you will have little trouble understanding this book.

## 2.2   Theorem on the Cone

The *Mori cone* $\overline{NE}\, X$ (see Figure 2.2.1) is probably the most profound and revolutionary of Mori's contributions to 3-folds. An $n$-dimensional projective variety $X$ over $\mathbb{C}$ is a $2n$-dimensional oriented compact topological space, and its second homology group $H_2(X, \mathbb{R})$ is a finite dimensional real vector space. Every algebraic curve $C \subset X$ can be triangulated and viewed as an oriented 2-cycle, and thus has a homology class $[C] \in H_2(X, \mathbb{R})$. Then by definition $\overline{NE}\, X$ is the closed convex cone in $H_2(X, \mathbb{R})$ generated by the classes

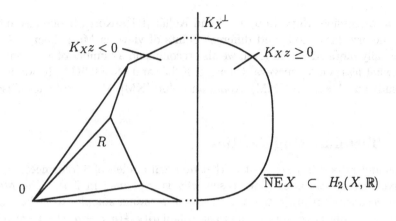

Figure 2.2.1: The Mori cone: $\overline{\text{NE}}\,X$ is locally rational polyhedral in $K_X z < 0$

$[C]$ of algebraic curves $C \subset X$. You can think of this as follows: $H_2(X, \mathbb{R})$ is a property of the topological space $X$, whereas the structure of $X$ as a projective algebraic variety provides the extra information of the Mori cone $\overline{\text{NE}}\,X \subset H_2(X, \mathbb{R})$.

The shape of $\overline{\text{NE}}\,X$ contains information about linear systems and embeddings $X \hookrightarrow \mathbb{P}^N$. Taking intersection number $D{\cdot}C$ with a divisor, or evaluating $\alpha \cap [C]$ with a cohomology class $\alpha \in H^2(X, \mathbb{R})$ (say, the first Chern class of a line bundle $L$) defines a linear form on $H_2(X, \mathbb{R})$. We say that $D$ or $\alpha$ is *nef* if this linear form is $\geq 0$ on $\overline{\text{NE}}\,X$; that is, a divisor $D$ is nef if $D \cdot C \geq 0$ for every curve $C \subset X$. Under an embedding, every algebraic curve must have positive degree; it is known that, under rather mild assumptions, $X$ is projective if and only if $\overline{\text{NE}}\,X$ is a genuine cone with a point.

To state Mori's theorem, we assume that the canonical divisor class $K_X$, (or equivalently, the first Chern class of the cotangent bundle) makes sense as a linear form on $H_2(X, \mathbb{R})$. This is a mild extra assumption on $X$, that certainly holds if $X$ is nonsingular or has at worst quotient singularities. The theorem on the cone then says that $\overline{\text{NE}}\,X$ is a rational polyhedral cone in the half-space of $H_2(X, \mathbb{R})$ on which $K_X$ is negative. This theorem is particularly powerful for Fano varieties, defined by the condition that $-K_X$ is ample: for these, the entire cone $\overline{\text{NE}}\,X$ is contained in $K_X z < 0$, so that $\overline{\text{NE}}\,X$ is a finite rational polyhedral cone.

## 2.3    Extremal rays and the contraction theorem

Mori theory applies mainly to varieties with $K_X$ not nef. This condition says that $K_X C < 0$ for some curve $C$, or that the part of the cone $\overline{\text{NE}}\,X$ in the half-space $K_X z < 0$ is nonempty. Since this part of the cone is locally rational polyhedral, it follows that, if $K_X$ is not nef, $\overline{\mathrm{NE}}\,X$ has at least one extremal ray $R$ with $K_X \cdot R < 0$. Here an *extremal ray* is just a half-line $R = \mathbb{R}_+ z \subset \overline{\mathrm{NE}}\,X$ that is extremal in the sense of convex geometry (that is,

$$z_1, z_2 \in \overline{\mathrm{NE}}\,X \text{ and } z_1 + z_2 \in R \implies z_1, z_2 \in R).$$

Let $R \subset \overline{\mathrm{NE}}\,X$ be an extremal ray with $K_X \cdot R < 0$. Then there exists a contraction morphism

$$f_R \colon X \to Y,$$

characterised by the property that a curve $C \subset X$ is mapped to a point if and only if $C \in R$ (more precisely, the class of $C$). The morphism $f_R \colon X \to Y$ is called a *Mori contraction* or an *extremal contraction*. It is determined by the extremal ray $R$, and has categorical properties such as $-K_X$ relatively ample and $\rho(X/Y) = 1$ that turn out to be surprisingly strong: for example $-K_X$ ample puts us in a position where vanishing results based on Kodaira vanishing kill almost all the cohomology.

The cone and contraction theorems are proved in Kollár and Mori [KM]; on the whole, we can get by without reference to the technicalities of the proof, and you may prefer to take these results on trust for now.

## 2.4    Types of extremal rays

The next step is the case division on the dimension of the image $Y$ and of the exceptional locus of the contraction morphism $f_R \colon X \to Y$, called the *classification of extremal rays* (or *rough classification*). The cases when the contraction $f_R \colon X \to Y$ has $\dim Y < \dim X$ lead to the definition of Mori fibre space and Fano varieties discussed in 2.6. In the other cases, we are dealing with birational modifications of $X$, and, as we see in 2.5, the aim is to proceed inductively towards a minimal model, as in the classical case of surfaces.

**Remark 2.4.1** Note the contrast with the classical case: for surfaces, the thing we contract is a geometric locus. We find a $-1$-curve $C$ and establish that it can be contracted in terms of a neighbourhood of $C$. In contrast, Mori theory in dimension $\geq 3$ works primarily in terms of *categorical definitions* and *existence theorems*: the thing to be contracted is an extremal ray $R$ of $\overline{\mathrm{NE}}\,X$ (the definition of which uses the totality of curves on $X$). The proof of the general theorems saying that $R$ is contractible by a morphism $f_R$ makes sophisticated use of numerical and cohomology vanishing properties of $X$.

The geometric nature of the contraction is only studied as a second step; even basic things such as the geometric locus that is contracted or even the dimension of the image cannot be anticipated. $f_R$ may be birational, a proper fibre space, or the constant morphism to a point. This curious inversion of thinking is another of Mori's characteristic contributions to the subject, and the logic still comes as a surprise to anyone knowing a traditional treatment of the classification of surfaces. After all, Castelnuovo and Enriques could scarcely have guessed that (i) contracting a $-1$-curve, (ii) projecting a geometrically ruled surface to its base curve, and (iii) the constant map of $\mathbb{P}^2$ to a point would find a unified treatment as extremal contractions, and that this idea, however outlandish it might appear at first sight, would lay the foundations of all future work in classification.

## 2.5   Birational modifications: divisorial contractions, flips and the minimal model program

The extremal contractions that are most similar to contracting a $-1$-curve on a nonsingular surface (Castelnuovo's criterion) are the *divisorial contractions*. Here the case assumption is that $f_R \colon X \to Y$ is birational, and contracts a divisor of $X$ to a locus of $Y$ of codimension $\geq 2$. The categorical properties of $f_R$ then guarantee automatically that the exceptional locus of $f_R$ is an *irreducible* divisor, and that $Y$ has terminal singularities. This is the point at which terminal singularities force themselves on our attention: even if $X$ is nonsingular, $Y$ may be singular. Because $Y$ is still in the Mori category, we can repeat the same game starting from $Y$.

The other birational case is when $f_R$ is *small*, that is, every component of $\mathrm{Exc}\, f_R$ has codimension $\geq 2$; in this case there cannot be any cohomology class in $H^2(X, \mathbb{R})$ that corresponds to the canonical divisor of $Y$, so that $Y$ can *never* have terminal singularities. (If such a class existed, its pullback to $X$ would coincide with $K_X$, which would then be numerically trivial on the fibres of $f_R$. This contradicts $-K_X$ ample, the defining property of a Mori extremal contraction.)

Because $Y$ is no longer in the Mori category, the minimal model program cannot just continue inductively from $Y$. The subject was stuck at this point for a few years in the 1980s, before Mori proved the 3-fold flip theorem: there is a *flip*

$$X \overset{t_R}{\dashrightarrow} X^+$$
$$\searrow \quad \swarrow$$
$$Y$$

$$(2.5.1)$$

where $X^+ \to Y$ is another birational map from a 3-fold $X^+$, characterised by the property that $K_{X^+}$ is ample over $Y$. In other words, the birational map $t_R \colon X \dashrightarrow X^+$ cuts out from $X$ a finite number of curves on which $K_X$ is negative, and in their place glues back into $X^+$ a finite number of curves on which $K_{X^+}$ is positive. The definition of flip may seem somewhat obscure, but many nice attributes of $X^+$ follow from it; in particular, the morphism $X^+ \to Y$ is also small, and $X^+$ again has terminal singularities, so is in the Mori category. In dimension $\geq 4$, the existence of the flip diagram (2.5.1) is called the *flip conjecture*; this seems to be one of the most intractable problems in the subject.

Divisorial contractions and Mori flips are the elementary steps in the Mori minimal model program. A sequence of these leads after a finite number of steps to a variety $X'$, which is either a *minimal model*, that is, a variety with $K_{X'}$ nef, or a Mori fibre space $f : X' \to S$.

## 2.6 The definition of Mori fibre space

We now discuss the remaining cases in the classification of extremal rays, when the contraction $f_R : X \to Y$ maps to a smaller dimensional variety, that is, $\dim Y < \dim X$. Then $f_R$ (or $X$ itself) is called a *Mori fibre space* (Mfs). Note that, following Iitaka and Ueno, we say *fibre space* to mean a morphism $f : X \to Y$, often assumed to have connected fibres and $Y$ normal, possibly with varying fibres, singular fibres, even fibres of different dimensions; this is not to be confused with the much stricter notion of fibre bundle.

The cases when $Y$ is a surface and $X \to Y$ is a conic bundle (that is, the general fibre is a conic) or when $Y$ is a curve and $X \to Y$ a fibre space of del Pezzo surfaces are the natural analogues of ruled surfaces. For the logical framework of Mori theory, we include in the definition of Mori fibre space the case that the contraction $f_R : X \to Y = \mathrm{pt.}$ is the constant map to a point: then the morphism $f_R$ is trivial, but its categorical properties include the fact that $-K_X$ is ample, and $\mathrm{Pic}\, X$ has rank 1. In this case $X$ is called a *Fano 3-fold*; in contrast to the classical terminology, we allow $X$ to be singular.

## 2.7 Biregular geometry versus birational geometry

The dividing line between biregular and birational geometry has changed through the generations, and is possibly still open to debate. The Italian school worked primarily in birational terms, and Zariski and Weil used birational ideas (at least in part) in setting up foundations for biregular geometry. The modern view, with scheme theory firmly established as the foundation, constructs birational geometry within this biregular framework. Thus, while the dichotomy between surfaces having nonvanishing plurigenera and ruled surfaces (or "adjunction terminates") is manifestly birational, we no longer think of it as the primary result of classification, but derive it from biregular results. This new view was instrumental in the success of Mori theory.

When we run a Mori minimal model program on a given 3-fold $V$, the end product is either a minimal model $X$ with $K_X$ nef, or a Mori fibre space, typically, a Fano 3-fold $X$ or a conic bundle over a surface $X \to S$. The properties that define the 3-fold $X$ are biregular in nature, so that we view $X$ as a biregular construction. From this point of view, the proof of classification should also be considered a biregular activity, since the point is to prove that a given minimal 3-fold $X$ has the right plurigenera and Kodaira dimension. Our conclusion is that birational geometry begins with the question of birational maps between different Mori fibre spaces.

# 3    What this book contains

This section discusses briefly the papers in this book, and their contribution to the above program of study. The papers are:

(1) K. Altmann: One-parameter families containing three-dimensional toric Gorenstein singularities

(2) J. Kollár: Nonrational covers of $\mathbb{P}^m \times \mathbb{P}^m$

(3) A. V. Pukhlikov: Essentials of the method of maximal singularities

(4) A. R. Iano-Fletcher: Working with weighted complete intersections

(5) A. Corti, A. Pukhlikov and M. Reid: Fano 3-fold hypersurfaces

(6) A. Corti: Singularities of linear systems and 3-fold birational geometry

(7) M. Reid: Twenty five years of 3-folds, an old person's view

Klaus Altmann's paper (1) is a study of the deformation theory of toric Gorenstein 3-fold singularities. It relates to the classification of 3-fold flips as follows: we know that any Mori flip diagram (2.5.1) can be obtained from a $\mathbb{C}^\times$ action on a 4-fold Gorenstein singularity $0 \in A$ by taking the quotient by the $\mathbb{C}^\times$ action in different interpretations – the so-called *variation of geometric invariant theory quotient*, see Dolgachev and Hu [DH], Reid [R2] and Section 5.3 below. Moreover, the general anticanonical divisor $S \in |-K_X|$ (the *general elephant*) is a surface with only Du Val singularities, according to Kollár and Mori [KM1], Theorem 1.7. Its inverse image in $A$ is a $\mathbb{C}^\times$ cover $B \to S$, and is a hyperplane section $B \subset A$, so that $A$ can be viewed as a 1-parameter deformation of $B$. It frequently happens that $S$ is of type $A_n$, and then $B$ is toric, so that Altmann's theory applies in many cases to give a classification of 3-fold flips. Altmann's previous work [Al] used the notion of Minkowski decomposition of polytopes to give a complete treatment of the deformation of isolated 3-fold toric Gorenstein singularities; in the present paper, he shows how to modify his method to the case of toric varieties having singularities in codimension 2.

János Kollár's paper (2) provides a new method of proving irrationality, adding to the known collection of rationally connected varieties that are not rational: finite covers of $\mathbb{P}^m \times \mathbb{P}^n$ with ramification divisor of large enough degree in one factor, and hypersurfaces in $\mathbb{P}^m \times \mathbb{P}^n$ of large enough degree. His technique involves reduction to characteristic $p$, and a rather clever and surprising analysis of the stability of the tangent bundle in characteristic $p$. In fact, he proves the slightly more general structural property that these varieties are not even ruled. In the case of conic bundles, these results are spectacularly close to the conjectural bound for rationality (compare, for example, paper (2), Remark 1.2.1.1 with Corti's paper (6), 4.10 and 4.11). This provides the strongest confirmation to date of the conjectures on conic bundles, in a numerical range that is inaccessible to all other methods.

The papers (3)–(6) form a connected suite of papers around the subject of *birational rigidity*. The notion, discussed in more detail in Section 4.5 below, originates in the famous result of Iskovskikh and Manin [IM] that a non-singular quartic 3-fold $X_4 \subset \mathbb{P}^4$ has no birational maps to Fano varieties (other than isomorphisms to itself). Pukhlikov's paper (3) describes his important simplification and elaboration of Iskovskikh and Manin's treatment. This paper is partly based on notes of lectures given at the 1995–96 Warwick algebraic geometry symposium and the preprint, with its clear treatment of the Russian methods, strongly stimulated our collaboration in the joint paper (5). The different approach in Pukhlikov's papers also offers a useful ideological and practical counterweight to the methods of Corti's paper (6).

Our long joint paper (5) is the real heart of this book. In it, we carry out a substantial portion of a program of research on birational rigidity, treating the *famous 95* families of Fano 3-fold weighted hypersurfaces. We refer to Section 4.5 and the introduction to paper (5) for further discussion of birational rigidity.

Anthony Iano-Fletcher's paper (4) is a well written tutorial introduction to weighted projective spaces and their subvarieties. This paper has been available for many years as a Max Planck Institute preprint, and is widely quoted in the literature; it contains many very useful results and methods of calculation, including one derivation of the list of the famous 95 hypersurfaces, and thus forms an essential prerequisite for paper (5).

Corti's paper (6) contains a detailed introduction to the Sarkisov program. It develops and applies powerful new methods to quantify and analyse the singularities of linear systems, clarifying and providing technical alternatives to the methods initiated by Iskovskikh and Manin based on the study of the resolution graph. The new ideas are based on the Shokurov connectedness principle in log birational geometry, and seem to provide the most powerful currently known technique to exclude birational maps between Mori fibre spaces. The results are applied to give rigidity criteria for Mori fibre spaces in a number of cases, and our joint paper (5) also appeals to them for one or two technical points.

Reid's historical paper (7) is a *Heldenleben* that needs no introduction.

# 4    Mori theory and birational geometry

Following our introductory remarks on Mori theory in Section 2, we now give a brief introduction to our view of birational geometry, including the Sarkisov program and birational rigidity.

## 4.1    Fano-style projections

Fano based his treatment of the 3-folds $V_{2g-2} \subset \mathbb{P}^{g+1}$ for $g \geq 7$ on the idea of constructing a birational map by projection from a suitably chosen centre. Typically, the double projection of $V$ from a line $L$ involves a diagram

$$
\begin{array}{ccc}
V' & \dashrightarrow & V'' \\
\downarrow & & \downarrow \\
V & & W,
\end{array}
\tag{4.1.1}
$$

where $V' \to V$ is the blowup of $L \subset V$, the map $V' \dashrightarrow V''$ flops the lines meeting $L$ (in good cases, finitely many lines with normal bundle of type $(-1, -1)$), and $V'' \to W$ contracts the surface $E \subset V''$ swept out by conics meeting $L$ to a curve $\Gamma \subset W$. Fano thought of the map $V \dashrightarrow W$ as the rational map defined by linear projection, and factoring it in biregular terms was not his primary concern.

For us, on the other hand, it is important to view Fano's projection as a general construction in the Mori category: $V' \to V$ is an extremal extraction, $V' \dashrightarrow V''$ a rational map that is an isomorphism in codimension 1 (in good cases, a composite of classic flops), and $V'' \to W$ the contraction of an extremal ray. All 4 of the varieties in (4.1.1) are in the Mori category, and the two morphisms are contractions of extremal rays.

**Remark 4.1.1** We take the opportunity to clear up a possible source of confusion that occurs throughout the subject: in Fano's case, the *single* projection from $L$ contracts the flopping lines by a morphism $V' \to \overline{V}$ to a variety $\overline{V}$ having (in good cases) only 3-fold ordinary double points; we think of $\overline{V}$ as the *midpoint* of the construction of the link $V \dashrightarrow W$. It is a Fano variety in some sense, since it has terminal singularities and $-K_V$ is ample, but it is *not in the Mori category*, because it is not $\mathbb{Q}$-factorial: the exceptional scroll over $L$ maps to a divisor in $\overline{V}$ that is not Cartier and not $\mathbb{Q}$-Cartier at the nodes.

## 4.2   Sarkisov links

Sarkisov links play the role of "elementary transformations" for birational maps between Mori fibre spaces. The *Sarkisov program* factors an arbitrary birational map $X \dashrightarrow Y$ between the total spaces of Mori fibre spaces $X \to S$ and $Y \to T$ as a chain of such links (see Corti's paper (6) and [Co] for details).

A general *Sarkisov link* is given by a diagram in the Mori category that is a variation on (4.1.1), but with general extremal contractions allowed as the morphisms. As discussed in much more detail in Corti [Co] and paper (6), if we start from a Mori fibre space $X \to S$, any link $(X/S) \dashrightarrow (Y/T)$ is given by one of the following constructions. First, we replace $X \to S$ by a new morphism $X_1 \to S_1$ having rank $N^1(X_1/S_1) = 2$: for this, either

(i) blow up $X$ by an extremal blowup $X_1 \to X$, and let $X_1 \to S_1$ be the composite $X_1 \to X \to S = S_1$; or

(ii) contract the base by an extremal contraction $S \to S_1$, and let $X_1 \to S_1$ be the composite $X_1 = X \to S \to S_1$.

In either case $N^1(X_1/S_1) = \mathbb{R}^2$, and $\overline{NE}(X_1/S_1)$ has an initial extremal ray corresponding to the given morphism $X_1 \to X$ or $X_1 \to S$. This sets up a restricted type of minimal model program called a *2-ray game*: because a cone in $\mathbb{R}^2$ is just a "wedge", it has a far side that is a (pseudo-) extremal ray (possibly not of Mori type). The case that leads to a link is when the minimal model program runs to completion in the Mori category: the far ray can be contracted, and possibly after a chain of inverse flips, flops and flips, the minimal model program ends with a divisorial contraction $X^{(n)} \to Y \to S_1 = T$ or a contraction of fibre type $X^{(n)} = Y \to T \to S_1$. There are two possible ways of starting the construction, and two ways of ending it, leading to Sarkisov links of Type I–IV.

Existence and uniqueness: the 2-ray game is entirely determined by the initial step $X_1/S_1$. It may happen that the initial step $X_1/S_1$ does not construct a link – either because an inverse flip demanded by the 2-ray game does not exist or has worse than terminal singularities, or because the final divisorial or fibre type contraction falls out of the Mori category. See paper (5), Section 5.5 and 7.6 for examples.

For surfaces over an algebraically closed field, links are the following familiar transformations: the blowup taking $\mathbb{P}^2$ to $\mathbb{F}_1$, its inverse contraction

$\mathbb{F}_1 \to \mathbb{P}^2$, the well known elementary transformations $\mathbb{F}_k \dashrightarrow \mathbb{F}_{k\pm 1}$ between scrolls over $\mathbb{P}^1$, and the "exchange of factors" of $\mathbb{F}_0 = \mathbb{P}^1 \times \mathbb{P}^1$ (in other words, the identity on $\mathbb{F}_0$, but viewed as exchanging its two projections). These are exactly the elementary steps in Castelnuovo's proof of Max Noether's theorem (discussed in 4.4 below).

## 4.3    The Sarkisov program

We explain in conceptual terms how to factor (or "untwist") a birational map $\varphi\colon X \dashrightarrow X'$ between Mori fibre spaces $X \to S$ and $X' \to S'$ as a chain of Sarkisov links. Untwisting is a constructive descending induction: let $\mathcal{H}'$ be a very ample complete linear system on $X'$, chosen at the outset and kept fixed throughout. Following the classical ideas of Cremona, Noether and Hudson, consider the linear system $\mathcal{H} = \varphi_*^{-1}\mathcal{H}'$ on $X$ obtained as the birational transform of $\mathcal{H}'$. Untwisting is the story of how we reduce the singularities and the degree of $\mathcal{H}$ to make $\varphi$ an isomorphism.

First we prove the *Noether–Fano–Iskovskikh inequalities*, that serve as a sensitive detector to locate the initial step $X_1/S_1$ of a Sarkisov link if $\varphi$ is not already an isomorphism: this is either a blowup of a maximal singularity of $\mathcal{H}$, or a way of viewing $X$ over a different base to make $\mathcal{H}$ look simpler. Next, when needed to factor a given map, the Sarkisov link $\psi\colon (X/S) \to (Y/T)$ with given initial step $X_1/S_1$ always exists.

At the start of the proof, we set up a discrete invariant of $\varphi$, its *Sarkisov degree* $\deg\varphi$. We only explain this for a Fano 3-fold $X$, when $-K_X$ is a $\mathbb{Q}$-basis of $\operatorname{Pic} X$. Then we set $\deg\varphi = n$, where $n$ is the positive rational number for which $\mathcal{H} \subset |-nK_X|$. We prove that the Sarkisov link $\psi\colon X \dashrightarrow Y$ provided by the NFI inequalities decreases the Sarkisov degree, in the sense that the composite map $\varphi\psi^{-1}\colon Y \dashrightarrow X'$ between $Y/T$ and $X'/S'$ has

$$\deg \varphi\psi^{-1} < \deg \varphi.$$

We say that $\varphi\psi^{-1}$ is an *untwisting* of $\varphi$ by $\psi$. The factorisation theorem then follows by descending induction on the Sarkisov degree. Of course, we are glossing over many subtle points, including the definition of Sarkisov degree for a strict Mori fibre space $X \to S$, the verification that untwisting by a link decreases $\deg\varphi$, and that a chain of untwistings must terminate. For the details, see paper (6).

## 4.4    A classical example

We content ourselves with illustrating how these ideas work in the most famous case of all, a birational map from $X = \mathbb{P}^2$ to $X' = \mathbb{P}^2$. Max Noether's inequality states that there are 3 points $P_1, P_2, P_3$ of $\mathbb{P}^2$ (possibly infinitely near), such that

$$m_1 + m_2 + m_3 > n = \deg \varphi, \quad \text{where } m_i = \operatorname{mult}_{P_i} \mathcal{H}. \qquad (4.4.1)$$

Here $\operatorname{mult}_{P_i} \mathcal{H}$ means the multiplicity of a general element of the linear system $\mathcal{H}$ at $P_i$. In the general case, when $P_1, P_2, P_3$ are distinct noncollinear points, we can choose coordinates so that these are the three coordinate points $(1, 0, 0)$, $(0, 1, 0)$, $(0, 0, 1)$, and it is easy to check that untwisting by the standard quadratic Cremona involution

$$\psi \colon (x_0 : x_1 : x_2) \dashrightarrow (x_1 x_2 : x_0 x_2 : x_1 x_2)$$

decreases the degree. Indeed, $\psi^{-1}(\text{line})$ is a conic passing through the $P_i$, so that after untwisting, the degree becomes

$$2n - \sum m_i < n.$$

The gap in this argument is that the points $P_i$ can be infinitely near. This can happen in two or three different ways, and in most of these cases, we can still construct a quadratic transformation centred on a suitably chosen coordinate triangle to untwist our map and decreases its degree. However, there are cases in which no single quadratic Cremona transformation decreases the degree.

This is the starting point of Castelnuovo's proof of Noether's theorem, a direct precursor of the Sarkisov program. Whatever infinitely near points there may be, there always exists one point $P \in \mathbb{P}^2$ with $m_P > \frac{n}{3}$. Of course, this follows from (4.4.1), but it is also easy to prove directly by an easy argument in the spirit of "termination of adjunction" (see paper (6), Theorem 2.4, where the inequality and the argument to prove it are generalised to any Mori fibre space). Blowing up this point by $\mathbb{F}_1 \to \mathbb{P}^1$ is the first link in a Sarkisov chain. It untwists because the Sarkisov degree measures divisors on $\mathbb{P}^2$ in terms of $-K_{\mathbb{P}^2} = \mathcal{O}(3)$, but measures relative divisors on $\mathbb{F}_1$ in terms of the relative $-K_{\mathbb{F}_1/\mathbb{P}^1} = \mathcal{O}(2)$ ([Co], 1.3 contains a more detailed description).

## 4.5   Birational rigidity

A Fano variety $X$ is *birationally rigid* if for every Mori fibre space $Y \to S$, the existence of a birational equivalence $X \dashrightarrow Y$ implies that $Y \cong X$. Once the Sarkisov program is established (that is, as yet only in dimension $\leq 3$), it is equivalent to say that there are either no Sarkisov links out of $X$, or only self-links $X \dashrightarrow X$.

The notion of birational rigidity for strict Mori fibre spaces is also well studied. However, the definition (paper (6), Definitions 1.2–3) is subtle and somewhat confusing, largely because it treats Mori fibre spaces up to *square birational equivalence*. It covers important results of Sarkisov on conic bundles and Pukhlikov on del Pezzo fibre spaces.

The main result of paper (5) states that a general member of any of the 95 families of Fano 3-fold hypersurfaces is rigid. We show that any birational map $\varphi\colon X \dashrightarrow Y$ factors as a chain of Sarkisov links $X \dashrightarrow X$, followed by an isomorphism $X \cong Y$. The paper is written to be essentially selfcontained. The Sarkisov program is introduced in a context where it is made simpler by various special circumstances. However, at one point, we rely on a technical statement (Theorem 5.3.3) that is proved in Corti's paper (6) (although it goes back essentially to Iskovskikh and Manin, and can be proved by the technique of Pukhlikov's paper (3)). A substantial part of paper (5) is devoted to the classification of links $X \dashrightarrow X$, which we also discuss in Section 5.5 below. It was rather surprising for us to discover that all the links can be described explicitly in terms of just two basic constructions in commutative algebra, which are natural generalisations of the classical Geiser and Bertini involutions of cubic surfaces.

## 4.6   Beyond rigidity

We now know a handful of varieties having precisely two models as Mori fibre space, either two Fano 3-folds (Corti and Mella [CM]), or two del Pezzo fibrations (Grinenko [Gr]), or one of each (in the case $X_{3,3} \subset \mathbb{P}(1,1,1,1,2,2)$ suggested by Grinenko and Pukhlikov). These varieties are not actually rigid, but nearly so. These and other examples suggest the following idea. Say that a birational map $\varphi\colon X \dashrightarrow X'$ between Mori fibre spaces $X \to S$ and $X' \to S'$ is a *square equivalence* if the following two conditions hold:

(1) There is a birational map $S \dashrightarrow S'$ making the obvious diagram commute. This condition is equivalent to saying that the generic fibre of $X \to S$ is birational to the generic fibre of $X' \to S'$ under $\varphi$.

(2) The birational map of (1), from the generic fibre of $X \to S$ to a generic fibre of $X' \to S'$, is in fact biregular.

We define the *pliability* of a Mori fibre space $X \to S$ as the set

$$\mathcal{P}(X/S) = \{\text{Mfs } Y \to T \mid X \text{ is birational to } Y\}/\text{square equivalence}.$$

Our general philosophy can be described as follows. We would like to describe the pliability $\mathcal{P}(X/S)$ of a Mori fibre space $X \to S$ in terms of its biregular geometry. There are reasons for thinking that $\mathcal{P}(X/S)$ will often have a reasonable description, say as a finite set or a finite union of algebraic varieties. A case division based on the various possibilities for the size of

$\mathcal{P}(X/S)$ can be used as a further birational classification of Mori fibre spaces. Note that our general philosophy can only be stated in the language of Mori theory.

Although the search for counterexamples to the Lüroth theorem played an important part in kick-starting the study of 3-folds around 1970, the last 25 years have seen little progress on criteria for rationality and unirationality, and these questions seem likely to remain intractable in the foreseeable future. The wealth of new examples in Mori theory must in any case cast doubt on the position of these problems as central issues in birational geometry; they date back after all to the golden days of innocence, when the classics (from Cremona through to Iskovskikh) had never really met the typical examples of birational geometry. Rationally connected seems to be the most useful modern replacement for (uni-)rationality, since it is robust on taking surjective image or under deformations, and there are good criteria for it. It is not that we are hostile to the rationality problem; rather, since we are committed to the classification of 3-folds, we need a theoretical framework capable of accomodating all varieties, with all their wealth of individual behaviour. Our suggested notion of pliability is a tentative step in this direction.

# 5    Some open problems

## 5.1    Explicit birational geometry

In a sense, we can define explicit birational geometry as the concrete study (including classification) of

(1) divisorial contractions, flips, and Mori fibre spaces;

(2) the way divisorial contractions and flips combine to form the links of the Sarkisov program.

The model is [YPG], Theorem 4.5, which classifies all 3-fold terminal singularities as a reasonably concrete, explicit and finite list of families. It seems reasonable to hope that Mori flips, divisorial contractions, Fano 3-folds and Sarkisov links between 3-fold Mori fibre spaces will eventually succumb to a similar treatment. The overall aim is to make everything else tractable in the same sense as the terminal singularities. This program can be expected to provide useful employment for algebraic geometers over several decades.

## 5.2    Divisorial contractions

**Problem 5.2.1** Fix a 3-fold terminal singularity $P \in Y$ (an analytic germ). Write down all 3-fold divisorial contractions $f\colon (E \subset X) \to (P \in Y)$ by explicit equations.

This seems a rather difficult problem in general. To do something useful with it, it is important to realise that, whereas you are free to pick your favourite singularity $P \in Y$, it is then your responsibility to classify all possible extremal blowups $X \to Y$ of $P$. The few known cases of this problem are $P \in Y = \frac{1}{r}(1, -1, a)$ treated by Kawamata [Ka], the ordinary node (see paper (6), Chapter 3), the singularity $xy = z^3 + w^3$ of Corti and Mella [CM]. Each of these results has significant applications to the Sarkisov program and 3-fold birational geometry; see paper (6), Section 6.3 and the forthcoming paper [CM]. See also [Ka] for more examples.

**Example 5.2.2** Let $P \in Y$ be a nonsingular point and $a, b$ coprime integers; then the weighted blowup $f \colon X \to Y$ with weights $(1, a, b)$ is an extremal divisorial contraction with $f(E) = P$. Corti conjectured in 1993 that this is the complete list.

## 5.3    What is a flip?

**Problem 5.3.1** Classify 3-fold flips $t \colon X \dashrightarrow X^+$

In a sense, this is done in the monumental paper of Kollár and Mori [KM1], but their description is not sufficiently explicit for some applications. The following, taken from [KM1], is the simplest example of flips (see Brown [Br] for many similar families of examples).

**Example 5.3.2** Let $f_{m-1}(x_1, x_2)$ be a homogeneous polynomial of degree $m - 1$ in 2 variables, and let $\mathbb{C}^\times$ act on $\mathbb{C}^5$ with weights $1, 1, m, -1, -1$. That is, the action is given by

$$x_1, x_2, x_3, y_1, y_2 \mapsto \lambda x_1, \lambda x_2, \lambda^m x_3, \lambda^{-1} y_1, \lambda^{-1} y_2.$$

Consider the $\mathbb{C}^\times$-invariant affine 4-fold $0 \in A \subset \mathbb{C}^5$ given by

$$x_4 y_1 = f_{m-1}(x_1, x_2).$$

An example of a flipping contraction $X \to Y$ and flip $X^+$ is obtained by taking the geometric invariant theory quotient (Spec of the ring of invariants) $Y = A /\!/ \mathbb{C}^\times$, and setting

$$X = \operatorname{Proj} \bigoplus_{n \leq 0} \mathcal{O}(nK_Y) \quad \text{and} \quad X^+ = \operatorname{Proj} \bigoplus_{n \geq 0} \mathcal{O}(nK_Y)$$

for the two sides of the flip.

Quite generally, all flips arise in this way from an affine Gorenstein 4-fold $0 \in A = \operatorname{Spec} \bigoplus_{n \in \mathbb{Z}} \mathcal{O}(nK_Y)$ with $\mathbb{C}^\times$ action (see [R2]), and the problem is to write manageable equations for $A$ in $\mathbb{C}^\times$-linearised coordinates. These considerations are the starting point of Klaus Altmann's paper (1).

## 5.4 Mori fibre spaces

**Problem 5.4.1** Classify Fano 3-folds (with $B_2 = 1$ and $\mathbb{Q}$-factorial terminal singularities) up to biregular equivalence.

An example, and a beginning of an answer to this problem, is the list of 95 Fano 3-fold (weighted) hypersurfaces of Reid and Iano-Fletcher (see paper (4), 16.6), but also the 86 codimension 2 (weighted) complete intersections (paper (4), 16.7), the 70 codimension 3 Pfaffian cases of Altınok [Al], Table 5.1, p. 69, etc. It is a feature of nonsingular Fano 3-folds that they all are linear sections of standard homogeneous varieties in their Plücker embedding. It is just possible that, when looked at from the correct angle, this beautiful and fundamentally simple structure will extend to singular Fano 3-folds.

## 5.5 The links of the Sarkisov program

It is in the study of the links of the Sarkisov program, the basis of its applicability, that explicit birational geometry really comes alive. Here we fix a Mori fibre space $X \to S$, and we ask to classify all links $X \dashrightarrow Y$, taking off from $X$ and landing at an arbitrary Mori fibre space $Y \to T$. For $X$ a general member of one of our famous 95 families of Fano 3-fold hypersurfaces, this program is carried out to completion in paper (5). In particular, at the end of the classification, we discover that $Y \cong X$. The links that occur fit into a very small number of known classes. On the other hand, in doing many concrete examples, it is our experience that each new case that we understand involves learning how to do computations (for example, in graded rings or in the geometry of projections), a process that can sometimes be rather tricky.

**Example 5.5.1** Consider the Fano variety $X = X_{2,2,2} \subset \mathbb{P}^6$ given as a complete intersection of three sufficiently general quadrics in $\mathbb{P}^6$, and choose a line $L \subset X$. Then there is a link $\tau_L : X \dashrightarrow X'$, to a conic bundle $X' \to \mathbb{P}^2$ with discriminant curve of degree 7.

The rational map $X \dashrightarrow \mathbb{P}^2$ is not too difficult to realise. For a point $x \in \mathbb{P}^6$, write $\Pi_x$ for the 2-plane spanned by $L$ and $x$. Write $\{Q_\lambda \mid \lambda \in \Lambda\}$ for the net of quadrics vanishing on $X$. Then the restriction $Q_{\lambda|\Pi_x} = L + \Gamma_\lambda$ is the union of $L$ plus a line $\Gamma_\lambda$, and it is easy to see that $x \notin X$ if and only if $\{\Gamma_\lambda \mid \lambda \in \Lambda\}$ is the whole of $\Pi_x^\vee$. Mapping $x \in X$ to the quadric $Q_\lambda$ containing all the $\Pi_x$ (which is unique, in general) gives a rational map $X \dashrightarrow \Lambda \cong \mathbb{P}^2$, whose fibres are conics.

The first point in seeing this process as a Sarkisov link is that we must understand it in explicit biregular terms, and factor the map $X \dashrightarrow \Lambda$ as a chain of flips, flops and divisorial contractions, followed by a Mori fibre space. In fact, if $x_0, \ldots, x_6$ are coordinates on $\mathbb{P}^6$ with $L : \{x_0 = x_1 = \cdots = x_4 = 0\}$, the equation of $X$ can be written as

$$M \begin{pmatrix} x_5 \\ x_6 \end{pmatrix} = \mathbf{q}$$

where $M$ is a $3 \times 2$ matrix of linear forms in the variables $x_0, \ldots, x_4$ and $\mathbf{q}$ is a 3-vector of quadric forms in the same variables. Write $\pi_L \colon X \to \overline{Y} \subset \mathbb{P}^4$ for the projection to $\mathbb{P}^4$. The image $\overline{Y} = \overline{Y}_4$ is the quartic 3-fold given by the equation

$$\det M\mathbf{q} = 0.$$

The singular locus of $\overline{Y}_4$ consists of the 44 ordinary nodes

$$\{\operatorname{rank} M\mathbf{q} = 1\}.$$

Consequently, letting $Y \to X$ be the blow up of $L \subset X$, the birational morphism $Y \to \overline{Y}$ contracts 44 lines with normal bundle $(-1, -1)$, and denoting the flop $t \colon Y \dashrightarrow Y'$, it is easy to check that the rational map $Y \dashrightarrow \Lambda$ described above becomes a morphism $Y' \to \Lambda$, which is in fact a Mori fibre space and a conic bundle. This explicit construction shows that the map $\tau_L \colon X \dashrightarrow Y'$ is a Sarkisov link of Type II.

Similarly, it is easy to construct self-links $\sigma_C \colon X \dashrightarrow X$ centred on conics and twisted cubics $C \subset X$.

The methods of paper (6), Section 6.6, should prove that the only links $X \dashrightarrow Y$, starting with a general $X = X_{2,2,2}$, are the $\tau_L$ and $\sigma_C$ just described.


# 6    Acknowledgments

This book originated in a 3-folds activity during three months Sep–Dec 1995, organised as part of the 1995–96 EPSRC Warwick algebraic geometry symposium. (Although it has grown almost beyond recognition in the intervening years.) As well as the funding from EPSRC, the 3-folds activity benefitted from a number of travel grants for Japanese visitors from the Royal Society/ Japan Society for the Promotion of Science administered by the Isaac Newton Institute, Cambridge and Kyoto Univ., RIMS, and from the support of AGE (Algebraic Geometry in Europe, EU HCM/TMR project, contract number ERBCHRXCT 940557).

It is our solemn duty, as editors, to apologise to the other authors for our slow pace in preparing this book, which has delayed by several years the appearance of the papers (1)–(4). This is especially reprehensible since the desire to understand their results and assimilate them into our own work has been a strong motivation for our research (this applies to Klaus Altmann's paper (1), and more especially to Sasha Pukhlikov's paper (3), which provided much of the stimulus for our papers (5) and (6)), and we are deeply conscious of the fact that our own work has improved partly as a result of the delay we have inflicted on theirs. We hope that the quality of the final product can go some way towards compensating for our transgressions.

# References

[Al]    S. Altınok, Graded rings corresponding to polarised K3 surfaces and Q-Fano 3-folds, Univ. of Warwick Ph. D. thesis, Sep. 1998, 93 + vii pp.

[Br]    G. Brown, Flips arising as quotients of hypersurfaces, Math. Proc. Camb. Phil. Soc. **127** (1999), 13–31

[Co]    A. Corti, Factoring birational maps of 3-folds after Sarkisov, J. Alg. Geom. **4** (1995), 223–254

[CM]    A. Corti and M. Mella, Terminal quartic 3-folds, in preparation

[CKM]    H. Clemens, J. Kollár and S. Mori, Higher dimensional complex geometry, Astérisque **166**, (1988)

[DH]    I. Dolgachev and Yi Hu, Variations of geometric invariant theory quotients, Publ. Math. IHES,

[Gr]    M. M. Grinenko, Birational automorphisms of a threefold double quadric with the simplest singularity, Mat. Sb. **189** (1998), 101–118; translation in Sb. Math. **189** (1998), 97–114

[IM]    V. A. Iskovskikh and Yu. I. Manin, Three-dimensional quartics and counterexamples to the Lüroth problem, Mat. Sb. (N.S.) **86** (128) (1971), 140–166; English translation, Math. USSR-Sb. **15** (1971), 141–166

[Ka]    Y. Kawamata, Divisorial contractions to 3-dimensional terminal quotient singularities, in Higher dimensional complex varieties (Trento 1994), 241–245, de Gruyter, Berlin, 1996

[Kol1]    J. Kollár, The structure of algebraic threefolds: an introduction to Mori's program, Bull. Amer. Math. Soc. **17** (1987), 211–273

[Kol2]    J. Kollár, Flips, flops, minimal models, etc. Surveys in differential geometry (Cambridge, MA, 1990), 113–199, Lehigh Univ., Bethlehem, PA, 1991

[KM]  J. Kollár and S. Mori, Birational geometry of algebraic varieties, CUP, 1998

[KM1]  J. Kollár and S. Mori, Classification of three dimensional flips, J. Amer. Math. Soc. **5** (1992), 533–703

[KMM]  Y. Kawamata, K. Matsuda and K. Matsuki, Introduction to the minimal model problem, in Algebraic geometry (Sendai, 1985), Adv. Stud. Pure Math. **10**, North-Holland, Amsterdam-New York, 1987, pp. 283–360

[M]  S. Mori, Flip theorem and the existence of minimal models for 3-folds. J. Amer. Math. Soc. **1** (1988), 117–253

[Mo]  S. Mori, Classification of higher-dimensional varieties, in Algebraic geometry (Bowdoin, 1985), Proc. Sympos. Pure Math., **46**, Part 1, AMS, 1987, pp. 269–331

[R1]  M. Reid, Tendencious survey of 3-folds, in Algebraic geometry (Bowdoin, 1985), Proc. Sympos. Pure Math., **46**, Part 1, AMS, 1987, pp. 333–344

[R2]  M. Reid, What is a flip?, Notes of colloquium talk, Utah, Dec 1992 and several subsequent lectures (in preparation as joint paper with Gavin Brown)

[YPG]  M. Reid, Young person's guide to canonical singularities, in Algebraic geometry (Bowdoin, 1985), Proc. Sympos. Pure Math., **46**, Part 1, AMS, 1987, pp. 345–414

[W]  P. M. H. Wilson, Towards birational classification of algebraic varieties, Bull. London Math. Soc. **19** (1987), 1–48

Alessio Corti
DPMMS, University of Cambridge,
Centre for Mathematical Sciences,
Wilberforce Road, Cambridge CB3 0WB, U.K.
e-mail: a.corti@dpmms.cam.ac.uk

Miles Reid,
Math Inst., Univ. of Warwick,
Coventry CV4 7AL, England
e-mail: miles@maths.warwick.ac.uk
web: www.maths.warwick.ac.uk/~miles

# One parameter families containing three dimensional toric Gorenstein singularities

Klaus Altmann

## Contents

1  Introduction                                                        21

2  Visualizing $T^1$                                                   23

3  Genuine deformations                                                29

4  Three dimensional toric Gorenstein singularities                    34

5  Work in progress and open problems                                  44

# 1  Introduction

## 1.1

Let $\sigma$ be a rational, polyhedral cone and $Y = Y_\sigma$ the affine toric variety it defines, which is normal but possibly singular; we study the deformation theory of $Y$. The infinitesimal deformation space $T_Y^1$ is a multigraded vector space, with homogeneous pieces determined by the combinatorial formulas of [Al 1].

If $Y_\sigma$ has only an isolated Gorenstein singularity, we can say even more (cf. [Al 2], [Al 3]): $T^1$ is concentrated in a single multidegree, and the corresponding homogeneous piece has an elementary geometric description in terms of Minkowski summands of a certain lattice polytope. We even obtain the entire versal deformation of $Y_\sigma$ (cf. [Al 4]).

## 1.2

Our first aim in the present paper is to interpret geometrically the formula for $T^1$ for arbitrary toric singularities in every multidegree. We again do this in terms of Minkowski summands of certain polyhedra; however, these polyhedra are not necessarily compact any longer, and their vertices are not necessarily lattice points (cf. 2.6).

In [Al 2] we studied so-called toric deformations. These only exist in negative multidegrees (that is, with multidegrees in $-\sigma^\vee$). They are genuine deformations with smooth parameter space, and are characterized by the fact that their total space is again toric. Now, armed with our new description of $T^1_Y$, we describe the Kodaira–Spencer map in these terms in Theorem 3.3. Moreover, in 3.5, we extend the construction of genuine deformations to non-negative degrees using a partial modification of our singularity $Y_\sigma$. Although the total space is no longer toric, we can still describe it and its Kodaira–Spencer map combinatorially.

## 1.3

After that, we focus on three dimensional Gorenstein toric singularities. As already mentioned, everything is known in the case of isolated singularities. However, as soon as $Y_\sigma$ has a one dimensional singular locus (necessarily of transversal type $A_k$), the situation changes drastically: $T^1_Y$ spreads in general into infinitely many multidegrees. In 4.3, using our geometric description of the graded pieces of $T^1_Y$, we detect all the nontrivial ones, and determine their dimension (which is usually 1). The easiest example of this kind is the cone over the weighted projective plane $\mathbb{P}(1,2,3)$ (cf. 4.4 and 4.8).

At present it seems impossible to describe the entire versal deformation, which is an infinite dimensional space. However, the infinitesimal deformations corresponding to the one dimensional homogeneous pieces of $T^1_Y$ are unobstructed, and we lift them in 4.5 to genuine one parameter families. Since the corresponding multidegrees are in general nonnegative, this can be done using the construction introduced in 3.5. Our treatment of the cone over $\mathbb{P}(1,2,3)$ continues in this spirit in 4.8.

These one parameter families can be thought of as a kind of skeleton, from which one hopes eventually to build the entire versal deformation. The most important open questions are the following:

1. which sets of one parameter families belong to a common irreducible component of the base space?

2. how can those families be combined to find a general fiber of this component (a smoothing of $Y_\sigma$)?

Answers to these questions would provide important information about three dimensional flips.

# 2  Visualizing $T^1$

## 2.1  Notation

As usual in toric geometry (see [Oda] for a detailed introduction), $N$ and $M$ denote dual lattices (that is, finitely generated, free Abelian groups) with a perfect pairing $\langle\,,\,\rangle : N \times M \to \mathbb{Z}$, and $N_{\mathbb{R}}$, $M_{\mathbb{R}}$ the corresponding $\mathbb{R}$-vector spaces obtained by extension of scalars. Let $\sigma \subset N_{\mathbb{R}}$ be the polyhedral cone with vertex at 0 spanned by fundamental generators $a^1, \ldots, a^M \in N$. We assume that the $a^i$ are primitive, that is, not proper multiples of other elements of $N$. We write $\sigma = \langle a^1, \ldots, a^M \rangle$.

The dual cone $\sigma^{\vee} := \{r \in M_{\mathbb{R}} \mid \langle \sigma, r \rangle \geq 0\}$ is given by the inequalities corresponding to $a^1, \ldots, a^M$. Intersecting $\sigma^{\vee}$ with the lattice $M$ yields a finitely generated semigroup $\sigma^{\vee} \cap M$. We denote by $E \subset \sigma^{\vee} \cap M$ its minimal set of generators, and call it the *Hilbert basis*. Then the affine toric variety $Y_{\sigma} := \mathrm{Spec}\,\mathbb{C}[\sigma^{\vee} \cap M] \subset \mathbb{C}^E$ has defining equations corresponding to the linear dependence relations among elements of $E$.

## 2.2

Most of the rings and modules relevant to the study of $Y_{\sigma}$ are $M$-graded (or multigraded); this applies to the modules $T_Y^i$, that describe infinitesimal deformations of $Y_{\sigma}$ and obstructions to extending them. Let $R \in M$. In [Al 1] and [Al 3], we defined the subsets

$$E_j^R := \left\{ r \in E \mid \langle a^j, r \rangle < \langle a^j, R \rangle \right\} \subset E$$

for $j = 1, \ldots, M$. These provide the main tool for building a complex $\mathrm{Span}(E^R)_{\bullet}$ of free Abelian groups, with

$$\mathrm{Span}(E^R)_{-k} := \bigoplus_{\substack{\tau \text{ a face of } \sigma \\ \text{with } \dim \tau = k}} \mathrm{Span}(E_{\tau}^R),$$

where

$$E_{\tau}^R := \bigcap_{a^j \in \tau} E_j^R \quad \text{for faces } \tau < \sigma, \quad \text{but} \quad E_0^R := \bigcup_{j=1}^{N} E_j^R,$$

and where the differentials are the obvious maps.

**Theorem 2.1 (cf. [Al 1], [Al 3])** *The homogeneous piece of $T_Y^1$ in degree $-R$ is given by*

$$T_Y^1(-R) = H^1\left(\mathrm{Span}(E^R)_{\bullet}^* \otimes_{\mathbb{Z}} \mathbb{C}\right).$$

If we assume in addition that $Y_\sigma$ is smooth in codimension two, the same thing holds for $T_Y^2$:

$$T_Y^2(-R) = H^2\Big(\text{Span}(E^R)_\bullet^* \otimes_\mathbb{Z} \mathbb{C}\Big).$$

In particular, to calculate $T_Y^1(-R)$, we need the vector spaces $\text{Span}_\mathbb{C} E_j^R$ and $\text{Span}_\mathbb{C} E_{jk}^R$, where $a^j, a^k$ span a two dimensional face of $\sigma$. The first of these is easily determined:

$$\text{Span}_\mathbb{C} E_j^R = \begin{cases} 0 & \text{if } \langle a^j, R \rangle \leq 0, \\ (a^j)^\perp & \text{if } \langle a^j, R \rangle = 1, \\ M_\mathbb{C} & \text{if } \langle a^j, R \rangle \geq 2. \end{cases}$$

The second is always contained in $(\text{Span}_\mathbb{C} E_j^R) \cap (\text{Span}_\mathbb{C} E_k^R)$ as a subspace of codimension $\leq 2$. As the following example shows, its actual size reflects the infinitesimal deformations of the two dimensional cyclic quotient singularity corresponding to the plane cone spanned by $a^j, a^k$. (These singularities are exactly the transversal types of the singularities of $Y_\sigma$ in codimension two.)

**Example 2.2** Write $Y_{n,q}$ for the two dimensional quotient of $\mathbb{C}^2$ by the action of $\mathbb{Z}/n\mathbb{Z}$ via $\left(\begin{smallmatrix} \xi & 0 \\ 0 & \xi^q \end{smallmatrix}\right)$, where $\xi$ is a primitive $n$th root of unity and $q$ is coprime to $n$. Then $Y_{n,q}$ is the toric variety given by the cone $\sigma = \langle (1,0); (-q,n) \rangle \subset \mathbb{R}^2$. The set $E \subset \sigma^\vee \cap \mathbb{Z}^2$ consists of the lattice points $r^0, \ldots, r^w$ along the compact faces of the boundary of the convex hull $\text{conv}\left((\sigma^\vee \setminus \{0\}) \cap \mathbb{Z}^2\right)$. There are integers $a_v \geq 2$ such that $r^{v-1} + r^{v+1} = a_v r^v$ for $v = 1, \ldots, w-1$. These may be obtained by expanding $n/(n-q)$ as a negative continued fraction (cf. [Oda], §1.6).

Assume $w \geq 2$ and let $a^1 = (1,0)$ and $a^2 = (-q,n)$. Then only two sets $E_1^R$ and $E_2^R$ are involved, and the previous theorem states that

$$T_Y^1(-R) = \left(\frac{\text{Span}_\mathbb{C} E_1^R \cap \text{Span}_\mathbb{C} E_2^R}{\text{Span}_\mathbb{C}(E_1^R \cap E_2^R)}\right)^*.$$

The only multidegrees $R \in \mathbb{Z}^2$ contributing to $T_Y^1$ are as follows:

(i) $R = r^1$ (and the case $R = r^{w-1}$ is similar): then $\text{Span}_\mathbb{C} E_1^R = (a^1)^\perp$, $\text{Span}_\mathbb{C} E_2^R = \mathbb{C}^2$ (or $(a^2)^\perp$ if $w = 2$), and $\text{Span}_\mathbb{C} E_{12}^R = 0$. Hence, $\dim T^1(-R) = 1$ (or 0 if $w = 2$).

(ii) $R = r^v$ with $2 \leq v \leq w-2$: $\text{Span}_\mathbb{C} E_1^R = \text{Span}_\mathbb{C} E_2^R = \mathbb{C}^2$, and $\text{Span}_\mathbb{C} E_{12}^R = 0$. Thus $\dim T^1(-R) = 2$.

(iii) $R = p \cdot r^v$ with $1 \leq v \leq w-1$, and $2 \leq p < a_v$ for $w \geq 3$; (or $v = 1 = w-1$ with $2 \leq p \leq a_1$ for $w = 2$): then $\text{Span}_\mathbb{C} E_1^R = \text{Span}_\mathbb{C} E_2^R = \mathbb{C}^2$, and $\text{Span}_\mathbb{C} E_{12}^R = \mathbb{C} \cdot R$. In particular, $\dim T^1(-R) = 1$.

## 2.3   Minkowski summands

**Definition 2.3** We define the *Minkowski sum* of two polyhedra $Q', Q'' \subset \mathbb{R}^n$ to be the polyhedron $Q' + Q'' := \{p' + p'' \mid p' \in Q', p'' \in Q''\}$. Obviously, this notion also makes sense for translation classes of polyhedra in arbitrary affine spaces.

Every polyhedron $Q$ decomposes as the Minkowski sum $Q = Q^c + Q^\infty$ of a (compact) polytope $Q^c$ and the *cone of unbounded directions* $Q^\infty$. The latter is uniquely determined by $Q$, whereas the compact summand is not. However, we can take $Q^c$ to be the minimal one – given as the convex hull of the vertices of $Q$ itself. If $Q$ was already compact, then $Q^c = Q$ and $Q^\infty = 0$.

*A polyhedron $Q'$ is called a* Minkowski summand *of $Q$ if there is a $Q''$ such that $Q = Q' + Q''$ and if, additionally, $(Q')^\infty = Q^\infty$.*

In particular, a Minkowski summand always has the same cone of unbounded directions as the original polyhedron, whereas its compact edges are a dilation of those of the original polyhedron (the dilation factor 0 is allowed).

## 2.4   Setup

Consider the cone $\sigma \subset N_{\mathbb{R}}$ and fix some element $R \in M$. Then

$$\mathbb{A}(R) := [R = 1] = \{a \in N_{\mathbb{R}} \mid \langle a, R \rangle = 1\} \subset N_{\mathbb{R}}$$

is an affine space; provided that $R$ is primitive, it comes with a lattice $\mathbb{L}(R) := [R = 1] \cap N$. The corresponding vector space is $\mathbb{A}_0(R) := [R = 0]$; this always has the lattice $\mathbb{L}_0(R) := [R = 0] \cap N$. We define the *cross-section* of $\sigma$ in degree $R$ to be the polyhedron

$$Q(R) := \sigma \cap [R = 1] \subset \mathbb{A}(R)$$

(here $Q$ stands for *Querschnitt*). This has cone of unbounded directions $Q(R)^\infty = \sigma \cap \mathbb{A}_0(R) \subset N_{\mathbb{R}}$. The compact part $Q(R)^c$ is generated by its vertices $\bar{a}^j := a^j / \langle a^j, R \rangle$ for $j$ satisfying $\langle a^j, R \rangle \geq 1$. A trivial but nevertheless important observation is the following: the vertex $\bar{a}^j$ is a lattice point (that is, $\bar{a}^j \in \mathbb{L}(R)$), if and only if $\langle a^j, R \rangle = 1$.

Fundamental generators of $\sigma$ contained in $R^\perp$ can still be "seen" as edges in $Q(R)^\infty$, but those with $\langle \bullet, R \rangle < 0$ are "invisible" in $Q(R)$. In particular, we can recover the cone $\sigma$ from $Q(R)$ if and only if $R \in \sigma^\vee$.

## 2.5

Write $d^1, \ldots, d^N \in R^\perp \subset N_\mathbb{R}$ for the compact edges of $Q(R)$. As in [Al 4], §2, for each compact 2-face $\varepsilon < Q(R)$ we define its *sign vector* $\underline{\varepsilon} \in \{0, \pm 1\}^N$ to be

$$\varepsilon_i := \begin{cases} \pm 1 & \text{if } d^i \text{ is an edge of } \varepsilon, \\ 0 & \text{otherwise,} \end{cases}$$

where the signs are chosen so that the oriented edges $\varepsilon_i \cdot d^i$ fit into a cycle along the boundary of $\varepsilon$. This determines $\underline{\varepsilon}$ up to sign (and either choice will do). In particular, $\sum_i \varepsilon_i d^i = 0$.

**Definition 2.4** For each $R \in M$ we define the vector spaces

$$V(R) := \left\{ (t_1, \ldots, t_N) \,\middle|\, \sum_i t_i \varepsilon_i d^i = 0 \text{ for every compact 2-face } \varepsilon < Q(R) \right\}$$

$$W(R) := \mathbb{R}^{\#\{\text{vertices of } Q(R) \text{ not in } N\}}.$$

The cone $C(R) := V(R) \cap \mathbb{R}^N_{\geq 0}$ measures the dilation of each compact edge, and therefore parametrizes exactly the Minkowski summands of positive multiples of $Q(R)$. For this reason, we call elements of $V(R)$ *generalized Minkowski summands*; they may have edges of negative length. (See [Al 4], Lemma 2.2 for a discussion of the compact case.) The vector space $W(R)$ provides coordinates $s_j$ for each vertex $\bar{a}^j \in Q(R) \setminus N$, that is, $\langle a^j, R \rangle \geq 2$.

## 2.6

Each compact edge $d^{jk} = \overline{\bar{a}^j \bar{a}^k}$ gives rise to a set of equations $G_{jk}$ relating elements $(\underline{t}, \underline{s}) \in V(R) \oplus W(R)$. These[1] sets are of one of the following three types:

(0) $G_{jk} = \emptyset$;

(1) $G_{jk} = \{s_j - s_k = 0\}$ whenever both coordinates exist in $W(R)$;

(2) $G_{jk} = \{t_{jk} - s_j = 0, t_{jk} - s_k = 0\}$, omitting equations that do not make sense.

---

[1] The edge corresponds to a codimension 2 singular locus of $Y_\sigma$, whose transversal type is the cyclic quotient singularity $\frac{1}{n}(1, q)$ described in Example 2.2. As we will see at the end of 2.7, the choice of $G_{jk}$ is essentially governed by the position of the monomial relative to the classification of Example 2.2 into cases (i), (ii), (iii).

Restricting $V(R) \oplus W(R)$ to the (at most) three coordinates $t_{jk}$, $s_j$, $s_k$, the actual choice of $G_{jk}$ is made such that these equations yield a subspace of dimension $1 + \dim T^1_{\langle a^j, a^k \rangle}(-R)$. Notice that the dimension of $T^1(-R)$ for the two dimensional quotient singularity assigned to the plane cone $\langle a^j, a^k \rangle$ can be obtained from Example 2.2.

**Theorem 2.5** *The infinitesimal deformations of $Y_\sigma$ in degree $-R$ equal*

$$T^1_Y(-R) = \left\{ (\underline{t}, \underline{s}) \in V_{\mathbb{C}}(R) \oplus W_{\mathbb{C}}(R) \,\middle|\, \begin{matrix} (\underline{t}, \underline{s}) \text{ satisfies the} \\ \text{equations } G_{jk} \text{ of } 2.6 \end{matrix} \right\} / \mathbb{C} \cdot (\underline{1}, \underline{1}).$$

The vector space $V(R)$ (encoding Minkowski summands) is, in a sense, the main tool to describe infinitesimal deformations. Depending on which of the above Types (0)–(2) the $G_{jk}$ belong to, the elements of $W(R)$ either provide additional parameters, or introduce conditions that exclude Minkowski summands not of some prescribed type.

If $Y$ is smooth in codimension two, then $G_{jk}$ is always of Type (2). In particular, the variables $\underline{s}$ are completely determined by the $\underline{t}$, and we obtain:

**Corollary 2.6** *If $Y$ is smooth in codimension two, then $T^1_Y(-R)$ is contained in $V_{\mathbb{C}}(R)/\mathbb{C} \cdot (\underline{1})$. It is made up of those $\underline{t}$ such that $t_{jk} = t_{kl}$ whenever $d^{jk}$, $d^{kl}$ are compact edges with a common non-lattice vertex $\bar{a}^k$ of $Q(R)$. Thus $T^1_Y(-R)$ equals the set of equivalence classes of those Minkowski summands of $\mathbb{R}_{\geq 0} \cdot Q(R)$ that preserve the stars of non-lattice vertices of $Q(R)$ up to homothety.*

## 2.7

**Proof of Theorem 2.5, Step 1**   From Theorem 2.1, we know that $T^1_Y(-R)$ equals the complexified cohomology of the complex

$$N_{\mathbb{R}} \to \bigoplus_j \left( \mathrm{Span}_{\mathbb{R}} \, E^R_j \right)^* \to \bigoplus_{\langle a^j, a^k \rangle < \sigma} \left( \mathrm{Span}_{\mathbb{R}} \, E^R_{jk} \right)^*.$$

According to 2.2, an element of $\bigoplus_j \left( \mathrm{Span}_{\mathbb{R}} \, E^R_j \right)^*$ can be represented by a family of elements

$$\begin{cases} b^j \in N_{\mathbb{R}} & \text{if } \langle a^j, R \rangle \geq 2, \\ b^j \in N_{\mathbb{R}} / \mathbb{R} \cdot a^j & \text{if } \langle a^j, R \rangle = 1. \end{cases}$$

Dividing by the image of $N_{\mathbb{R}}$ means shifting this family by a common vector $b \in N_{\mathbb{R}}$. On the other hand, the family $\{b^j\}$ must map to 0 in the complex;

that is, for each compact edge $\overline{a^j, a^k} < Q$ the functions $b^j$ and $b^k$ must be equal on $\mathrm{Span}_{\mathbb{R}} E_{jk}^R$. Since

$$(a^j, a^k)^\perp \subset \mathrm{Span}_{\mathbb{R}} E_{jk}^R \subset (\mathrm{Span}_{\mathbb{R}} E_j^R) \cap (\mathrm{Span}_{\mathbb{R}} E_k^R),$$

we immediately obtain the necessary condition $b^j - b^k \in \mathbb{R}a^j + \mathbb{R}a^k$. However, the actual behavior of $\mathrm{Span}_{\mathbb{R}} E_{jk}^R$ will require a closer look in Step 3 below.

**Step 2**   We introduce new "coordinates":

$\overline{b}^j := b^j - \langle b^j, R \rangle\, \overline{a}^j \in R^\perp$, which is defined even in the case $\langle a^j, R \rangle = 1$;

$s_j := - \langle b^j, R \rangle$ for $j$ meeting $\langle a^j, R \rangle \geq 2$ (inducing an element of $W(R)$).

The shift of the $b^j$ by an element $b \in N_{\mathbb{R}}$ (that is, $(b^j)' = b^j + b$) appears in these new coordinates as

$$
\begin{aligned}
(\overline{b}^j)' &= (b^j)' - \langle (b^j)', R \rangle\, \overline{a}^j = b^j + b - \langle b^j, R \rangle\, \overline{a}^j - \langle b, R \rangle\, \overline{a}^j \\
&= \overline{b}^j + b - \langle b, R \rangle\, \overline{a}^j, \\
s_j' &= - \langle (b^j)', R \rangle = s_j - \langle b, R \rangle.
\end{aligned}
$$

In particular, an element $b \in R^\perp$ does not change the $s_j$, but shifts the points $\overline{b}^j$ inside the hyperplane $R^\perp$. Hence, the set of the $\overline{b}^j$ should be considered modulo translation inside $R^\perp$ only. On the other hand, the condition $b^j - b^k \in \mathbb{R}a^j + \mathbb{R}a^k$ changes into $\overline{b}^j - \overline{b}^k \in \mathbb{R}\overline{a}^j + \mathbb{R}\overline{a}^k$ or even $\overline{b}^j - \overline{b}^k \in \mathbb{R}(\overline{a}^j - \overline{a}^k)$ (consider the values of $R$). Hence, the $\overline{b}^j$ form the vertices of a Minkowski summand of $Q(R)$, or at least a generalized Minkowski summand. Modulo translation, this summand is completely described by the dilation factors $t_{jk}$ obtained from

$$\overline{b}^j - \overline{b}^k = t_{jk} \cdot (\overline{a}^j - \overline{a}^k).$$

Now, the remaining part of the action of $b \in N_{\mathbb{R}}$ comes down to an action of $\langle b, R \rangle \in \mathbb{R}$ only:

$$
\begin{aligned}
t_{jk}' &= t_{jk} - \langle b, R \rangle, \quad \text{and} \\
s_j' &= s_j - \langle b, R \rangle, \quad \text{as we already know.}
\end{aligned}
$$

Up to now, we have found that $T_Y^1(-R) \subset V_{\mathbb{C}}(R) \oplus W_{\mathbb{C}}(R)/(\underline{1}, \underline{1})$.

**Step 3**   Actually, the elements $b^j$ and $b^k$ must coincide on $\mathrm{Span}_{\mathbb{R}}\, E_{jk}^R$, which may be larger than just $(a^j, a^k)^\perp$. To measure the difference, consider the quotient space $\mathrm{Span}_{\mathbb{R}}\, E_{jk}^R / (a^j, a^k)^\perp$ contained in the two dimensional vector space $M_{\mathbb{R}}/(a^j, a^k)^\perp = \mathrm{Span}_{\mathbb{R}}(a^j, a^k)^*$. Since this quotient space coincides with the set $\mathrm{Span}_{\mathbb{R}}\, E_{jk}^{\overline{R}}$ corresponding to the two dimensional cone $\langle a^j, a^k \rangle \subset \mathrm{Span}_{\mathbb{R}}(a^j, a^k)$, where $\overline{R}$ denotes the image of $R$ in $\mathrm{Span}_{\mathbb{R}}(a^j, a^k)^*$, we may assume that $\sigma = \langle a^1, a^2 \rangle$ (that is, $j = 1$, $k = 2$) represents a two dimensional cyclic quotient singularity. In particular, we only need to discuss the three cases (i)–(iii) of Example 2.2:

In (i) and (ii) we have $\mathrm{Span}_{\mathbb{R}}\, E_{12}^R = 0$, that is, no additional equation is needed. This means $G_{12} = \emptyset$ is of Type (0) (see 2.6). On the other hand, if $T_Y^1 = 0$, then the vector space $\mathbb{R}^3_{(t_{12}, s_1, s_2)}/\mathbb{R} \cdot (\underline{1})$ has to be killed by identifying the three variables $t_{12}$, $s_1$ and $s_2$; we obtain Type (2).

Case (iii) provides $\mathrm{Span}_{\mathbb{R}}\, E_{12}^R = \mathbb{R} \cdot R$. Hence, as an additional condition we obtain that $b^1$ and $b^2$ have to be equal on $R$. By definition of $s_j$, this means that $s_1 = s_2$, and $G_{12}$ must be of Type (1).   $\square$

# 3   Genuine deformations

## 3.1

In [Al 2] we studied the so-called *toric deformations* in a given multidegree $-R \in M$. These are genuine deformations, in the sense that they are defined over smooth parameter spaces; they are characterized by the fact that the total space together with the embedding of the special fiber still belongs to the toric category. Despite the fact they look so special, it seems that toric deformations cover a big part of the versal deformation of $Y_\sigma$. They only exist in negative degrees (that is, $R \in \sigma^\vee \cap M$), but here they form a kind of skeleton. If $Y_\sigma$ is an isolated toric Gorenstein singularity, then toric deformations even provide all irreducible components of the versal deformation (cf. [Al 4]).

After briefly recalling the idea of this construction, we show how the new formula for $T_Y^1$ of Theorem 2.1 can be used to describe the Kodaira–Spencer map of toric deformations. We follow this by the study of nonnegative degrees: if $R \notin \sigma^\vee \cap M$, then we are still able to construct genuine deformations of $Y_\sigma$; however, these are no longer toric.

## 3.2

Let $R \in \sigma^\vee \cap M$. Then, as in [Al 2], §3, an $m$-parameter toric deformation of $Y_\sigma$ in degree $-R$ corresponds to a splitting of $Q(R)$ as a Minkowski sum

$$Q(R) = Q_0 + Q_1 + \cdots + Q_m$$

satisfying the following conditions:

(i) $Q_0 \subset \mathbb{A}(R)$ and $Q_1, \ldots, Q_m \in \mathbb{A}_0(R)$ are polyhedra with common cone of unbounded directions $Q(R)^\infty$.

(ii) Each supporting hyperplane $t$ of $Q(R)$ defines a face $F(Q_i, t)$ of each of the polyhedra $Q_0, Q_1, \ldots, Q_m$, and the Minkowski sum of these faces equals $F(Q(R), t)$. With at most one exception (which can depend on $t$), these faces contain lattice vertices (points of $N$).

**Remark 3.1** In [Al 2] we distinguished between the case of primitive and nonprimitive elements $R \in M$: if $R$ is a multiple of some element of $M$, then $\mathbb{A}(R)$ does not contain lattice points at all. In particular, condition (ii) just means that $Q_1, \ldots, Q_m$ must be lattice polyhedra.

On the other hand, for primitive $R$, the $(m + 1)$ summands $Q_i$ appear on an equal footing and may be put into the same space $\mathbb{A}(R)$. Their Minkowski sum must then be interpreted inside this affine space.

**Given a Minkowski decomposition, how to obtain the corresponding toric deformation?**

Defining $\widetilde{N} := N \oplus \mathbb{Z}^m$ (and $\widetilde{M} := M \oplus \mathbb{Z}^m$), we have to embed the summands as $(Q_0, 0), (Q_1, e^1), \ldots, (Q_m, e^m)$ into the vector space $\widetilde{N}_\mathbb{R}$; here $\{e^1, \ldots, e^m\}$ denotes the standard basis of $\mathbb{Z}^m$. Together with $(Q(R)^\infty, 0)$, these polyhedra generate a cone $\widetilde{\sigma} \subset \widetilde{N}$ containing $\sigma$ via the inclusion

$$N \hookrightarrow \widetilde{N} \quad \text{defined by} \quad a \mapsto \left(a; \langle a, R \rangle, \ldots, \langle a, R \rangle\right).$$

Actually, $\sigma$ equals $\widetilde{\sigma} \cap N_\mathbb{R}$, and we obtain an inclusion $Y_\sigma \hookrightarrow X_{\widetilde{\sigma}}$ between the associated toric varieties.

On the other hand, $[R, 0] \colon \widetilde{N} \to \mathbb{Z}$ and $\mathrm{pr}_{\mathbb{Z}^m} \colon \widetilde{N} \to \mathbb{Z}^m$ induce regular functions $f \colon X_{\widetilde{\sigma}} \to \mathbb{C}$ and $(f^1, \ldots, f^m) \colon X_{\widetilde{\sigma}} \to \mathbb{C}^m$, respectively. The resulting map $(f^1 - f, \ldots, f^m - f) \colon X_{\widetilde{\sigma}} \to \mathbb{C}^m$ is flat and has $Y_\sigma \hookrightarrow X_{\widetilde{\sigma}}$ as special fiber.

## 3.3

Let $R \in \sigma^\vee \cap M$ and $Q(R) = Q_0 + \cdots + Q_m$ be a decomposition satisfying conditions (i) and (ii) of 3.2. Denote by $(\overline{a}^j)_i$ the vertex of $Q_i$ induced from $\overline{a}^j \in Q(R)$, so that $\overline{a}^j = (\overline{a}^j)_0 + \cdots + (\overline{a}^j)_m$.

**Theorem 3.2** *The Kodaira-Spencer map of the corresponding toric deformation $X_{\widetilde{\sigma}} \to \mathbb{C}^m$ is the map*

$$\varrho \colon \mathbb{C}^m = T_{\mathbb{C}^m, 0} \longrightarrow T_Y^1(-R) \subset V_\mathbb{C}(R) \oplus W_\mathbb{C}(R) / \mathbb{C} \cdot (\underline{1}, \underline{1})$$

*sending $e^i$ to the pair $[Q_i, \underline{s}^i] \in V(R) \oplus W(R)$ (for $i = 1, \dots, m$) with*

$$s^i_j := \begin{cases} 0 & \text{if the vertex } (\overline{a}^j)_i \text{ of } Q_i \text{ belongs to the lattice } N, \\ 1 & \text{if } (\overline{a}^j)_i \text{ is not a lattice point.} \end{cases}$$

**Remark 3.3** Setting $e^0 := -(e^1 + \dots + e^m)$, we obtain $\varrho(e^0) = [Q_0, \underline{s}^0]$ where $\underline{s}^0$ is defined in the same way as $\underline{s}^i$ above.

## 3.4   Proof of Theorem 3.2

We to derive the formula for the Kodaira–Spencer map from the more technical result of [Al 2], Theorem 5.3. If we use in addition [Al 3], Theorem 6.1, this theorem describes $\varrho(e^i) \in T^1_Y(-R) = H^1\big(\mathrm{Span}_{\mathbb{C}}(E^R)^*_\bullet\big)$ as follows:

Let $E = \{r^0, \dots, r^w\} \subset \sigma^{\vee} \cap M$. Its elements may be lifted via $\widetilde{M} \twoheadrightarrow M$ to $\tilde{r}^v \in \tilde{\sigma}^{\vee} \cap \widetilde{M}$ (for $v = 0, \dots, w$); denote their $i$th entry of the $\mathbb{Z}^m$-part by $\hat{\eta}^v_i$. Then, given elements $v^j \in \mathrm{Span}\, E^R_j$, we may represent them as $v^j = \sum_v q^j_v r^v$ with $q^j \in \mathbb{Z}^{E^R_j}$, and $\varrho(e^i)$ sends $v^j$ to the integer $-\sum_v q^j_v \eta^v_i$. Using the notation of 2.7 for $\varrho(e^i)$, this means that $b^j$ sends elements $r^v \in E^R_j$ onto $-\eta^v_i \in \mathbb{Z}$.

By construction of $\tilde{\sigma}$, we have the inequalities

$$\big\langle ((\overline{a}^j)_0, 0), \tilde{r}^v \big\rangle \geq 0 \quad \text{and} \quad \big\langle ((\overline{a}^j)_i, e^i), \tilde{r}^v \big\rangle \geq 0 \quad \text{for } i = 1, \dots, m,$$

which sum up to $\langle \overline{a}^j, r^v \rangle = \big\langle (\overline{a}^j, \underline{1}), \tilde{r}^v \big\rangle \geq 0$. On the other hand, $r^v \in E^R_j$ is equivalent to $\langle \overline{a}^j, r^v \rangle < 1$. Hence, for $i = 0, \dots, m$, whenever $(\overline{a}^j)_i \in Q_i$ belongs to the lattice, the corresponding inequality becomes an equality. With at most one exception, this must always happen. Hence for $i = 1, \dots, m$,

$$\langle (\overline{a}^j)_i, r^v \rangle + \eta^v_i = \begin{cases} 0 & \text{if } (\overline{a}^j)_i \in N, \\ \langle \overline{a}^j, r^v \rangle & \text{if } (\overline{a}^j)_i \notin N, \end{cases}$$

meaning that $b^j = (\overline{a}^j)_i$ or $b^j = (\overline{a}^j)_i - \overline{a}^j$ respectively. By definition of $\overline{b}^j$ and $s_j$ given in 2.7, we are done.   $\square$

## 3.5

We now treat the case of nonnegative degrees; let $R \in M \setminus \sigma^{\vee}$. It often happens that the easiest way to solve a problem is to change the question until there is no problem left. We can do this here by changing our cone $\sigma$ into some $\tau^R$ such that the degree $-R$ becomes negative. We define

$$\tau := \tau^R := \sigma \cap [R \geq 0], \quad \text{that is,} \quad \tau^{\vee} = \sigma^{\vee} + \mathbb{R}_{\geq 0} \cdot R.$$

The cone $\tau$ defines an affine toric variety $Y_\tau$. Since $\tau \subset \sigma$, it comes with a map $g \colon Y_\tau \to Y_\sigma$; in other words, $Y_\tau$ is an open part of a modification of $Y_\sigma$. The important observation is

$$\tau \cap [R=0] \;=\; \sigma \cap [R=0] \;=\; Q(R)^\infty, \quad \text{and}$$
$$\tau \cap [R=1] \;=\; \sigma \cap [R=1] \;=\; Q(R),$$

implying $T^1_{Y_\tau}(-R) = T^1_{Y_\sigma}(-R)$ by Theorem 2.6. Moreover, even the genuine toric deformations $X_{\tilde\tau} \to \mathbb{C}^m$ of $Y_\tau$ carry over to $m$-parameter (nontoric) deformations $X \to \mathbb{C}^m$ of $Y_\sigma$:

**Theorem 3.4** *Each Minkowski decomposition $Q(R) = Q_0 + Q_1 + \cdots + Q_m$ satisfying (i) and (ii) of 3.2 provides an $m$-parameter deformation $X \to \mathbb{C}^m$ of $Y_\sigma$. Via some birational map $\tilde g \colon X_{\tilde\tau} \to X$, it is compatible with the toric deformation $X_{\tilde\tau} \to \mathbb{C}^m$ of $Y_\tau$ presented in 3.2.*

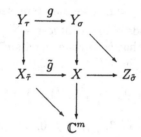

The total space $X$ is no longer toric, but it sits via birational maps between $X_{\tilde\tau}$ and some affine toric variety $Z_{\tilde\sigma}$ that still contains $Y_\sigma$ as a closed subset.

## 3.6   Proof

We first construct $\tilde N$, $\tilde M$ and $\tilde\tau \subset \tilde N_\mathbb{R}$ by the recipe of 3.2. In particular, $N \subset \tilde N$, and the projection $\pi \colon \tilde M \to M$ sends $[r; g_1, \ldots, g_m]$ onto $r + (\sum_i g_i)R$. Defining $\tilde\sigma := \tilde\tau + \sigma$ (hence $\tilde\sigma^\vee = \tilde\tau^\vee \cap \pi^{-1}(\sigma^\vee)$), we obtain the commutative diagram

$$
\begin{array}{ccc}
\mathbb{C}[\tilde\tau^\vee \cap \tilde M] & \longleftarrow & \mathbb{C}[\tilde\sigma^\vee \cap \tilde M] \\
\pi \downarrow & & \downarrow \pi \\
\mathbb{C}[\tau^\vee \cap M] & \longleftarrow & \mathbb{C}[\sigma^\vee \cap M]
\end{array}
$$

with surjective vertical maps. The canonical elements $e_1, \ldots, e_m \in \mathbb{Z}^m \subset \tilde M$ together with $[R; 0] \in \tilde M$ are preimages of $R \in M$. Hence, the corresponding monomials $x^{e_1}, \ldots, x^{e_m}, x^{[R,0]}$ in the semigroup algebra $\mathbb{C}[\tilde\tau^\vee \cap \tilde M]$ (these were called $f^1, \ldots, f^m, f$ in 3.2) map onto $x^R \in \mathbb{C}[\tau^\vee \cap M]$, which is not regular on $Y_\sigma$. We define $Z_{\tilde{\sigma}}$ as the affine toric variety corresponding to $\tilde{\sigma}$, and $X$ as

$$X := \operatorname{Spec} B, \quad \text{where} \quad B := \mathbb{C}[\tilde{\sigma}^\vee \cap \widetilde{M}][f^1 - f, \ldots, f^m - f] \subset \mathbb{C}[\tilde{\tau}^\vee \cap \widetilde{M}].$$

This means that $X$ is obtained from $X_{\tilde{\tau}}$ by eliminating all the variables except those lifted from $Y_\sigma$ and the deformation parameters themselves. By construction of $B$, the vertical algebra homomorphisms $\pi$ induce a surjection $B \twoheadrightarrow \mathbb{C}[\sigma^\vee \cap M]$.

**Lemma 3.5** *Any element of $\mathbb{C}[\tilde{\tau}^\vee \cap \widetilde{M}]$ can be written in a unique way as a sum*

$$\sum_{(v_1, \ldots, v_m) \in \mathbb{N}^m} c_{v_1, \ldots, v_m} \cdot (f^1 - f)^{v_1} \cdots (f^m - f)^{v_m},$$

*with $c_{v_1, \ldots, v_m} \in \mathbb{C}[\tilde{\tau}^\vee \cap \widetilde{M}]$ such that $s - e_i \notin \tilde{\tau}^\vee$ (for $i = 1, \ldots, m$) for any of its monomial terms $x^s$. Moreover, these sums belong to the subalgebra $B$ if and only if their coefficients $c_{v_1, \ldots, v_m}$ do so.*

**Proof** *(a) Existence:* Let $s - e_i \in \tilde{\tau}^\vee$ for some $s, i$. Then, with $s' := s - e_i + [R, 0]$ we obtain

$$x^s = x^{s'} + x^{s-e_i}(x^{e_i} - x^{[R,0]}) = x^{s'} + x^{s-e_i}(f^i - f).$$

Since $e_i = 1$ and $[R, 0] = 0$ if evaluated on $(Q_i, e^i) \subset \tilde{\tau}$, this process eventually stops.

*(b) Membership of $B$:* For the previous reduction step, we have to show that if $s \in \mathbb{C}[\tilde{\sigma}^\vee \cap \widetilde{M}]$, then the same holds for $s'$ and $s - e_i$. Since $\pi(s') = \pi(s) \in \sigma^\vee$, this is clear for $s'$. It remains to check that $\pi(s - e_i) \in \sigma^\vee$. Let $a \in \sigma$ be an arbitrary test element; we distinguish two cases:

Case 1: $\langle a, R \rangle \geq 0$. Then $a$ belongs to the subcone $\tau$, and $\pi(s - e_i) \in \tau^\vee$ yields $\langle a, \pi(s - e_i) \rangle \geq 0$.

Case 2: $\langle a, R \rangle \leq 0$. This implies

$$\langle a, \pi(s - e_i) \rangle = \langle a, s \rangle - \langle a, R \rangle \geq \langle a, s \rangle \geq 0.$$

*(c) Uniqueness:* Let $p := \sum c_{v_1, \ldots, v_m} \cdot (f^1 - f)^{v_1} \cdots (f^m - f)^{v_m}$ (satisfying the above conditions) be equal to 0 in $\mathbb{C}[\tilde{\tau}^\vee \cap \widetilde{M}]$. Using the projection $\pi \colon \widetilde{M} \to M$ makes everything $M$-graded. Since the factors $(f^i - f)$ are homogeneous (of degree $R$), we may assume that the same holds for $p$, hence also for its coefficients $c_{v_1, \ldots, v_m}$.

**Claim 3.6** *These coefficients are just monomials.*

Indeed, if $s, s' \in \tilde{\tau}^\vee$ had the same image under $\pi$, we could assume that some $e_i$-coordinate of $s'$ is smaller than that of $s$. Hence, $s - e_i$ would still be equal to $s$ on $(Q_0, 0)$ and on any $(Q_j, e^j)$ (for $j \neq i$), but even greater than or equal to $s'$ on $(Q_i, e^i)$. This would imply $s - e_i \in \tilde{\tau}^\vee$, contradicting our assumption for $p$.

Say $c_{v_1,\dots,v_m} = \lambda_{v_1,\dots,v_m} x^\bullet$; we use the projection $\widetilde{M} \to \mathbb{Z}^m$ to take $p$ into the ring $\mathbb{C}[\mathbb{Z}^m] = \mathbb{C}[y_1^{\pm 1}, \dots, y_m^{\pm 1}]$. The elements $x^\bullet$, $f^i$, $f$ map onto $y^\bullet$, $y_i$, and 1, respectively. Hence, $p$ turns into

$$\bar{p} = \sum_{(v_1,\dots,v_m) \in \mathbb{N}^m} \lambda_{v_1,\dots,v_m} \cdot y^\bullet \cdot (y_1 - 1)^{v_1} \cdots (y_m - 1)^{v_m}.$$

By induction through $\mathbb{N}^m$, we obtain that vanishing of $\bar{p}$ implies the vanishing of its coefficients: replace $y_i - 1$ by $z_i$, and take partial derivatives. This proves the claim.

Now, we can easily see that $X \to \mathbb{C}^m$ is flat and has $Y_\sigma$ as special fiber: Lemma 3.5 means that for $k = 0, \dots, m$, we have inclusions

$$B/(f^1 - f, \dots, f^k - f) \hookrightarrow \mathbb{C}[\tilde{\tau}^\vee \cap \widetilde{M}]/(f^1 - f, \dots, f^k - f).$$

The values $k < m$ yield that $(f^1 - f, \dots, f^m - f)$ forms a regular sequence even in the subring $B$, meaning that $X \to \mathbb{C}^m$ is flat. With $k = m$ we obtain that the surjective map $B/(f^1 - f, \dots, f^m - f) \to \mathbb{C}[\sigma^\vee \cap M]$ is also injective. $\square$

# 4    Three dimensional toric Gorenstein singularities

## 4.1

By [Ish], Theorem 7.7, toric Gorenstein singularities always arise from the following construction: assume given a *lattice polytope* $P \subset \mathbb{R}^n$. We embed the whole space (containing $P$) into the "height one" slice of $N_\mathbb{R} := \mathbb{R}^n \oplus \mathbb{R}$, and consider the cone $\sigma$ generated by $P$; write $M_\mathbb{R} := (\mathbb{R}^n)^* \oplus \mathbb{R}$ for the dual space and $N, M$ for the natural lattices. Our polytope $P$ may be recovered from $\sigma$ as

$$P = Q(R^*) \subset \mathbb{A}(R^*) \quad \text{with } R^* := [\underline{0}, 1] \in M.$$

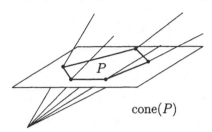

cone($P$)

The fundamental generators $a^1, \ldots, a^M \in \mathbb{L}(R^*)$ of $\sigma$ coincide with the vertices of $P$. (This involves a slight abuse of notation; we use the same symbol $a^j$ for both $a^j \in \mathbb{Z}^n$ and $(a^j, 1) \in M$.)

If $\overline{a^j a^k}$ is an edge of $P$, we denote by $\ell(j, k) \in \mathbb{Z}$ its "length" induced from the lattice $\mathbb{Z}^n \subset \mathbb{R}^n$. Every edge provides a codimension two singular stratum of $Y_\sigma$ with transversal type $A_{\ell(j,k)-1}$. In particular, $Y_\sigma$ is smooth in codimension two if and only if every edge of $P$ is primitive, that is, of length $\ell = 1$.

## 4.2

As usual, we fix some element $R \in M$. We know the vector spaces $V(R)$ and $W(R)$ from 2.5; we introduce the subspace

$$V'(R) := \{\underline{t} \in V(R) \mid t_{jk} \neq 0 \Rightarrow 1 \leq \langle a^j, R \rangle = \langle a^k, R \rangle \leq \ell(j, k)\} \subset V(R).$$

This represents the Minkowski summands of $Q(R)$ that contract to a point any compact edge *not* satisfying $\langle a^j, R \rangle = \langle a^k, R \rangle \leq \ell(j, k)$.

**Theorem 4.1** *For $T_Y^1(-R)$, we distinguish two different types of $R \in M$:*

  *(i) If $R \leq 1$ on $P$ (or equivalently $\langle a^j, R \rangle \leq 1$ for $j = 1, \ldots, M$), then $T_Y^1(-R) = V_{\mathbb{C}}(R)/(\underline{1})$. Moreover, concerning Minkowski summands, we may replace the polyhedron $Q(R)$ by its compact part $P \cap [R = 1]$ (a face of $P$).*

  *(ii) If $R$ does not satisfy the previous condition, then $T_Y^1(-R) = V'(R)$.*

**Proof** The first case follows from Theorem 2.6 simply because $W(R) = 0$. For (ii), we assume that there are vertices $a^j$ contained in the affine halfspace $[R \geq 2]$. These vertices can be connected to one another inside this halfspace via paths along edges of $P$.

The two dimensional cyclic quotient singularities corresponding to the edges $\overline{a^j a^k}$ of $P$ are themselves Gorenstein. In the language of Example 2.2, this means that $w = 2$, and we obtain

$$
\dim T^1_{\langle a^j, a^k \rangle}(-R) = \begin{cases} 1 & \text{if } \langle a^j, R \rangle = \langle a^k, R \rangle = 2, \ldots, \ell(j,k) \\ & \text{(case (iii) in 2.2)}, \\ 0 & \text{otherwise.} \end{cases}
$$

In particular, $T^1_{\langle a^j, a^k \rangle}(-R)$ cannot be two dimensional, and (in the notation of 2.6) the equations $s_j - s_k = 0$ belong to $G_{jk}$ whenever $\langle a^j, R \rangle, \langle a^k, R \rangle \geq 2$. This means that for elements of

$$
T^1_Y \subset \left( V_{\mathbb{C}}(R) \oplus W_{\mathbb{C}}(R) \right) / \mathbb{C} \cdot (\underline{1}, \underline{1}),
$$

all entries of the $W_{\mathbb{C}}(R)$-part must be equal, or even zero, after dividing by $\mathbb{C} \cdot (\underline{1}, \underline{1})$. Moreover, if not both $\langle a^j, R \rangle$ and $\langle a^k, R \rangle$ equal one, the vanishing of $T^1_{\langle a^j, a^k \rangle}(-R)$ implies that $G_{jk}$ also contains the equation $t_{jk} - s_\bullet = 0$.  $\square$

**Corollary 4.2** *For toric Gorenstein singularities, the condition 3.2, (ii) that allow us to build genuine deformations simplifies: $Q_1, \ldots, Q_m$ just have to be lattice polyhedra.*

**Proof** If $R \leq 1$ on $P$, then $Q(R)$ is itself a lattice polyhedron. Hence, condition (ii) automatically comes down to this simpler form.

In the second case, $T^1_Y(-R)$ involves some $W(R)$ part. On the one hand, via the Kodaira–Spencer map, it indicates which vertices of which polyhedron $Q_i$ belong to the lattice. On the other, we have observed in the previous proof that the entries of $W(R)$ are all equal. This exactly implies our claim.  $\square$

## 4.3

To treat three dimensional toric Gorenstein singularities, we now focus on *plane lattice polygons* $P \subset \mathbb{R}^2$. The vertices $a^1, \ldots, a^M$ are arranged in a cycle. For $j \in \mathbb{Z}/M\mathbb{Z}$, we write $d^j := a^{j+1} - a^j \in \mathbb{L}_0(R^*)$ for the edge from $a^j$ to $a^{j+1}$ (see 2.4), and $\ell(j) := \ell(j, j+1)$ for its length.

Let $s^1, \ldots, s^M$ be the fundamental generators of the dual cone $\sigma^\vee$, labelled so that $\sigma \cap (s^j)^\perp$ equals the face spanned by $a^j, a^{j+1} \in \sigma$. In particular, skipping the last coordinate of $s^j$ yields the (primitive) inner normal vector at the edge $d^j$ of $P$.

**Remark 4.3** For the convenience of those who prefer to live in $M$ rather than $N$, we show how to see the integers $\ell(j)$ in the dual world: choose a fundamental generator $s^j$ and denote by $r, r' \in M$ the elements of the Hilbert basis closest to $s^j$ along the two adjacent faces of $\sigma^{\vee}$, respectively (see Figure 1 in 4.7). Then, $\{R^*, s^j\}$ together with either $r$ or $r'$ form a basis of the lattice $M$, and $(r + r') - \ell(j)R^*$ is a multiple of $s^j$.

In the very special case of plane lattice polygons (that is, three dimensional toric Gorenstein singularities), we can describe $T_Y^1$ and the genuine deformations (for fixed $R \in M$) explicitly. First, we can easily spot the degrees carrying infinitesimal deformations:

**Theorem 4.4** *In general (with the exceptions (4–5)), $T_Y^1(-R)$ is only nontrivial in the following cases:*

(1) $R = R^*$ with $\dim T_Y^1(-R) = M - 3$;

(2) $R = qR^*$ (for $q \geq 2$) with $\dim T_Y^1(-R) = \max\{0, \#\{j \mid q \leq \ell(j)\} - 2\}$;

(3) $R = qR^* - ps^j$ with $2 \leq q \leq \ell(j)$ and $p \in \mathbb{Z}$ sufficiently large such that $R \notin \mathrm{int}(\sigma^{\vee})$. In this case, $T_Y^1(-R)$ is one dimensional.

*Additional degrees exist only in the following two (overlapping) exceptional cases:*

(4) $P$ contains a pair of parallel edges $d^j, d^k$, both longer than every other edge. Then $\dim T_Y^1(-qR^*) = 1$ for $q$ in the range

$$\max\{\ell(l) \mid l \neq j, k\} < q \leq \min\{\ell(j), \ell(k)\}.$$

(5) $P$ contains a pair of parallel edges $d^j, d^k$ with distance $d$ $(d := \langle a^j, s^k \rangle = \langle a^k, s^j \rangle)$. If $\ell(k) > d \geq \max\{\ell(l) \mid l \neq j, k\}$, then $\dim T_Y^1(-R) = 1$ for $R = qR^* + ps^j$ with $1 \leq q \leq \ell(j)$ and $1 \leq p \leq (\ell(k) - q)/d$.

The cases (1), (2), (4), and (5) yield at most finitely many (negative) degrees in $T_Y^1$. Type (3) consists of $\ell(j) - 1$ infinite series corresponding to any vertex $a^j \in P$; these series contain only nonnegative degrees, except possibly for the leading elements ($R$ might sit on $\partial \sigma^{\vee}$).

**Proof** These assertions are straightforward consequences of Theorem 4.1, so that the following brief remark should be sufficient: the condition $\langle a^j, R \rangle = \langle a^{j+1}, R \rangle$ means $d^j \in R^{\perp}$. Moreover, if $R \notin \mathbb{Z} \cdot R^*$, there is at most one edge (or a pair of parallel edges) having this property. $\square$

## 4.4

**Example 4.5** A typical example of a nonisolated, three dimensional toric Gorenstein singularity is the cone over the weighted projective space $\mathbb{P}(1,2,3)$. We use it to illustrate our calculations of $T^1$ as well as the forthcoming construction of genuine one parameter families. $P$ has the vertices $(-1,-1)$, $(2,-1)$, $(-1,1)$, so that $\sigma$ is the cone generated by

$$a^1 = (-1,-1;1), \quad a^2 = (2,-1;1), \quad a^3 = (-1,1;1).$$

Since our singularity is a cone over a projective variety, $\sigma^\vee$ also appears as a cone over some lattice polygon. In this example, $\sigma$ and $\sigma^\vee$ happen to be isomorphic. We obtain

$$\sigma^\vee = \langle s^1, s^2, s^3 \rangle \quad \text{with} \quad s^1 = [0,1;1], s^2 = [-2,-3;1], s^3 = [1,0;1].$$

The Hilbert basis $E \subset \sigma^\vee \cap \mathbb{Z}^3$ consists of these three fundamental generators together with

$$R^* = [0,0;1], \quad v^1 = [-1,-2;1], \quad v^2 = [0,-1;1], \quad w = [-1,-1;1].$$

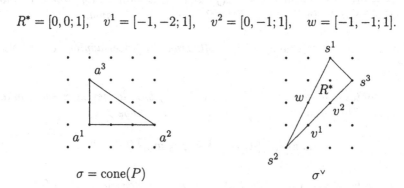

$$\sigma = \text{cone}(P) \qquad\qquad \sigma^\vee$$

In particular, $Y_\sigma$ has embedding dimension 7. The edges of $P$ have length $\ell(1) = 3$, $\ell(2) = 1$, and $\ell(3) = 2$. Thus $Y_\sigma$ has one dimensional singularities of transversal type $A_2$ and $A_1$. According to Theorem 4.4, $Y_\sigma$ admits only infinitesimal deformations of type (3). Their degrees come in three series:

($\alpha$) $2R^* - p_\alpha s^3$ with $p_\alpha \geq 1$. Even the leading element $R^\alpha = [-1,0,1]$ is not contained in $\sigma^\vee$.

($\beta$) $2R^* - p_\beta s^1$ with $p_\beta \geq 1$. The leading element equals $R^\beta = v^2 = [0,-1,1]$ and sits on the boundary of $\sigma^\vee$.

($\gamma$) $3R^* - p_\gamma s^1$ with $p_\gamma \geq 2$. The leading element is $R^\gamma = [0,-2,1] \notin \sigma^\vee$.

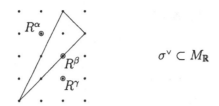

$$\sigma^\vee \subset M_\mathbb{R}$$

## 4.5

Each degree belonging to type (3) (that is, $R = qR^* - ps^j$ with $2 \le q \le \ell(j)$) provides an infinitesimal deformation. To show that they are unobstructed by describing how they lift to genuine one parameter deformations should be no problem: just split the polygon $Q(R)$ into a Minkowski sum satisfying conditions (i) and (ii) of 3.2, then construct $\tilde{\tau}$, $\tilde{\sigma}$ and $(f^1 - f)$ as in 3.2 and 3.5.

However, we prefer to present the result for our special case all at once using new coordinates. Let $P \subset \mathbb{A}(R^*) = \mathbb{R}^2 \times \{1\} \subset \mathbb{R}^3 = N_\mathbb{R}$ be a lattice polygon as in 4.3, and $R = qR^* - ps^j$ as just mentioned. Then $\sigma, \tau \subset N_\mathbb{R}$ are the cones over $P$ and $P \cap [R \ge 0]$ respectively, and the one parameter family in degree $-R$ is obtained as follows:

**Proposition 4.6** *The cone $\tilde{\tau} \subset N_\mathbb{R} \oplus \mathbb{R} = \mathbb{R}^4$ is generated by the elements:*

(i) $(a, 0) - \langle a, R \rangle (\underline{0}, 1)$ *as* $a \in P \cap [R \ge 0]$ *runs through the vertices from the $R^\perp$-line to $a^j$,*

(ii) $(a, 0) - \langle a, R \rangle (d^j/\ell(j), 1)$ *as* $a \in P \cap [R \ge 0]$ *runs from $a^{j+1}$ back up to the line $R^\perp$, and*

(iii) $(\underline{0}, 1)$ *and* $(d^j/\ell(j), 1)$.

*The vector space $N_\mathbb{R}$ containing $\sigma$ sits in $N_\mathbb{R} \oplus \mathbb{R}$ as $N_\mathbb{R} \times \{0\}$. Via this embedding, we have as usual $\tilde{\sigma} = \tilde{\tau} + \sigma$. The monomials $f$ and $f^1$ are given by their exponents $[R, 0]$ and $[R, 1] \in M \oplus \mathbb{Z}$ respectively.*

Geometrically, one can think of $\tilde{\tau}$ as generated by the interval $I$ with vertices as in (iii) and by the polygon $P'$ obtained as follows: "squeeze" $P \cap [R \ge 0]$ along $R^\perp$ by a cone with base $q/\ell(j) \cdot \overline{a^j a^{j+1}}$ and some vertex on the line $R^\perp$; take $-\langle \bullet, R \rangle$ as an additional, fourth coordinate. Then, $[R^*, 0]$ is still 1 on $P'$ and equals 0 on $I$. Moreover, $[R, 0]$ vanishes on $I$ and on the $R^\perp$-edge of $P'$; $[R, 1]$ vanishes on the whole of $P'$.

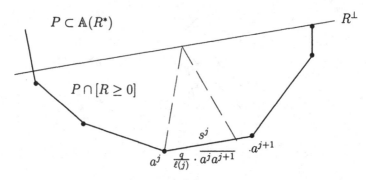

Figure 1: Parallel lines $R^\perp$ and $a^j a^{j+1}$ at distance $q/p$ apart in $\mathbb{A}(R^*)$

**Proof**  We change coordinates. If $g := \gcd(p,q)$ denotes the "length" of $R$, then we can find an $s \in M$ such that $\{s, R/g\}$ forms a basis of $M \cap (d^j)^\perp$. Adding some $r \in M$ with $\langle d^j/\ell(j), r\rangle = 1$ (the $r$ of Remark 4.3 will do) yields a $\mathbb{Z}$-basis for the whole lattice $M$. We consider the following commutative diagram:

$$
\begin{array}{ccc}
N & \xrightarrow[\sim]{(s,r,R/g)} & \mathbb{Z}^3 \\
{\scriptstyle(\mathrm{id},0)}\Big\downarrow & & \Big\downarrow{\scriptstyle(\mathrm{id},\,g\,\cdot\,\mathrm{pr}_3)} \\
N \oplus \mathbb{Z} & \xrightarrow[\sim]{([s,0],[r,0],[R/g,0],[R,1])} & \mathbb{Z}^3 \oplus \mathbb{Z}
\end{array}
$$

The left hand side contains the data relevant for our proposition. Carrying them over to the right yields:

- $[0,0,g] \in (\mathbb{Z}^3)^*$ as the image of $R$;

- $[0,0,g,0], [0,0,0,1] \in (\mathbb{Z}^4)^*$ as the images of $[R,0]$ and $[R,1]$ respectively;

- $\tau$ becomes a cone with affine cross-section

$$
Q([0,0,g]) =
$$
$$
\mathrm{conv}\left(\left(\,\langle a,s\rangle \,/\, \langle a,R\rangle\,;\, \langle a,r\rangle \,/\, \langle a,R\rangle\,;\, 1/g\right) \,\Big|\, a \in P \cap [R \geq 0]\right);
$$

- $I$ changes into the unit interval $(Q_1, 1)$ reaching from $(0,0,0,1)$ to $(0,1,0,1)$;

- finally, cone$(P')$ maps onto the cone spanned by the convex hull $(Q_0, 0)$ of the points $(\langle a, s \rangle / \langle a, R \rangle; \langle a, r \rangle / \langle a, R \rangle; 1/g; 0)$ for $a \in P \cap [R \geq 0]$ on the $a^j$-side and $(\langle a, s \rangle / \langle a, R \rangle; \langle a, r \rangle / \langle a, R \rangle - 1; 1/g; 0)$ for $a$ on the $a^{j+1}$-side respectively.

Since $Q([0, 0, g])$ equals the Minkowski sum of the interval $Q_1 \subset \mathbb{A}_0([0, 0, g])$ and the polygon $Q_0 \subset \mathbb{A}([0, 0, g])$, we are done by 3.2.  □

## 4.6

To see how the original equations of the singularity $Y_\sigma$ are perturbed, it is useful to study first the dual cones $\tilde{\tau}^\vee$ or $\tilde{\sigma}^\vee = \tilde{\tau}^\vee \cap \pi^{-1}(\sigma^\vee)$:

**Proposition 4.7** *If $s \in \sigma^\vee \cap M$, the element of $M \oplus \mathbb{Z}$ given by*

$$S := \begin{cases} [s, 0] & \text{if } \langle d^j, s \rangle \geq 0 \\ [s, - \langle d^j / \ell(j), s \rangle] & \text{if } \langle d^j, s \rangle \leq 0 \end{cases}$$

*is a lift of $s$ to $\tilde{\sigma}^\vee \cap (M \oplus \mathbb{Z})$. (Notice that it does not depend on $p, q$, but only on $j$.) Moreover, if $s^v$ runs through the edges of $P \cap [R \geq 0]$, the elements $S^v$ together with $[R, 0]$ and $[R, 1]$ form the fundamental generators of $\tilde{\tau}^\vee$.*

**Proof** Since we know $\tilde{\tau}$ from the previous proposition, the calculations are straightforward and we omit them.  □

## 4.7

Recall from 2.1 that $E$ denotes the minimal set generating the semigroup $\sigma^\vee \cap M$. Any $s \in E$ has an associated variable $z_s$, and $Y_\sigma \subset \mathbb{C}^E$ is given by binomial equations arising from linear relations among elements of $E$. Everything will be clear by considering an example: a linear relation such as $s^1 + 2s^3 = s^2 + s^4$ transforms into $z_1 z_3^2 = z_2 z_4$.

The fact that $\sigma$ defines a Gorenstein variety (that is, $\sigma$ is a cone over a lattice polytope) implies that $E$ consists only of $R^*$ and elements of $\partial \sigma^\vee$ including the fundamental generators $s^v$. If $E \cap \partial \sigma^\vee$ is ordered clockwise, then any two adjacent elements together with $R^*$ form a $\mathbb{Z}$-basis of the three dimensional lattice $M$.

In particular, any three consecutive elements of $E \cap \partial \sigma^\vee$ provide a unique linear relation among them and $R^*$. (We have already met this fact in Remark 4.3, where $r, s^j, r'$ were consecutive elements.) The resulting "boundary" equations do not generate the ideal of $Y_\sigma \subset \mathbb{C}^E$. Nevertheless, to describe a deformation of $Y_\sigma$, it is sufficient to know only how this set of equations is perturbed. Moreover, if one has to avoid boundary equations "overlapping" a certain spot on $\partial\sigma^\vee$, then it will even be possible to drop up to two of them from the list.

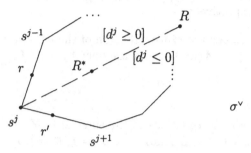

**Theorem 4.8** *The one parameter deformation of $Y_\sigma$ in degree $-(qR^* - ps^j)$ is completely determined by the following perturbations:*

(i) *(Boundary) equations involving only variables induced from $[d^j \geq 0] \subset \sigma^\vee$ remain unchanged. The same statement holds for $[d^j \leq 0]$.*

(ii) *The boundary equation $z_r z_{r'} - z_{R^*}^{\ell(j)} z_{s^j}^k = 0$ corresponding to the triple $\{r, s^j, r'\}$ is perturbed to $\left(z_r z_{r'} - z_{R^*}^{\ell(j)} z_{s^j}^k\right) - t z_{R^*}^{\ell(j)-q} z_{s^j}^{k+p} = 0$. Divide everything by $z_{s^j}^k$ if $k < 0$.*

**Proof**  Restricted to either $[d^j \geq 0]$ or $[d^j \leq 0]$, the map $s \mapsto S$ lifting elements of $E$ to $\tilde{\sigma} \cap (M \oplus \mathbb{Z})$ is linear. Hence, any linear relation remains true, and part (i) is proved.

For the second part, we consider the boundary relation $r + r' = \ell(j)R^* + ks^j$ with a suitable $k \in \mathbb{Z}$. By Proposition 4.6, the summands involved lift to the elements $[r, 0]$, $[r', 1]$, $[R^*, 0]$, and $[s^j, 0]$ respectively. In particular, the relation breaks down and has to be replaced by

$$[r, 0] + [r', 1] = [R, 1] + \big(\ell(j) - q\big)[R^*, 0] + (k + p)[s^j, 0], \quad \text{and}$$

$$\ell(j)[R^*, 0] + k[s^j, 0] = [R, 0] + \big(\ell(j) - q\big)[R^*, 0] + (k + p)[s^j, 0].$$

The monomials corresponding to $[R, 1]$ and $[R, 0]$ are $f^1$ and $f$ respectively. They are *not* regular on the total space $X$, but their difference $t := f^1 - f$ is. Hence, the difference of the monomial versions of both equations yields the result.

Finally, we should remark that (i) and (ii) cover all boundary equations except those overlapping the intersection of $\partial\sigma^\vee$ with $\overline{R^*R}$.  $\square$

## 4.8

We return to Example 4.4 and discuss the one parameter deformations in degree $-R^{\alpha}$, $-R^{\beta}$, and $-R^{\gamma}$ respectively:

**Case $\alpha$:** $R^{\alpha} = [-1, 0, 1] = 2R^* - s^3$ means $j = 3$, $q = \ell(3) = 2$, and $p = 1$. Hence, the line $R^{\perp}$ has distance $q/p = 2$ from its parallel through $a^3$ and $a^1$. In particular, $\tau = \langle a^1, c^1, c^3, a^3 \rangle$ with $c^1 = (1, -1, 1)$ and $c^3 = (3, -1, 3)$.

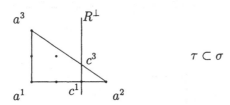

$$\tau \subset \sigma$$

We construct the generators of $\tilde{\tau}$ by the recipe of Proposition 4.6: $a^3$ treated via (i) and $a^1$ treated via (ii) yield the same element $A := (-1, 1, 1, -2)$; from the $R^{\perp}$-line we obtain $C^1 := (1, -1, 1, 0)$ and $C^3 := (3, -1, 3, 0)$; finally (iii) provides $X := (0, 0, 0, 1)$ and $Y := (0, -1, 0, 1)$. Hence, $\tilde{\tau}$ is the cone over the pyramid with plane base $XYC^1C^3$ and $A$ as top. (The relation between the vertices of the quadrangle equals $3C^1 + 2X = C^3 + 2Y$.) Moreover, $\tilde{\sigma}$ equals $\tilde{\sigma} = \tilde{\tau} + \mathbb{R}_{\geq 0}a^2$ with $a^2 := (a^2, 0)$. Since $A + 2X + 2a^2 = C^3$ and $A + 2Y + 2a^2 = 3C^1$, $\tilde{\sigma}$ is a simplex generated by $A$, $X$, $Y$, and $a^2$.

Denoting the variables assigned to $s^1, s^2, s^3, R^*, v^1, v^2, w \in E \subset \sigma^{\vee} \cap M$ by $Z_1, Z_2, Z_3, U, V_1, V_2, W$ respectively, there are six boundary equations:

$$Z_3 W Z_1 - U^3 = Z_1 Z_2 - W^2 = 0,$$
$$W V_1 - U Z_2 = Z_2 V_2 - V_1^2 = V_1 Z_3 - V_2^2 = V_2 Z_1 - U^2 = 0.$$

Only the latter four are covered by Theorem 4.8. They are perturbed to

$$W V_1 - U Z_2 = Z_2 V_2 - V_1^2 = V_1 Z_3 - V_2^2 = V_2 Z_1 - U^2 - t_{\alpha} Z_3 = 0.$$

**Case $\beta$:** $R^{\beta} = [0, -1, 1] = 2R^* - s^1$ means $j = 1$, $\ell(1) = 3$, $q = 2$, and $p = 1$. Hence, $R^{\perp}$ still has distance 2, but now from the line $a^1 a^2$.

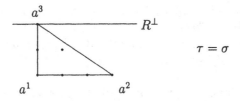

$$\tau = \sigma$$

We obtain

$$\widetilde{\tau} = \langle (-1,-1,1,-2); (0,-1,1,-2); (-1,1,1,0); (0,0,0,1); (1,0,0,1) \rangle.$$

The boundary equation $Z_3 W Z_1 - U^3 = 0$ corresponds to Theorem 4.8, (ii); it is perturbed to $Z_3 W Z_1 - U^3 - t_\beta U Z_1 = 0$.

**Case $\gamma$:**   $R^\gamma = [0,-2,1] = 3R^* - 2s^1$ means $j = 1$, $q = \ell(1) = 3$, and $p = 2$.

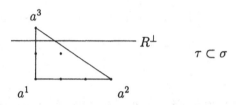

Here we have

$$\widetilde{\tau} = \langle (-1,-1,1,-3); (-2,1,2,0); (-1,2,4,0); (0,0,0,1); (1,0,0,1) \rangle,$$

and the previous boundary equation provides $Z_3 W Z_1 - U^3 - t_\gamma Z_1^2 = 0$.

# 5   Work in progress and open problems

## 5.1

For a fixed $j$, denote by $t_{q,p}$ the parameter corresponding to $R = qR^* - ps^j$ and take them as coefficients of formal power series $t_q(z) := \sum_{p \gg 0} t_{q,p} z^p$ for $2 \le q \le \ell(j)$.

**Conjecture 5.1** *The one parameter families corresponding to a fixed $j$ fit into a common huge family defined over a smooth, infinite dimensional parameter space. As in Theorem 4.8, the boundary equations sitting completely inside $[d^j \ge 0]$ or $[d^j \le 0]$ remain unchanged. The equation $z_r z_{r'} - z_{R^*}^{\ell(j)} z_{s^j}^k = 0$ turns into $\left( z_r z_{r'} - z_{R^*}^{\ell(j)} z_{s^j}^k \right) - \sum_{2 \le q \le \ell(j)} t_q(z_{s^j}) z_{R^*}^{\ell(j)-q} z_{s^j}^k = 0$.*

If this conjecture is true, then it should be possible to prove it directly by lifting equations and relations, without using Minkowski decompositions.

| | Equations | Perturbations |
|---|---|---|
| (0) | $Z_3 W Z_1 - U^3$ | $-(t_\alpha + Z_3\xi_\alpha)U Z_3 - (t_\beta + Z_1\xi_\beta)U Z_1 - (t_\gamma + Z_1\xi_\gamma)Z_1^2$ |
| (1) | $Z_1 Z_2 - W^2$ | $-t_\alpha(t_\alpha + Z_3\xi_\alpha)\xi_\beta - t_\alpha V_1 - \xi_\alpha\xi_\beta(V_2 + t_\beta)Z_1$ |
| | | $-\xi_\alpha(V_2 + t_\beta)^2$ |
| (2) | $W V_1 - U Z_2$ | $-(t_\alpha + Z_3\xi_\alpha)t_\alpha\xi_\gamma - \xi_\alpha(t_\gamma + Z_1\xi_\gamma)(V_2 + t_\beta)$ |
| (3) | $Z_2 V_2 - V_1^2$ | $-(t_\alpha + Z_3\xi_\alpha)\xi_\beta V_1 - t_\alpha\xi_\gamma W - \xi_\alpha\xi_\beta(t_\gamma + Z_1\xi_\gamma)U$ |
| | | $-\xi_\alpha(t_\gamma + Z_1\xi_\gamma)^2$ |
| (4) | $V_1 Z_3 - V_2^2$ | $+(t_\alpha + Z_3\xi_\alpha)\xi_\beta Z_3 - (t_\beta + Z_1\xi_\beta)V_2 - (t_\gamma + Z_1\xi_\gamma)U$ |
| (5) | $V_2 Z_1 - U^2$ | $-(t_\alpha + Z_3\xi_\alpha)Z_3$ |
| (6) | $V_2 W - U V_1$ | $-(t_\alpha + Z_3\xi_\alpha)(t_\gamma + Z_1\xi_\gamma + U\xi_\beta)$ |
| (7) | $Z_3 W - U V_2$ | $-(t_\beta + Z_1\xi_\beta)U - (t_\gamma + Z_1\xi_\gamma)Z_1$ |
| (8) | $Z_2 Z_3 - V_1 V_2$ | $-t_\alpha\xi_\gamma U - (t_\beta + Z_1\xi_\beta)V_1 - (t_\gamma + Z_1\xi_\gamma)W$ |
| (9) | $Z_1 V_1 - U W$ | $-(t_\alpha + Z_3\xi_\alpha)(V_2 + t_\beta)$ |

Table 5.1: Deformation of the cone over $\mathbb{P}(1,2,3)$. The base space is defined by $t_\alpha t_\beta = t_\alpha t_\gamma = 0$.

## 5.2

Jan Stevens has used the computer algebra system Macaulay to compute the versal deformation of our example $\text{Cone}(\mathbb{P}(1,2,3))$. In the list of perturbed equations given in Table 5.1, we denote the deformation parameters by $t_\alpha, t_\beta, t_\gamma, \xi_\alpha, \xi_\beta$ and $\xi_\gamma$. The first ones are the same as used in 4.8 covering $R^\alpha, R^\beta, R^\gamma$, respectively. The latter three parameters are formal power series $\xi_\alpha = \xi_\alpha(Z_3)$, $\xi_\beta = \xi_\beta(Z_1)$, and $\xi_\gamma = \xi_\gamma(Z_1)$; their coefficients, including that of $Z^0$ are the actual deformation parameters. Up to shift of degrees, the notation is similar to that of the previous conjecture; here we have $\deg_M \xi_\alpha = R^\alpha - s^3$, $\deg_M \xi_\beta = R^\beta - s^1$, and $\deg_M \xi_\gamma = R^\gamma - s^1$, respectively.

Equation (0) is not needed as a generator of the ideal of $Y_\sigma \subset \mathbb{C}^7$. It was put into the list since (0)–(5) are the boundary equations mentioned in 4.8. I'm afraid that the plus sign in (4) seems to be no mistake since, for instance, the original relation between equations (4), (6), and (7) lifts to $U \cdot \boxed{4} + Z_3 \cdot \boxed{6} - V_2 \cdot \boxed{7} - (t_\gamma + Z_1\xi_\gamma) \cdot \boxed{5} = 0$.

The versal base space is given by $t_\alpha t_\beta = t_\alpha t_\gamma = 0$, that is, it consists of two (infinite dimensional) smooth components. Restricting the deformation to the subspaces $[t_\alpha = \xi_\alpha = 0]$ and $[t_\beta = t_\gamma = \xi_\beta = \xi_\gamma = 0]$ yields two smooth families corresponding to $j = 1$ and $j = 3$, respectively. They equal exactly the deformations predicted in 5.1 and contain the 3 one parameter families shown in 4.8.

## 5.3

Let us focus on the smooth two parameter family involving only $t_\alpha$ and $\xi_\beta :=$ $\xi_\beta(0)$, setting the remaining parameters to zero. From the above list, only the following perturbation terms survive:

(0) $\left(Z_3 W Z_1 - U^3\right) - t_\alpha U Z_3 - \xi_\beta U Z_1^2$    (5) $\left(V_2 Z_1 - U^2\right) - t_\alpha Z_3$

(1) $\left(Z_1 Z_2 - W^2\right) - t_\alpha^2 \xi_\beta - t_\alpha V_1$    (6) $\left(V_2 W - U V_1\right) - t_\alpha \xi_\beta U$

(2) $\left(W V_1 - U Z_2\right)$    (7) $\left(Z_3 W - U V_2\right) - \xi_\beta Z_1 U$

(3) $\left(Z_2 V_2 - V_1^2\right) - t_\alpha \xi_\beta V_1$    (8) $\left(Z_2 Z_3 - V_1 V_2\right) - \xi_\beta Z_1 V_1$

(4) $\left(V_1 Z_3 - V_2^2\right) + t_\alpha \xi_\beta Z_3 - \xi_\beta Z_1 V_2$    (9) $\left(Z_1 V_1 - U W\right) - t_\alpha V_2$

Equation (3) turns into a binomial by restricting to the subfamily $[\xi_\gamma = 0]$ or $[t_\alpha = 0]$. This makes it difficult to hope that our two parameter family has a total space which is at least close to being toric (for instance, by some partial modifications).

## 5.4

Nevertheless, we try to obtain those "mixed" deformations involving degrees $-(qR^* - ps^j)$ with more than one $j$ by the methods developed so far. We begin with the $t_\alpha$-family introduced in 4.8 and try to deform varieties close to its total space. We recall the relevant cones:

$$\sigma = \langle a^1, a^2, a^3 \rangle \text{ and } \tau = \langle a^1, c^1, c^3, a^3 \rangle, \text{ with}$$

$$\begin{aligned} a^1 &= (-1,-1,1), \\ a^2 &= (2,-1,1), \\ a^3 &= (-1,1,1) \end{aligned} \quad \text{and} \quad \begin{aligned} c^1 &= (1,-1,1), \\ c^3 &= (3,-1,3); \end{aligned}$$

$$\tilde{\sigma} = \langle A, X, Y, a^2 \rangle \text{ and } \tilde{\tau} = \langle A, C^1, C^3, X, Y \rangle, \text{ with}$$

$$\begin{aligned} A &= (-1,1,1,-2), \\ C^1 &= (1,-1,1,0), \\ C^3 &= (3,-1,3,0) \end{aligned} \quad \text{and} \quad \begin{aligned} X &= (0,0,0,1), \\ Y &= (0,-1,0,1), \\ a^2 &= (a^2,0). \end{aligned}$$

The two dimensional faces ("edges") of the simplicial cone $\tilde{\sigma}$ are smooth. Hence, by 2.6 the variety $Z_{\tilde{\sigma}}$ is rigid.

That means we have to focus on infinitesimal deformations of $X_{\tilde{\tau}}$. We begin at the level of special fibers with a quick view of $T_{\tilde{\tau}}^1$ in comparison with $T_\sigma^1$: again from 2.6, we see that the vector space $T_\tau^1(-R)$ is nontrivial only for $R = [0,-1,0]$ and for the four series

$$R_\lambda^{(1)} = R^\alpha - \lambda s^3; \quad R_\lambda^{(2)} = R^\beta - \lambda s^1; \quad R_\lambda^{(3)} = v^1 - \lambda R^\alpha; \quad R_\lambda^{(4)} = s^1 - \lambda s^2$$

with $\lambda \geq 0$.

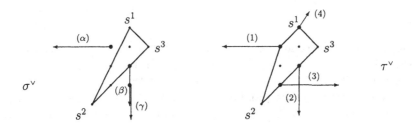

$T^1(-R)$ is always one dimensional, except that $\dim T^1_\tau(-[0,-2,0]) = 2$; here the series (2) and (3) meet (with $\lambda = 1$). In particular, we have lost the $\gamma$-deformations in $Y_\tau$. They are not compatible with the partial modification $Y_\tau \to Y_\sigma$.

As already mentioned in 4.8, $\tilde\tau$ is the cone over the pyramid with plane base $XYC^1C^3$ and $A$ as its top. The edges of the triangular face $AC^1C^3$ correspond to $A_1$ singularities; the remaining edges are smooth. The Hilbert basis $E$ of $\sigma^\vee \cap M \subset \tau^\vee \cap M$ lifts uniquely to $\tilde\tau^\vee \cap \mathbb{Z}^4$; together with $[R^\alpha, 0]$ and $[R^\alpha, 1]$ we obtain the Hilbert basis of the latter semigroup:

$$
\begin{array}{lll}
S^1 = [0,1,1,1] & V^1 = [-1,-2,1,0] & R^* = [0,0,1,0] \\
S^2 = [-2,-3,1,0] & V^2 = [0,-1,1,0] & [R^\alpha,0] = [-1,0,1,0] \\
S^3 = [1,0,1,0] & W = [-1,-1,1,0] & [R^\alpha,1] = [-1,0,1,1]
\end{array}
$$

Using our formula of 2.6 for $T^1$ again, we see that $T^1_{\tilde\tau}(-R)$ is at most one dimensional. $X_{\tilde\tau}$ admits infinitesimal deformations in the following degrees:

(a) $R^{(a)}_\lambda = [-1,-2,1,-1] - \lambda[R^\alpha,1]$ for $\lambda \geq 1$.

Here, $R$ equals 2 on the vertices $A$, $C^1$, $C^3$, and it is nonpositive otherwise. In particular, $Q(R) = \tilde\tau \cap [R=1]$ is a triangle, and we obtain $V(R) = W(R) = \mathbb{R}$. Projecting $\mathbb{Z}^4 \to M$ maps this series onto $R^{(3)}_\lambda$ ($\lambda \geq 1$) which was mentioned above.

(b) $R^{(b)}_{\lambda,\mu} = [-1,-2,1,0] - \lambda[R^\alpha,0] - \mu[R^\alpha,1]$ for $\lambda, \mu \geq 0$.

The compact part of $Q(R)$ equals the interval reaching from $C^1/2$ to $C^3/2$. Again, this series projects onto the third one for $\tau$.

(c) $R^{(c)}_{\lambda,\mu} = [0,-1,1,-1] - \lambda S^1 - \mu[R^\alpha,1]$ for $\lambda \geq 2$; $\mu \geq 0$.

$Q(R)$ now has $\overline{AC^1}/2$ as compact part. Mapping the degrees into $M$ yields the $\beta$-series $R^{(2)}_\lambda$ with $\lambda \geq 2$.

(d) $R^{(d)}_{\lambda,\mu} = [0,1,1,0] - \lambda S^2 - \mu[R^\alpha,1]$ for $\lambda, \mu \geq 0$.

This final series corresponds to the edge $\overline{AC^3}$, and it maps onto $R^{(4)}_\lambda$ ($\lambda \geq 0$).

## 5.5

From the previous section we know that there is no infinitesimal deformation either of $X_{\tilde{7}}$ or $Z_{\tilde{\sigma}}$ in any lift of degree $-R_1^{(2)}$. In particular, our attempt to reconstruct the $(t_\alpha, \xi_\beta)$-deformation of 5.3 has failed.

However, since we are just interested in mixing $\alpha$- and $\beta$-deformations, we may try degree $T := R_{2,0}^{(c)} = [0, -3, -1, -3]$ from the series (c) above. It is a lift of $R_2^{(2)} \in M$ corresponding to the deformation parameter arising as the coefficient of $Z$ in the formal power series $\xi_\beta(Z)$. By abuse of notation, we will call this parameter $\xi_\beta$ again. The corresponding family may be obtained from 5.3 via the substitution $\xi_\beta := \xi_\beta Z_1$.

Now we may construct the genuine deformation of $X_{\tilde{7}}$ occurring in degree $-T$. The usual approach (splitting $Q(R)$ into a Minkowski sum) yields the following result:

- On the partial modification of the new total space we have variables $Z_1, Z_2, Z_3, U, V_1, V_2, W$ arising as lifts from $\sigma^\vee$. Moreover, we call $R_0, R_1, T_0, T_1$ the variables corresponding to the lifts of $[R^\alpha, 0], [R^\alpha, 1]$ or to the two lifts of $T$, respectively. In particular, $t_\alpha = R_1 - R_0$ and $\xi_\beta = T_1 - T_0$ are the deformation parameters.

- The modified total space corresponds to a cone

$$\Delta = \langle B^1, B^2, B^3, C^3, Y, Z \rangle,$$

where $B^1, B^2, B^3$ have arisen from the Minkowski splitting $\overline{AC^1} = \{B^1\} + \overline{B^2B^3}$ inside the affine plane $[T = 1]$. The variables mentioned above correspond to the equally named points of $\Delta^\vee \cap \mathbb{Z}^5$. The Hilbert basis consists of $\{Z_1, Z_2, Z_3, V_1, V_2, R_0, R_1, T_0, T_1\}$ and an additional element $S$ (with $U = S + Z_1$, $W = S + R_1$). It might be useful to know $\Delta$ and its dual cone in detail. Therefore, we write down the following table indicating the pairing between the mutually dual cones:

|       | $Z_1$ | $Z_2$ | $Z_3$ | $U$ | $V_1$ | $V_2$ | $W$ | $R_0$ | $R_1$ | $T_0$ | $T_1$ | $S$ |
|-------|-------|-------|-------|-----|-------|-------|-----|-------|-------|-------|-------|-----|
| $B^1$ | 0     | 0     | 0     | 1   | 0     | 0     | 1   | 0     | 0     | 2     | 0     | 1   |
| $B^2$ | 0     | 0     | 0     | 0   | 0     | 0     | 0   | 1     | 0     | 0     | 1     | 0   |
| $B^3$ | 0     | 1     | 1     | 0   | 1     | 1     | 0   | 0     | 0     | 0     | 1     | 0   |
| $C^3$ | 2     | 0     | 6     | 3   | 2     | 4     | 1   | 0     | 0     | 0     | 0     | 1   |
| $Y$   | 0     | 3     | 0     | 0   | 2     | 1     | 1   | 0     | 1     | 0     | 0     | 0   |
| $Z$   | 2     | 0     | 0     | 3   | 0     | 0     | 3   | 6     | 2     | 0     | 0     | 0   |

- Proceeding as in the proof of Theorem 4.8, the ten original equations defining $Y_\sigma$ (cf. the second column of Table 5.1) induce binomial equations for the modified total space:

  (0) and (5) provide $WZ_1 - R_1U = U^2 - Z_1^3T_0 = R_0Z_3 - Z_1^3T_1 = V_2Z_1 - R_1Z_3 = 0$;

  (1) provides $Z_1Z_2 - R_1V_1 = W^2 - Z_1R_1^2T_0 = R_0V_1 - Z_1R_1^2T_1 = 0$;

  (9) provides $Z_1V_1 - V_2R_1 = UW - T_0R_1Z_1^2 = V_2R_0 - T_1R_1Z_1^2 = 0$;

  the equations (2), (3), (4), (6), (7), and (8) remain unchanged.

- To describe the deformation of $X_{\tilde\tau}$ we have to eliminate $T_0$ and $T_1$; just their difference $\xi_\gamma = T_1 - T_0$ may occur. There is nothing to do with equations (2), (3), (4), (6), (7), (8); for the remaining ones we obtain

  (0;5)    $WZ_1 - R_1U = U^2 - R_0Z_3 + \xi_\beta Z_1^3 = V_2Z_1 - R_1Z_3 = 0$

  (1)      $Z_1Z_2 - R_1V_1 = W^2 - R_0V_1 + \xi_\beta R_1^2 Z_1$

  (9)      $Z_1V_1 - V_2R_1 = UW - V_2R_0 + \xi_\beta R_1Z_1^2.$

- Finally, we should remember that we are eventually interested in deformations of $X$ rather than $X_{\tilde\tau}$. Hence, we have to do the same job again with $t_\alpha = R_1 - R_0$. The best result for equations (0), (1), (5), and (9) is

  (0)      $(Z_3WZ_1 - U^3) - U(t_\alpha Z_3 + \xi_\beta Z_1^3)$

  (1)      $(Z_1Z_2 - W^2) - t_\alpha V_1 - \xi_\beta R_1^2 Z_1$

  (5)      $(V_2Z_1 - U^2) - t_\alpha Z_3 - \xi_\beta Z_1^3$

  (9)      $(Z_1V_1 - UW) - t_\alpha V_2 - \xi_\beta R_1 Z_1^2.$

In particular, we have not been able to eliminate $R_1$ from equations (1) and (9). It indicates that our deformation of $X_{\tilde\tau}$ *does not blow down* onto the $X$ level.

## 5.6

We have used $\tau \subset \sigma$ to obtain a partial modification $Y_\tau \to Y_\sigma$. Deformations of $Y_\sigma$ have been obtained by blowing down deformations of $Y_\tau$.

Is it possible, or even better, to consider the whole modification $Y_\Sigma \to Y_\sigma$ instead of the open chart only? $\Sigma$ is the fan consisting of the two cones $\tau = \sigma \cap [R \geq 0]$ and $\tau' := \sigma \cap [R \leq 0]$. Apart from the fact that this approach looks more natural, there might be another advantage: the lack of certain degrees providing infinitesimal deformations of $Y_\tau$ may be overcome by blowing down local trivial deformations of $Y_\Sigma$.

# References

[Al 1]   Altmann, K.: Computation of the vector space $T^1$ for affine toric varieties. J. Pure Appl. Algebra **95** (1994), 239-259.

[Al 2]   Altmann, K.: Minkowski sums and homogeneous deformations of toric varieties. Tôhoku Math. J. **47** (1995), 151-184.

[Al 3]   Altmann, K.: Infinitesimal deformations and obstructions for toric singularities. J. Pure Appl. Algebra **119** (1997), 211-235.

[Al 4]   Altmann, K.: The versal deformation of an isolated, toric Gorenstein singularity. Invent. math. **128** (1997), 443-479.

[Ish]   Ishida, M.-N.: Torus embeddings and dualizing complexes. Tôhoku Math. J. **32**, 111-146 (1980).

[Oda]   Oda, T.: Convex bodies and algebraic geometry. Ergebnisse der Mathematik und ihrer Grenzgebiete (3/15), Springer-Verlag, 1988.

[Re]   Reid, M.: Some flips. Talk at the conferences in Santa Cruz (July 1995) and Warwick (September 1995).

Klaus Altmann
Institut für reine Mathematik, Humboldt-Universität zu Berlin
Ziegelstr. 13A,
D–10099 Berlin, Germany.
e-mail: altmann@mathematik.hu-berlin.de

# Nonrational covers of $\mathbb{CP}^m \times \mathbb{CP}^n$

## János Kollár

# 1 Introduction

This article aims to provide further examples of higher dimensional varieties that are rationally connected, but not rational, and not even ruled. The original methods of [Iskovskikh–Manin71, Clemens–Griffiths72] have been further developed by many authors (for example, [Beauville77, Iskovskikh80, Bardelli84]), and give a fairly complete picture in dimension three. On the other hand, in higher dimension, only special examples were known until recently [Artin–Mumford72, Sarkisov81, Sarkisov82, Pukhlikov87, CTO89]. The rationality question for hypersurfaces in $\mathbb{P}^n$ was considered in [Kollár95]. There I proved the following:

**Theorem 1.1 [Kollár95]** *Let $X_d \subset \mathbb{CP}^{n+1}$ be a very general hypersurface of degree $d$. Assume that*

$$\frac{2}{3}n + 3 \leq d \leq n + 1. \tag{1.1.1}$$

*Then $X_d$ is not rational.* $\square$

Here *very general* means that the result holds for hypersurfaces corresponding to a point in the complement of countably many closed subvarieties in the space of all hypersurfaces.

The method applies to hypersurfaces in $\mathbb{P}^m \times \mathbb{P}^{n+1}$. Let $X_{c,d}$ be such a hypersurface of bidegree $(c, d)$. Via the projection $X_{c,d} \to \mathbb{P}^m$, it can be viewed as a family of degree $d$ hypersurfaces in $\mathbb{P}^{n+1}$ parametrized by $\mathbb{P}^m$. $X_{c,d}$ is rationally connected if $d \leq n + 1$, no matter what $c$ is (cf. [Kollár96, IV.6.5]).

A straightforward application of the method of [Kollár95] provides an analog of Theorem 1.1 for hypersurfaces when $c \geq m + 3$ and $d, n$ satisfy the inequalities (1.1.1). However, it is more interesting to study some cases where the fibers of $X_{c,d} \to \mathbb{P}^m$ are rational. Here I propose to work out two cases: conic bundles and families of cubic surfaces.

The method of [Kollár95] works naturally for cyclic covers, and it is easier to formulate the results that way.

51

Fix a prime $p$ and let $X_{ap,bp} \to \mathbb{CP}^m \times \mathbb{CP}^n$ be a degree $p$ cyclic cover ramified along a very general hypersurface of bidegree $(ap, bp)$. [Kollár95] shows that $X_{ap,bp}$ is not rational and not even ruled if $ap > m + 1$ and $bp > n + 1$. In this paper, I study the case when $bp = n + 1$ and $n = 1, 2$. The main results are the following:

**Theorem 1.2** *Let* $X_{2a,2} \to \mathbb{CP}^m \times \mathbb{CP}^1$ *be a double cover ramified along a very general hypersurface of bidegree* $(2a, 2)$, *where* $m \geq 2$. *Then* $X_{2a,2}$ *is not rational if* $2a > m + 1$.

*More precisely, if* $Y$ *is any variety of dimension* $m$ *and* $\varphi\colon Y \times \mathbb{P}^1 \dashrightarrow X_{2a,2}$ *a dominant map then* $2 \mid \deg \varphi$.

**Remark 1.2.1**

**1.2.1.1** $X_{2a,2} \to \mathbb{CP}^n$ is a conic bundle. For conic bundles we have the very strong results of [Sarkisov81, Sarkisov82], saying that a conic bundle is not rational if the locus of singular fibers plus 4 times the canonical class of the base is effective. In our case the locus of singular fibers is a divisor in $\mathbb{P}^m$ of degree $4a > 2m + 2$. This is lower than Sarkisov's bound $4m + 4$. Sarkisov's normal crossing assumptions are also not satisfied by $X_{2a,2} \to \mathbb{CP}^n$. Thus some of our cases are not covered by Sarkisov's results. This suggests the possibility that the bounds of [Sarkisov81, Sarkisov82] can be improved considerably. This is not clear even for conic bundles over surfaces.[1]

**1.2.1.2** It is natural to ask whether the projection $X_{2a,2} \to \mathbb{CP}^m$ is the only conic bundle structure of $X_{2a,2}$. The proof gives such examples in characteristic 2, but not over $\mathbb{C}$.

For families of cubic surfaces the result is weaker:

**Theorem 1.3** *Let* $X_{3a,3} \to \mathbb{CP}^m \times \mathbb{CP}^2$ *be a cyclic triple cover ramified along a very general hypersurface of bidegree* $(3a, 3)$, *where* $m \geq 1$. *Then* $X_{3a,3}$ *is not rational and not even ruled if* $3a > m + 1$.

**Remark 1.3.1** Similar results for $m = 1$ were proved by [Bardelli84].

For hypersurfaces in $\mathbb{CP}^m \times \mathbb{CP}^{n+1}$ these imply the following:

**Theorem 1.4**

**1.4.1** *Let* $X_{c,2} \subset \mathbb{CP}^m \times \mathbb{CP}^2$ *be a very general hypersurface of bidegree* $(c, 2)$, *where* $m \geq 2$. *Then* $X_{c,2}$ *is not rational if* $c \geq m + 3$.

**1.4.2** *Let* $X_{c,3} \subset \mathbb{CP}^m \times \mathbb{CP}^3$ *be a very general hypersurface of bidegree* $(c, 3)$, *where* $m \geq 1$. *Then* $X_{c,3}$ *is not rational if* $c \geq m + 4$.

---

[1] Compare Corti's article, Section 4.2.

**Proof** By [Kollár96, V.5.12–13], it is sufficient to find an algebraically closed field $k$ and a single example of a nonruled hypersurface $X_{c',2} \subset \mathbb{P}^m \times \mathbb{P}^2$, respectively $X_{c',3} \subset \mathbb{P}^m \times \mathbb{P}^3$ over $k$ for some $c' \leq c$. We proceed to construct such examples in characteristic two for conic bundles and in characteristic three for families of cubic surfaces.

Consider $\mathbb{A}^m \times \mathbb{A}^n$ with coordinates $(u_1, \ldots, u_m; v_1, \ldots, v_n)$. Let $f(u,v)$ be a polynomial of bidegree $(c, n+1)$. We have a hypersurface

$$Z := (y^{n+1} - f(u,v) = 0) \subset V := \mathbb{A}^1 \times \mathbb{A}^m \times \mathbb{A}^n,$$

where $y$ is a coordinate on $\mathbb{A}^1$. There are several natural ways to associate a projective variety to $Z$:

**1.4.3.1** We can view $V$ as a coordinate chart in $\mathbb{P}^m \times \mathbb{P}^{n+1}$. The closure $\overline{Z}_1$ of $Z$ is a hypersurface of bidegree $(c, n+1)$. For $n = 1, 2$ this gives our examples $X_{c,2}$ and $X_{c,3}$.

**1.4.3.2** We can compactify $\mathbb{A}^m \times \mathbb{A}^n$ to $\mathbb{P}^m \times \mathbb{P}^n$ and view $f$ as a section of the line bundle $\mathcal{O}_{\mathbb{P}^m \times \mathbb{P}^n}(c, n+1)$. Assume that $(n+1) \mid c$ and let $L = \mathcal{O}(\frac{c}{n+1}, 1)$. The corresponding cyclic cover $\overline{Z}_2 = \mathbb{P}^m \times \mathbb{P}^n[\sqrt[n+1]{f}]$ (see Definition 2.3) gives another compactification of $Z$.

Theorems 1.2 and 1.3 show, using the second representation, that $\overline{Z}_2$ is not ruled if $f$ is very general and $c > n + 1$. This implies Theorem 1.4. $\square$

## 1.5   Generalizations

It is clear from the proof that it applies to many different cases. The main point is to have a family of conics or cubic surfaces whose branch divisor is sufficiently large, but I found it hard to write down a reasonably general statement.

## 1.6   Outline of the proof of Theorems 1.2–1.3

By [Kollár96, V.5.12–13] it is sufficient to find an algebraically closed field $k$ and a single example of a nonruled cyclic cover $X_{2a',2} \to \mathbb{P}^m \times \mathbb{P}^1$ (respectively, $X_{3a',3} \to \mathbb{P}^m \times \mathbb{P}^2$) for some $a' \leq a$. We proceed to construct such examples in characteristic two for conic bundles and in characteristic three for families of cubic surfaces.

Fix an algebraically closed field $k$ of characteristic $p$. Let $\pi\colon X_{ap,bp} \to \mathbb{P}^m \times \mathbb{P}^n$ be a degree $p$ cyclic cover corresponding to a very general hypersurface of bidegree $(ap, bp)$ (cf. Definition 2.3). Since $\pi$ is purely inseparable, $X_{ap,bp}$ is (purely inseparably) unirational. We intend to show that it is frequently not ruled.

Section 2 is a review of the machinery of inseparable cyclic covers and their applications to nonrationality problems. $X_{ap,bp}$ has isolated singularities by (2.1.4) and (2.2.3). These can be resolved; if $\pi_Y \colon Y \to X_{ap,bp} \to \mathbb{P}^m \times \mathbb{P}^n$ is a resolution, then by Corollary 2.5 there is a nonzero map

$$\pi_Y^* \mathcal{O}(ap - m - 1, bp - n - 1) \to \overset{m+n-1}{\bigwedge} \Omega_Y^1 \cong \Omega_Y^{m+n-1}.$$

Thus [Kollár95] shows that $X_{ap,bp}$ is not rational and not even ruled if $ap > m + 1$ and $bp > n + 1$. In this paper I study the case when $ap > m + 1$ and $bp = n + 1$.

Section 3 contains a nonruledness criterion. Let $f \colon Y \to \mathbb{P}^m$ be the composite of $\pi_Y$ with the first projection. Corollary 3.2 shows that if $ap > m + 1$ and $bp = n + 1$, then $Y$ is ruled if and only if the generic fiber of $f$ is ruled over the function field $k(\mathbb{P}^m)$. This is the main technical departure from [Kollár95]. There I used varieties in positive characteristic which were shown to be not even separably uniruled. The varieties here are separably uniruled, but we are able to get a description of all separable unirulings.

Rationality or ruledness over nonclosed fields is a very interesting question. Unfortunately I cannot say much, except for $n = 1, 2$.

When $n = 1$, the generic fiber of $f$ is a plane conic. Conics over nonclosed fields are considered in Section 4; Theorem 4.1 gives a complete description of their unirulings. Theorem 1.2 is implied by Corollary 2.5, Theorem 3.1 and Proposition 4.2.

If $n = 2$, the generic fiber of $f$ is a cubic surface $S$, given by an equation $u^3 = f_3(x, y, z)$. Since $X_{ap,bp}$ has only isolated singularities, $S$ is regular over $k(\mathbb{P}^m)$, though we will see that it is not smooth (Remark 5.2). Thus we are led in Section 5 to investigate regular Del Pezzo surfaces over arbitrary fields. We are forced to study the situation over nonperfect fields; this introduces several new features. The main result is Theorem 5.7 which generalizes results of Segre and Manin to nonperfect fields. Together with Corollary 2.5 and Theorem 3.1, this result implies Theorem 1.3.

## 1.7   Terminology

I follow the terminology of [Kollár96].

If a scheme $X$ is defined over a field $F$, and $E$ is a field extension of $F$, then $X_E$ denotes the scheme obtained by base extension.

For a field $F$, $\overline{F}$ denotes an algebraic closure.

$X_F$ is called *rational* if it is birational to $\mathbb{P}_F^n$, and the birational map is defined over $F$. Sometimes for emphasis I say that $X_F$ is rational over $F$. The same convention applies to other notions, such as irreducible, etc.

We say that $X_F$ is ruled (respectively uniruled) if there is a scheme $Y_F$ with $\dim Y_F = \dim X_F - 1$ and a birational (respectively dominant) map $Y_F \times \mathbb{P}^1 \dashrightarrow X_F$. In the uniruled case, it may happen that $Y_F$ is reducible over $\overline{F}$.

If $X_{\overline{F}}$ is rational then we say that $X_F$ is geometrically rational. Similarly for other notions (ruled, uniruled, irreducible etc.).

Following standard terminology, we say that a scheme is *regular* (or *nonsingular*) if all of its local rings are regular. Over nonperfect fields there are schemes that are nonsingular but not smooth.

## 1.8   Acknowledgements

Partial financial support was provided by the NSF under grant number DMS-9102866. I thank M. Grinenko and M. Reid for several useful comments.

# 2   Inseparable cyclic covers

We first recall the definitions and basic properties of critical points of sections of line bundles in positive characteristic. For proofs, see [Kollár95] or [Kollár96, V.5].

**Definition 2.1** Let $X$ be a smooth variety over an algebraically closed field $k$ and $f$ a function on $X$. Let $x \in X$ be a closed point and assume that $f$ has a critical point at $x$. Choose local coordinates $x_1, \ldots, x_n$ at $x$.

**2.1.1** $f$ has a *nondegenerate critical point* at $x$ if $\partial f / \partial x_1, \ldots, \partial f / \partial x_n$ generate the maximal ideal of the local ring $\mathcal{O}_{x,X}$. This notion is independent of the local coordinates chosen.

If $\operatorname{char} k \neq 2$ or $\operatorname{char} k = 2$ and $n$ is even then $f$ has a nondegenerate critical point at $x$ if and only if $f$ can be written in suitable local coordinates as

$$f = c + \begin{cases} x_1 x_2 + x_3 x_4 + \cdots + x_{n-1} x_n + f_3 & \text{if } n \text{ is even,} \\ x_1^2 + x_2 x_3 + \cdots + x_{n-1} x_n + f_3 & \text{if } n \text{ is odd,} \end{cases}$$

with $f_3 \in m_x^3$.

**2.1.2** If $\operatorname{char} k = 2$ and $\dim X$ is odd, then every critical point is degenerate.

**2.1.3** Assume that $\operatorname{char} k = 2$ and $\dim X$ is odd. A critical point of $f$ is called *almost nondegenerate* if $\operatorname{length} \mathcal{O}_{x,X}/(\partial f / \partial x_1, \ldots, \partial f / \partial x_n) = 2$. Equivalently, $f$ can be written in suitable local coordinates as

$$f = c + a x_1^2 + x_2 x_3 + \cdots + x_{n-1} x_n + b x_1^3 + f_3, \quad \text{where } b \neq 0.$$

**2.1.4** Assume that char $k \mid d$. Then the hypersurface

$$Z = (y^d - f(x_1, \ldots, x_n) = 0)$$

is singular at the point $(y, x) \in Z$ if and only if $x \in X$ is a critical point of $f$.

**2.1.5** Let $L$ be a line bundle on $X$ and $s \in H^0(X, L^d)$ a section. Let $U \subset X$ be an affine open subset such that $L_{|U} \cong \mathcal{O}_U$. Choose such an isomorphism. Then $s_{|U}$ can be viewed as section of $\mathcal{O}_U^{\otimes d} \cong \mathcal{O}_U$. Thus it makes sense to talk about its critical points. If char $k \mid d$ then this is independent of the choice of $U$ and of the trivialization $L_{|U} \cong \mathcal{O}_U$. (This fails if the characteristic does not divide $d$.)

The usual Morse lemma can be generalized to positive characteristic. We use it in a somewhat technical form.

**Proposition 2.2** *Let $X$ be a smooth variety over a field of characteristic $p$, and $L$ a line bundle on $X$. Let $d$ be an integer divisible by $p$ and $W \subset H^0(X, L^d)$ a finite dimensional vector subspace. Let $m_x$ denote the ideal sheaf of $x \in X$. Assume that:*

**2.2.1** *For every closed point $x \in X$ the restriction map $W \to (\mathcal{O}_X/m_x^2) \otimes L^k$ is surjective,*

**2.2.2** *For every closed point $x \in X$ there is an $f_x \in W$ which has an (almost) nondegenerate critical point at $x$.*

*Then a general section $f \in W$ has only (almost) nondegenerate critical points.*

**Proof**   This is a simple dimension count. Fix $x \in X$ and let $W_x \subset W$ be the set of functions with a critical point at $x$. By (2.2.1), $W_x$ has codimension $n$. In $W_x$ the functions with an (almost) nondegenerate critical point at $x$ form an open subset $W_x^0$ which is nonempty by (2.2.2). Thus the set of functions with a degenerate critical point is $\bigcup_x (W_x - W_x^0)$ and it has codimension at least one in $W$.   $\square$

**Lemma 2.2.3** *Let $X_1, X_2$ be smooth varieties over a field of characteristic $p$ and $L_i$ very ample line bundles on $X_i$. Let $L = p_1^* L_1 \otimes p_2^* L_2$ be the corresponding line bundle on $X = X_1 \times X_2$. If $p \mid d$ then a general section $f \in H^0(X, L)$ has only (almost) nondegenerate critical points.*

**Proof** Pick a point $x = (x_1, x_2)$. The condition (2.2.1) is clearly satisfied. To check (2.2.2), choose global sections $u_i \in H^0(X_1, L_1)$ and $v_j \in H^0(X_2, L_2)$ that give local coordinates at $x_1$, respectively $x_2$.

If $p \neq 2$ then $\sum u_i^2 + \sum v_j^2$ gives a section of $L^d$ with a nondegenerate critical point at $x$.

If $p = 2$ we need to consider a few cases. Set $n_i = \dim X_i$. We plan to use the function

$$g = \sum_{i=1}^{n_i/2} u_{2i-1} u_{2i} + \sum_{j=1}^{n_2/2} v_{2j-1} v_{2j}.$$

If both of the $n_i$ are even, then we can take $f = g$. If both of the $n_i$ are odd, then we can use $f = g + u_{n_1} v_{n_2}$. Otherwise we may assume that $n_1$ is odd and $n_2$ is even. Then we use

$$f = g + u_{n_1} v_{n_2 - 1} + u_{n_1}^2 v_{n_2}.$$

Explicit computation shows that $f$ has an almost nondegenerate critical point. $\square$

**Definition 2.3 (Cyclic covers)** Let $X$ be a scheme, $L$ a line bundle on $X$ and $s \in H^0(X, L^d)$ a section. Assume for simplicity that its divisor of zeros $(s = 0)$ is reduced. The cyclic cover of $X$ obtained by taking a $d$th root of $s$, denoted by $X[\sqrt[d]{s}]$ is the scheme locally constructed as follows:

Let $U \subset X$ be an open set such that $L_{|U} \cong \mathcal{O}_U$. Then $s_{|U}$ can be identified with a function $f \in H^0(U, \mathcal{O}_U)$. Let $V \subset \mathbb{A}^1 \times U$ be the closed subset defined by the equation $y^d - f = 0$ where $y$ is the coordinate on $\mathbb{A}^1$. The resulting schemes can be patched together in a natural way to get a scheme $X[\sqrt[d]{s}]$; cf. [Kollár96, II.6.1]. We are only interested in it up to birational equivalence, so the precise definitions are unimportant.

The only result about cyclic covers we need is the following special case of [Kollár96, V.5.10]:

**Proposition 2.4** *Let $X$ be a smooth variety of dimension $n$ over a field $k$ of characteristic $p$, $L$ a line bundle on $X$ and $d$ an integer divisible by $p$. Suppose that $s \in H^0(X, L^d)$ is a section with (almost) nondegenerate critical points, and let $\pi: Y \to X$ be a smooth projective model of $X[\sqrt[d]{s}]$ (Y always exists).*

*Then there is a nonzero map*

$$\pi^*(K_X \otimes L^d) \to \bigwedge^{n-1} \Omega_Y^1 \cong \Omega_Y^{n-1}. \quad \square$$

Applied to the cyclic covers $\overline{Z}_2 = \mathbb{P}^m \times \mathbb{P}^n[\sqrt[2]{s}]$ from the introduction (see 1.4.3.2), we get the following:

**Corollary 2.5** *Fix a prime $p$ and let $k$ be an algebraically closed field of characteristic $p$. Let $s \in H^0(\mathbb{P}^m \times \mathbb{P}^n, \mathcal{O}(a,b)^{\otimes p})$ be a general section and $q\colon Y \to \mathbb{P}^m \times \mathbb{P}^n$ a smooth projective model of $\mathbb{P}^m \times \mathbb{P}^n[\sqrt[p]{s}]$.*
*Then there is a nonzero map*

$$q^*\mathcal{O}(ap-m-1, bp-n-1) \to \bigwedge^{m+n-1} \Omega_Y^1 \cong \Omega_Y^{m+n-1}. \quad \Box$$

# 3   A nonruledness criterion

In this section we prove the following generalization of [Kollár96, V.5.11].

**Theorem 3.1** *Let $X, Y$ be smooth proper varieties and $f\colon Y \to X$ a surjective morphism, where $n = \dim Y$. Let $M$ be a big line bundle on $X$ and assume that for some $i > 0$ there is a nonzero map*

$$h\colon f^*M \to \bigwedge^i \Omega_Y^1.$$

**3.1.1** *Let $Z$ be an affine variety of dimension $n-1$ and $\varphi\colon Z \times \mathbb{P}^1 \to Y$ a dominant and separable morphism. Then there is a morphism $\psi\colon Z \to X$ which fits in the commutative diagram*

$$
\begin{array}{ccc}
Z \times \mathbb{P}^1 & \xrightarrow{\ \varphi\ } & Y \\
\downarrow & & \downarrow{\scriptstyle f} \\
Z & \xrightarrow{\ \psi\ } & X.
\end{array}
$$

**3.1.2** *Let $F = k(X)$ be the field of rational functions on $X$ and $Y_F$ the generic fiber of $f$. There is a one-to-one correspondence*

$$\left\{ \begin{array}{c} \text{degree } d \text{ separable} \\ \text{unirulings of } Y \end{array} \right\} \longleftrightarrow \left\{ \begin{array}{c} \text{degree } d \text{ separable} \\ \text{unirulings of } Y_F \end{array} \right\}.$$

*In particular, $Y$ is ruled if and only if $Y_F$ is ruled over $F$.*

**3.1.3** *Assume that for any two general points $y_1, y_2 \in Y_{\overline{F}}$ there is a morphism $f = f(y_1, y_2)\colon \mathbb{P}^1 \to Y_{\overline{F}}$ such that $y_1, y_2 \in \operatorname{im} f$, $Y_{\overline{F}}$ is smooth along $\operatorname{im} f$ and $f^*T_{Y_{\overline{F}}}$ is semipositive. Then any birational selfmap of $Y$ preserves $f$. Hence there is an exact sequence*

$$1 \to \operatorname{Bir}(Y_F) \to \operatorname{Bir}(Y) \to \operatorname{Bir}(X).$$

**Remark 3.1.4** The assumption (3.1.3) is satisfied if $Y_{\overline{F}}$ is separably rationally connected (cf. [Kollár96, IV.3.2]). More generally, it also holds for the cyclic covers of $\mathbb{P}^n$ that we are considering [Kollár96, V.5.19].

**Proof** $M$ is big, hence there is an open set $U \subset X$ such that sections of $M^k$ separate points of $U$ for $k \gg 1$. In particular, if $g: C \to X$ is a nonconstant morphism from a smooth proper curve to $X$ whose image intersects $U$, then $\deg g^* M > 0$.

Let $g: C \to Y$ be a morphism such that $g^* \Omega_Y^1$ is seminegative. We have a map

$$g^* h: g^* f^* M \to \overset{i}{\bigwedge} g^* \Omega_Y^1.$$

Thus either $(f \circ g)(C) \subset X \setminus U$ or $(f \circ g)(C)$ is a single point. This will allow us to identify the fibers of $f$.

To prove (3.1.1), pick a general point $z \in Z$ and let $\varphi_z: \mathbb{P}^1 \to Y$ be the restriction of $\varphi$ to $\{z\} \times \mathbb{P}^1$. Then

$$\Omega_{Z \times \mathbb{P}^1 | \{z\} \times \mathbb{P}^1}^1 \cong \mathcal{O}_{\mathbb{P}^1}^{n-1} + \mathcal{O}_{\mathbb{P}^1}(-2),$$

and $\varphi$ gives a map

$$\Phi: (f \circ \varphi_z)^* M \xrightarrow{\varphi_z^* h} \varphi_z^* \overset{i}{\bigwedge} \Omega_Y^1 \xrightarrow{\wedge^i d\varphi} \overset{i}{\bigwedge} \left( \mathcal{O}_{\mathbb{P}^1}^{n-1} + \mathcal{O}_{\mathbb{P}^1}(-2) \right),$$

which is nonzero for general $z$ since $\varphi$ is separable. Thus $\deg(f \circ \varphi_z)^* M \le 0$. By the above remarks, this implies that $f \circ \varphi_z$ is a constant morphism.

Pick a point $0 \in \mathbb{P}^1$ and define $\psi: Z \to X$ by $\psi(z) := f \circ \varphi(z, 0)$. This proves (3.1.1).

Let $Z_F$ be the generic fiber of $\psi$. We obtain a dominant $F$-morphism $\varphi_F: Z_F \times \mathbb{P}^1 \to Y_F$ which is birational (respectively, separable) if and only if $\varphi$ is birational (respectively, separable).

Conversely, if $W_F$ is any variety and $W_F \to Y_F$ a morphism, then it extends to a map $W \dashrightarrow Y$ of the same degree. This proves (3.1.2).

Finally assume (3.1.3). Then there is an open set $Y^0 \subset Y$ with the property that if $y_1, y_2 \in Y^0$ and $f(y_1) = f(y_2)$, then there is a morphism $f = f_{(y_1, y_2)}: \mathbb{P}^1 \to Y$ such that $y_1, y_2 \in \operatorname{im} f$, $Y$ is smooth along $\operatorname{im} f$ and $f^* T_Y$ is semipositive.

Let $\varphi: Y \dashrightarrow Y$ be a birational selfmap of $Y$; $\varphi$ is defined outside a codimension 2 set $Z \subset Y$. By [Kollár96, II.3.7], the image of the general $f_{(y_1, y_2)}$ is disjoint from $Z$. Thus we have an injection

$$f_{(y_1, y_2)}^* T_Y \hookrightarrow (\varphi \circ f_{(y_1, y_2)})^* T_Y,$$

which shows that the latter is also semipositive. Thus $\varphi(y_1)$ and $\varphi(y_2)$ are in the same fiber of $f$. Therefore $\varphi$ preserves $f$, which gives the exact sequence

$$1 \to \mathrm{Bir}(Y_F) \to \mathrm{Bir}(Y) \to \mathrm{Bir}(X). \quad \Box$$

Applying Theorem 3.1 to the projection $Y \to \mathbb{P}^m \times \mathbb{P}^n \to \mathbb{P}^m$ of Corollary 2.5 we obtain:

**Corollary 3.2** *Fix a prime $p$ and let $k$ be an algebraically closed field of characteristic $p$. Let $s \in H^0(\mathbb{P}^m \times \mathbb{P}^n, \mathcal{O}(a,b)^{\otimes p})$ be a general section. Assume that $ap > m + 10$ and $bp = n + 1$.*

*Then $Y' := \mathbb{P}^m \times \mathbb{P}^n[\sqrt[p]{s}]$ is ruled if and only if the generic fiber of $Y' \to \mathbb{P}^m$ is ruled over the field $k(x_1, \ldots, x_m)$.*

*Furthermore, $Y'$ has a degree $d$ separable uniruling if and only if the generic fiber of $Y' \to \mathbb{P}^m$ has a degree $d$ separable uniruling over the field $k(x_1, \ldots, x_m)$.* $\Box$

Corollary 3.2 naturally leads to the following:

**Question 3.3** Let $X_F$ be a variety over a field. When is $X_F$ ruled over $F$?

The problem is mainly interesting when $X_F$ is geometrically ruled. I cannot say much in general, so I consider only two simple examples: conics and Del Pezzo surfaces. One should bear in mind that in our applications, $F$ is the function field of a variety in positive characteristic, so is not perfect. Also, we need these results in characteristic 2 for conics and in characteristic 3 for cubic surfaces. These are the most unusual cases.

**Example 3.3.1** If $X_F$ is an arbitrary variety having no $F$-points, then $X_F$ is not rational, but it can happen that it is ruled. For instance, if $Y_F$ has no $F$-points then $Y \times \mathbb{P}^1$ is ruled and has no $F$-points.

This can happen even for a quadric in $\mathbb{P}^3$. For example, choose $a, b \in F$ such that $C = (x_0^2 + ax_1^2 + bx_2^2 = 0)$ has no $F$-points (say $F = \mathbb{R}$ and $a = b = 1$). Then $C \times \mathbb{P}^1$ is birational to the quadric $Q = (y_0^2 + ay_1^2 + by_2^2 + aby_3^2 = 0)$ via the map

$$\varphi \colon (x_0 : x_1 : x_2, s : t) \mapsto (sx_0 + atx_1 : sx_1 - tx_0 : sx_2 : tx_2);$$

but $Q$ clearly has no $F$-points.

# 4    Conics over nonclosed fields

The aim of this section is to study conics over arbitrary fields. We study when they are ruled or uniruled. The main result is Theorem 4.1, but for the applications we also need Proposition 4.2.

**Theorem 4.1** *Let $F$ be a field and $C_F \subset \mathbb{P}_F^2$ an irreducible and reduced conic (which may be reducible or nonreduced over $\overline{F}$). The following are equivalent:*

**4.1.1** *$C_F$ has a point in $F$.*

**4.1.2** *$C_F$ is ruled.*

**4.1.3** *$C_F$ has an odd degree uniruling.*

*If $C_F$ is smooth then these are also equivalent to*

**4.1.4** *$C_F \cong \mathbb{P}_F^1$.*

**Proof** Let $P$ be an $F$-point of $C$. If $C$ is geometrically irreducible, then projecting it from $P$ gives a birational map $C \to \mathbb{P}_F^1$; hence $C$ is ruled. Otherwise, projection exhibits $C$ as a cone over a length two subscheme of $\mathbb{P}_F^1$, thus again $C$ is ruled. This shows that $(4.1.1) \Rightarrow (4.1.2)$ while $(4.1.2) \Rightarrow (4.1.3)$ is clear.

The proof of $(4.1.3) \Rightarrow (4.1.1)$ is longer. Let $A$ be a zero dimensional $F$-scheme and $\varphi_F : A \times_F \mathbb{P}^1 \to C_F$ an odd degree uniruling. Let $A_i \subset A$ be the irreducible components. $\deg \varphi_F = \sum_i \deg(\varphi_{F|A_i \times \mathbb{P}^1})$, thus one of the $\deg(\varphi_{F|A_i \times \mathbb{P}^1})$ is odd. Thus we may assume that $A$ is irreducible. $\deg(\varphi_{F|\mathrm{red}\,A \times \mathbb{P}^1})$ divides $\deg \varphi_F$, hence we may also assume that $A = \mathrm{Spec}_F F'$ where $F' \supset F$ is a field extension.

By base change to $F'$ we obtain

$$\mathbb{P}_{F'}^1 \hookrightarrow \mathrm{Spec}_{F'}(F' \otimes_F F') \times \mathbb{P}_F^1 \to C_{F'}.$$

Thus by the Lüroth theorem, every irreducible component of $\mathrm{red}\,C_{F'}$ is birational to $\mathbb{P}_{F'}^1$. Passing to the algebraic closure we obtain

$$\varphi_{\overline{F}} : \mathrm{Spec}_{\overline{F}}(\overline{F} \otimes_F F') \times \mathbb{P}^1 \to C_{\overline{F}}.$$

The left hand side may have several irreducible components, conjugate to each other. Let $\overline{\varphi} : \mathbb{P}^1 \to \mathbb{P}^1$ be the induced map between any of the irreducible and reduced components. By counting degrees we obtain the following:

**Claim 4.1.5** *Notation as above. Then*

**4.1.5.1** $\deg \varphi_F = \deg(F'/F) \deg(\overline{\varphi})$ *if $C_{\overline{F}}$ is a smooth conic, and*

**4.1.5.2** $2 \deg \varphi_F = \deg(F'/F) \deg(\overline{\varphi})$ *if $C_{\overline{F}}$ is a singular conic.*

*Thus $\deg(F'/F)$ is odd in the first case, and not divisible by 4 in the second.* □

Assume first that $C_{\overline{F}}$ is a smooth conic. $A \times_F \operatorname{Spec} F'$ has a closed point, thus we get a morphism $\mathbb{P}^1_{F'} \to C_{F'}$. Hence $C_{F'}$ has a point in $F'$. This in turn gives an odd degree point on $C_F$; let $L$ be the corresponding line bundle. The restriction of $\mathcal{O}_{\mathbb{P}^2}(1)$ to $C_F$ is a line bundle of degree 2. We conclude that $C_F$ has a line bundle of degree 1. Any of its sections gives an $F$-point on $C_F$.

If $C_{\overline{F}}$ is a pair of intersecting lines, then the intersection point is defined over $F$.

Finally consider the case when $C_{\overline{F}}$ is a double line; this can happen only in characteristic 2. The equation of $C_F$ is $\sum b_i x_i^2 = 0$.

**Claim 4.1.6** *Let $C_F = (\sum b_i x_i^2 = 0)$ be an irreducible conic over a field of characteristic 2. Let $E/F$ be a separable extension. Then any $E$-point of $C_F$ is an $F$-point.*

**Proof** We may assume that $E/F$ is Galois. Assume that $P$ is an $E$-point which is not an $F$-point. Conjugates of $P$ over $F$ also give $E$-points, thus we obtain that red $C_{\overline{F}}$ is defined over $E$. red $C_{\overline{F}}$ is also defined over the purely inseparable extension $F^i = F(\sqrt{b_0}, \sqrt{b_1}, \sqrt{b_2})$, hence also over the intersection $F'' \cap F^i = F$ (cf. [Kollár96, I.3.5]). This is a contradiction. $\square$

As in the irreducible case, we know that $C_{F'}$ has an $F'$-point. Thus by Claim 4.1.6, $F'/F$ is not separable. In view of (4.1.5.2) we conclude that there is a subextension $F' \supset F'' \supset F$ such that $F'/F$ has odd degree (hence is separable) and $F' = F''(\sqrt{s})$ for some $s \in F''$. By Claim 4.1.6 it is enough to show that $C$ has an $F''$-point. As we mentioned earlier, red $C_{F'}$ is birational to $\mathbb{P}^1_{F'}$, thus red $C_{F'}$ is a line in $\mathbb{P}^2$. Therefore $C_{F'}$ is a double line with equation $(\sum a_i x_i)^2 = 0$ where $a_i \in F'$. $a_i^2 \in F''$, thus the equation of $C$ over $F''$ is $\sum a_i^2 x_i^2 = 0$. Write $a_i = c_i + s d_i$ where $c_i, d_i \in F''$. The equation of $C$ is

$$\sum a_i^2 x_i^2 = (\sum c_i x_i)^2 + s^2 (\sum d_i x_i)^2 = 0.$$

The solution of $\sum c_i x_i = \sum d_i x_i = 0$ gives an $F''$-point on $C$.

The equivalence with (4.1.4) was established in the course of the proof. $\square$

**Remark 4.1.7** If $C_{\overline{F}}$ is a smooth conic, then $C_F$ has a degree 2 separable uniruling. Take any general line in $\mathbb{P}^2$. Its intersection points with $C_F$ are in a separable extension $E \supset F$ of degree 2. Thus $C_E \cong \mathbb{P}^1_E$.

It remains to establish that the generic fibers appearing in Theorem 3.1 are not ruled. There should be a general result about conics over function fields, but I could not find a simple proof. For our applications the following is sufficient:

**Proposition 4.2** *Let $k$ be a field of characteristic 2 and $F = k(x_1, \ldots, x_m)$ the field of rational functions in $m \geq 2$ variables. Fix an even integer $d \geq 2$ and let $a, b, c \in k[x_1, \ldots, x_m]$ be general (inhomogeneous) polynomials of degree $d$. Then the conic*

$$C = (y_0^2 = ay_1^2 + by_1y_2 + cy_2^2) \subset \mathbb{P}_F^2$$

*is not ruled (over $F$). Moreover $C$ does not have any odd degree uniruling.*

**Comments 4.2.1** It is worth remarking that for Proposition 4.2 to hold, it is essential that $k(x_1, \ldots, x_m)$ is not perfect. A point on $C$ is given by $P = (\sqrt{a}, 1, 0)$. If $F$ is perfect of characteristic 2, then $P$ is an $F$-point and $C$ is rational.

**Proof** By Theorem 4.1, it is sufficient to establish that $C$ has no $F$-points. We can identify $ay_1^2 + by_1y_2 + cy_2^2$ with a section $s$ of $\mathcal{O}_{\mathbb{P}^m \times \mathbb{P}^1}(d, 2)$. Let $Y := \mathbb{P}^m \times \mathbb{P}^1[\sqrt{s}]$ be the corresponding double cover. The generic fiber of $\pi \colon Y \to \mathbb{P}^m$ is $C$; thus it is sufficient to prove that $\pi$ has no sections. By Lemma 2.2.3, $Y$ has only isolated singularities. The fibers of $\pi$ over $b = 0$ are double lines. This shows that $\pi$ does not even have local sections at the generic point of $b = 0$. $\square$

# 5    Del Pezzo surfaces over a nonclosed field

## 5.1    Del Pezzo surfaces

To complete the proof of Theorem 1.3, we need to study the Del Pezzo surfaces:

**5.1.1** $S_F$ with equation $u^3 = f_3(x, y, z)$ in characteristic 3.

Although we do not need it, it is very natural to study also the surfaces:

**5.1.2** $T_F$ with equation $u^2 = f_4(x, y, z)$ in characteristic 2.

To get an idea of the geometry of these surfaces, we study them first over perfect fields.

**Remark 5.2 (The case of perfect fields)** Assume first that our base field is algebraically closed. It is easy to see that we can write

$$f_3 = l_1l_2l_3 + l_4^3, \quad \text{and} \quad f_4 = l_1l_2l_3l_4 + q^2$$

where the $l_i$ are linear and $q$ quadratic. Thus we can make coordinate changes $u = u - l_4$ (respectively, $u = u - q$) and then a suitable coordinate change among the $x, y, z$ to reduce the equations to the form

$$u^3 = xyz, \quad \text{and} \quad u^2 = xyz(x + y + z).$$

From this we see that the cubic $S$ has three singular points of type $A_2$. The degree two Del Pezzo $T$ has seven singular points of type $A_1$ at the seven points of $\mathbb{F}_2\mathbb{P}^2$.

Consider next the case when the base field $F$ is perfect. Then these singular points are defined over $F$. For the cubic $S_F$ resolve the singular points, and contract the birational transforms of the 3 coordinate axes to get a Del Pezzo surface of degree 6. Over a perfect field $F$, a Del Pezzo surface of degree 6 is rational if and only if it has an $F$-point [Manin72, IV.7.8]. Our surface $S_F$ does have $F$-points over perfect fields of characteristic 3, for example $P = (1, 0, 0, \sqrt[3]{f_3(1, 0, 0)})$.

For the surface $T_F$, resolve the singular points, and contract the birational transforms of the seven lines in $\mathbb{F}_2\mathbb{P}^2$ to get a Brauer–Severi variety. It has a point in a degree 7 extension, hence it is isomorphic to $\mathbb{P}^2$. Thus our surfaces $S_F$ and $T_F$ are rational over any perfect field.

In our case the field $F$ is not perfect, the surfaces $S_F$ and $T_F$ are regular over $F$ but they are not smooth. In order to show that they are not ruled, we have to understand how the presence of regular but nonsmooth points effects the birational geometry of a surface.

The crucial result is the following:

**Theorem 5.3** *Let $F$ be a field and $S, T$ regular and proper surfaces over $F$. Assume that $T$ is smooth except possibly at finitely many points. Let $f: S \dashrightarrow T$ be a birational map.*

*Then $f$ is defined at all nonsmooth points of $S$.*

**Proof**  There is a sequence $p: S' \to S$ of blowups of closed points such that $f \circ p: S' \to T$ is a morphism. Let $P \in S$ be a closed nonsmooth point, and assume that $f$ is not defined at $P$. Then $p^{-1}(P)$ is 1-dimensional and there is an irreducible component $E \subset p^{-1}(P)$ such that $p \circ f$ is a local isomorphism at the generic point of $E$. This implies that $S'$ is smooth at the generic point of $E$. This contradicts the following lemma.  $\square$

**Lemma 5.3.1** *Let $F$ be a field and $S, S'$ regular surfaces over $F$. Suppose that $p: S' \to S$ is a proper and birational morphism and $P \in S$ a closed nonsmooth point.*

*Then $S$ is not smooth at all points of $p^{-1}(P)$.*

**Proof** By induction, it is sufficient to consider the case when $p$ is the blowup of $P$. We may also assume that $S$ is an affine neighborhood of $P$ such that the maximal ideal $m_P$ is generated by two global sections $u, v \in m_P \subset \mathcal{O}_S$. The blowup can be described by two affine charts, one of which is

$$S_1' := (u - vs = 0) \subset S \times \mathbf{A}^1, \quad \text{where } s \text{ is a global coordinate on } \mathbf{A}^1.$$

By assumption $S$ is not smooth at $P$, so $S \times \mathbf{A}^1$ is not smooth along $P \times \mathbf{A}^1$. $S_1'$ is a Cartier divisor on $S \times \mathbf{A}^1$, thus is it also not smooth along $P \times \mathbf{A}^1$. This is what we had to prove. $\square$

In the course of the proof we used the following results about birational transformations of regular surfaces. See, for example, [Zariski58] for a proof.

**Proposition 5.3.2** *Let $S, T$ be proper and regular surfaces over a field $F$ and $\varphi \colon S \dashrightarrow T$ a birational map. Then $\varphi$ is a composite of blowups and blowdowns (of closed points). In particular, $h^i(S, \mathcal{O}_S) = h^i(T, \mathcal{O}_T)$.* $\square$

**Corollary 5.3.3** *Let $F$ be a field and $S$ a regular and proper surface over $F$. Assume that $S$ is smooth except possibly at finitely many points. Let $f \colon S \dashrightarrow S$ be a birational map.*
*Then $f$ is a local isomorphism at all nonsmooth points of $S$.*

**Proof** Factor $f$ as

$$f \colon S \xleftarrow{\ p\ } S' \xrightarrow{\ p'\ } S$$

where $p, p'$ are birational morphisms. By Theorem 5.3 we see that $p$ and $p'$ are both local isomorphisms at the nonsmooth points of $S$. $\square$

This gives the following rationality criterion:

**Corollary 5.4** *Let $F$ be a field and $S$ a proper regular surface over $F$. Assume that $S$ is generically smooth. The following are equivalent:*

**5.4.1** *$S$ is rational (over $F$).*

**5.4.2** *There is a two dimensional linear system $L = |C|$ on $S$ with (infinitely near) base points $P_i$ of multiplicity $m_i$ such that*

**5.4.2.1** *a general $C \in L$ is birational to $\mathbb{P}^1$;*

**5.4.2.2** *$S$ is smooth along a general $C \in L$;*

**5.4.2.3** *$C \cdot K_S + \sum m_i = -3$ and $C^2 - \sum m_i^2 = 1$.*

**Proof** Assume that there is a birational map $f \colon S \dashrightarrow \mathbb{P}^2_F$. Let $Z \subset S$ be the locus of nonsmooth points. By Theorem 5.3, $f$ is defined along $Z$ and $f(Z)$ is zero dimensional. Set $L = f_*^{-1}|\mathcal{O}_{\mathbb{P}^2}(1)|$. In (5.4.2.1–2) are clear, and (5.4.2.3) is the usual equalities (5.4.3.1).

Conversely, assume (5.4.2). Resolve the base points of $L$ to obtain a base point free linear system $L'$ on $S'$. From (5.4.2.3) we obtain that

$$C' \cdot K_{S'} = -3 \quad \text{and} \quad C'^2 = 1 \quad \text{for } C' \in L'.$$

Thus the linear system $L'$ maps $S'$ birationally to $\mathbb{P}^2_F$. $\square$

What we really need is a ruledness criterion. If $S$ is ruled, it can be birational to a surface which is the product of $\mathbb{P}^1$ with a conic which has no $F$-points. Thus the natural linear system obtained on $S$ is two dimensional and its general member is geometrically reducible. I found it clearer to concentrate instead on a single curve, which may not be defined over $F$. We do not get an equivalence any longer, but for our applications this does not matter.

**Remark 5.4.3** For linear systems we considered base points with multiplicities. If we look at a general member, the corresponding notion is a curve with assigned multiplicities. All infinitely near multiple points are assigned with their multiplicity, but we may also have some smooth points assigned with multiplicity one.

If $C \subset S$ is a curve on a smooth surface with assigned multiplicities $m_i$ at the points $P_i$ and $\varphi \colon S \to S'$ is a birational map, there is a natural birational transform $C' \subset S'$ with assigned multiplicities $m_i'$ at the points $P_i'$. To define this, we need only the case when $\varphi$ is the blowup of a point or its inverse, where the definition is the obvious one. The values of the expressions

$$C \cdot K_S + \sum m_i \quad \text{and} \quad C^2 - \sum m_i^2 \qquad (5.4.3.1)$$

are birational invariants, cf. [Hudson27, p. 5].

**Corollary 5.5** *Let $F$ be a field and $S$ a regular and proper surface over $F$ such that $S$ is generically smooth, geometrically irreducible and $h^1(S, \mathcal{O}_S) = 0$. Assume that $S$ is ruled (over $F$).*

*Then there is a rational curve $C \subset S_{\bar{F}}$ with assigned (infinitely near) multiple points $P_i$ of multiplicity $m_i$ such that*

**5.5.1** $S_{\bar{F}}$ *is smooth along $C$.*

**5.5.2** $C \cdot K_S + \sum m_i = -2$ *and* $C^2 - \sum m_i^2 = 0$.

**5.5.3** $\mathcal{O}_{S_{\bar{F}}}(2C) \in \operatorname{Pic} S$.

**Proof** By assumption there is a regular, geometrically integral curve $D$ and a birational map $f: S \dashrightarrow D \times \mathbb{P}^1$. By Proposition 5.3.2,

$$0 = h^1(S, \mathcal{O}_S) = h^1(D \times \mathbb{P}^1, \mathcal{O}_{D \times \mathbb{P}^1}) = h^1(D, \mathcal{O}_D).$$

Thus $D$ is isomorphic to a smooth conic.

Let $d \in D_{\overline{F}}$ be a general point, $C' = d \times \mathbb{P}^1$ and $C = f_*^{-1}(C') \subset S_{\overline{F}}$ the corresponding birational transform. $S$ is smooth along $C$ by Theorem 5.3, and (5.5.2) is the usual equalities (5.4.3.1).

Finally, $D$ has a degree 2 point defined over $F$, thus the line bundle $\mathcal{O}_{D \times \mathbb{P}^1}(2C')$ is defined over $F$. Therefore $\mathcal{O}_{S_{\overline{F}}}(2C)$ is also defined over $F$. $\square$

The main result of this section is Theorem 5.7. Over a perfect field it is a special case of more general results of Segre and Manin (see [Manin72]). The proofs in [Manin72] use the structure of the Picard group of smooth Del Pezzo surfaces to compute the action of certain involutions. In our case the Picard groups are small by Lemma 5.6, and it is easier to use the geometric ideas of [Segre43, Segre51] to analyze the involutions. Theorem 5.3 essentially says that all the relevant geometry takes place inside the smooth locus, where the geometric description of the involutions works well.

**Lemma 5.6** *Let $S_F$ denote an integral cubic surface $u^3 = f_3(x, y, z) \subset \mathbb{P}^3$ for char $F = 3$, or an integral double plane with equation $u^2 = f_4(x, y, z)$ for char $F = 2$.*

*Then $\mathrm{Pic}(S_F) = \mathbb{Z}[-K_S]$.*

**Proof** Let $\pi: S_F \to \mathbb{P}^2$ be the projection to the $(x, y, z)$-plane. $\pi$ is purely inseparable, thus if $C \subset S_F$ is any divisor then $\pi^*(\pi_*(C)) = (\deg \pi)C$. Therefore $-K_S = \pi^*(\mathcal{O}_{\mathbb{P}^2}(1))$ generates $\mathrm{Pic}(S_F) \otimes \mathbb{Q}$. Since $K_S^2 \leq 3$, we get that $K_S$ is not divisible in $\mathrm{Pic}(S_F)$, hence $-K_S$ generates $\mathrm{Pic}(S_F)/(\text{torsion})$. Thus we are left to prove that $\mathrm{Pic}\, S$ has no torsion.

Let $[C] \in \mathrm{Pic}\, S$ be a numerically trivial Cartier divisor. By Riemann-Roch,

$$h^0(\mathcal{O}_S(C)) + h^0(\mathcal{O}_S(K_S - C)) \geq \chi(\mathcal{O}_S) = 1.$$

$-K_S$ is ample, so $h^0(\mathcal{O}_S(K_S - C)) = 0$. Thus $\mathcal{O}_S(C)$ has a section and $\mathcal{O}_S(C) \cong \mathcal{O}_S$. $\square$

**Theorem 5.7** *Let $F$ be a field and $S$ a regular Del Pezzo surface over $F$. Assume that $S_{\overline{F}}$ is integral, $\chi(\mathcal{O}_S) = 1$, $\mathrm{Pic}\, S = \mathbb{Z}[-K_S]$ and $K_S^2 = 1, 2, 3$.*

*Then $S$ is not ruled (over $F$).*

**Remark 5.7.1** Let $S$ be a regular Del Pezzo surface over $F$ such that $S_{\overline{F}}$ is integral, $\chi(\mathcal{O}_S) = 1$, and $K_S^2 = 1, 2, 3$. The structure theory of Del Pezzo surfaces shows that $S$ is a cubic surface for $K_S^2 = 3$, a double plane for $K_S^2 = 2$ and as expected for $K_S^2 = 1$ (cf. [Kollár96, III.3]).

[Reid94] contains examples of nonnormal Del Pezzo surfaces for which $\chi(\mathcal{O}_S) < 1$. Some of these may have regular models over nonperfect fields. I have not checked if this indeed happens or whether anything can be said about their arithmetic properties.

**Proof**   Assuming that $S$ is ruled, we derive a contradiction. Corollary 5.5 guarantees the existence of a rational curve $C \subset S_{\overline{F}}$ satisfying the properties (5.5.1–3). By (5.5.3) $\mathcal{O}(2C)$ is in Pic $S$. Since $K_S^2 \leq 3$, $K_S$ is not divisible by 2 in $\operatorname{Pic}(S_{\overline{F}})$ and therefore $\mathcal{O}(C)$ is in Pic $S$. This implies that $C \in |-dK_{S_{\overline{F}}}|$ for some $d$. We show that there cannot be such a curve:

**Proposition 5.8** *Let $S$ be an integral Del Pezzo surface over an algebraically closed field such that $\chi(\mathcal{O}_S) = 1$, and $K_S^2 = 1, 2, 3$. Then $S$ does not contain any curve $C$ satisfying the following properties:*

**5.8.1** *$C$ is birational to $\mathbb{P}^1$;*

**5.8.2** *$S$ is smooth along $C$;*

**5.8.3** *$C \in |-dK_S|$ for some $d$.*

**5.8.4** *$\sum_i m_i = dK_S^2 - 2$ and $\sum_i m_i^2 = d^2 K_S^2$, where $P_i$ are the (assigned and infinitely near) multiple points of $C$ with multiplicity $m_i$.*

**Proof**   If $m_i \leq d$ for every $i$ then

$$\sum_i m_i^2 \leq d \sum_i m_i = d^2 K_S^2 - 2d < d^2 K_S^2,$$

a contradiction. Thus there is a point $P = P_1$ such that $m = m_1 > d$.

Let $P \in L \subset S$ be a line (that is, $-K_S \cdot L = 1$). Then $m \leq (C \cdot L) = d$. Therefore $P$ cannot lie on any line.

If $K_S^2 = 1$ then any member of $|-K_S|$ is a line, hence there is a line through any point. Hence we are done if $K_S^2 = 1$.

Next consider the case when $K_S^2 = 3$. That is, $S \subset \mathbb{P}^3$ is a cubic surface.

Let $D \subset S$ be the intersection of $S$ with the tangent plane at $P$. Since there is no line through $P$, $D$ is an irreducible cubic whose unique singular point is at $P$. In particular, $D$ is contained in the smooth locus of $S$.

The point $P$ determines a birational selfmap $\tau$ of $S$ as follows. Take any point $Q \in S$, connect $P, Q$ with a line and let $\tau(Q)$ be the third intersection point of the line with $S$. $\tau$ is an automorphism of $S - D$. Another way of describing $\tau$ is the following. Projecting $S$ from $P$ gives a diagram

$$
\begin{array}{ccc}
B_P S & \xrightarrow{\;q\;} & \mathbb{P}^2 \\
{\scriptstyle p}\downarrow & & \\
S & &
\end{array}
$$

Let $E \subset B_P S$ be the exceptional curve; $C', D' \subset B_P S$ the birational transforms of $C, D$. Then $q$ is a degree two morphism and $\tau$ is the involution interchanging the two sheets. (Computing with the local equation at $P$ shows that $q$ is separable in characteristic 2 if $D$ is irreducible, thus $\tau$ always exists.) Furthermore, $\tau(E) = D'$ and $p^* \mathcal{O}_S(1)(-E) = q^* \mathcal{O}_{\mathbb{P}^2}(1)$. Thus

$$
C' + (m - d)E \in |q^* \mathcal{O}_{\mathbb{P}^2}(d)|, \quad \text{hence} \quad \tau(C' + (m - d)E) \in |q^* \mathcal{O}_{\mathbb{P}^2}(d)|.
$$

Pushing this down to $S$ we obtain that

$$
\tau(C) + (m - d)D \in |\mathcal{O}_S(d)|, \quad \text{hence} \quad \tau(C) \in |\mathcal{O}_S(d - (m - d))|.
$$

Thus $\tau(C)$ satisfies all the properties (5.8.1–4) and its degree is lower than the degree of $C$. We obtain a contradiction by induction on $d$.

A similar argument works if $K_S^2 = 2$, but the details are a little more complicated. I just outline the arguments, leaving out some simple details.

We already proved that $P$ is not on any line, and a similar argument shows that $P$ cannot be a singular point of a member of $|-K_S|$.

Since $h^0(S, -2K_S) = 7$, there is a curve $D \in |-2K_S|$ which has a triple point at $P$. In fact, $D$ is unique, it is a rational curve and $P$ is its only singular point. Thus $S$ is smooth along $D$. As before we look at the blowup diagram

$$
\begin{array}{ccc}
B_P S & \xrightarrow{\;q\;} & Q \subset \mathbb{P}^3 \\
{\scriptstyle p}\downarrow & & \\
S & &
\end{array}
$$

where $Q$ is a quadric cone, the image of $B_P S$ by the linear system $|-2K_{B_P S}|$. Now $q$ is a degree two morphism and $\tau$ is the involution interchanging the two sheets. (Again one can see that $q$ is separable in characteristic 2 if $|-K_S|$ does not have a member which is singular at $P$.) Let $E \subset B_P S$ be the exceptional curve; $C', D' \subset B_P S$ the birational transforms of $C, D$. $\tau(E) = D'$ and $p^* \mathcal{O}_S(2)(-2E) = q^* \mathcal{O}_{\mathbb{P}^3}(1)$. As before we obtain that

$$
\tau(C) \in |\mathcal{O}_S(d - 2(m - d))|.
$$

We obtain a contradiction by induction on $d$. $\quad\square$

# References

[Artin–Mumford72] M. Artin and D. Mumford, Some elementary examples of uniruled varieties which are not rational, Proc. London. Math. Soc. **25** (1972) 75–95

[Bardelli84] F. Bardelli, Polarized mixed Hodge structures: on irrationality of threefolds via degeneration, Annali di Math. pura e appl. **137** (1984) 287–369

[Beauville77] A. Beauville, Variétés de Prym et jacobiennes intermédiaires, Ann. Sci. E. N. S. **10** (1977) 309–391

[Clemens–Griffiths72] H. Clemens and P. Griffiths, The intermediate Jacobian of the cubic threefold, Ann. Math. **95** (1972) 281–356

[CTO89] J.-L. Colliot-Thélène and M. Ojanguren, Variétés unirationelles non rationelles: au-del de l'exemple d'Artin et Mumford, Inv. Math. (1989) 141–158

[Hudson27] H. Hudson, Cremona transformations, Cambridge Univ. Press, 1927

[Iskovskikh80] V. A. Iskovskikh, Birational automorphisms of three dimensional algebraic varieties, J. Soviet Math **13** (1980) 815–868

[Iskovskikh–Manin71] V. A. Iskovskikh and Yu. I. Manin, Three-dimensional quartics and counterexamples to the Lüroth problem, Mat. Sb. **86** (1971) 140–166 = Math. USSR Sbornik **15** (1971) 141–166

[Kollár95] J. Kollár, Nonrational hypersurfaces, J. Amer. Math. Soc. **8** (1995) 241–249

[Kollár96] J. Kollár, Rational curves on algebraic varieties, Springer Verlag, Ergebnisse der Math. **32**, 1996

[Manin66] Yu. I. Manin, Rational surfaces over perfect fields (in Russian) Publ. Math. IHES **30** (1966), 55–114

[Manin72] Yu. I. Manin, Cubic forms (in Russian), Nauka 1972 = Cubic forms. Algebra, geometry, arithmetic. Second edition. North-Holland, Amsterdam–New York, 1986

[Nishimura55] H. Nishimura, Some remarks on rational points, Mem. Coll. Sci. Univ. Kyoto **29** (1955) 189–192

[Pukhlikov87] A. V. Pukhlikov, Birational isomorphisms of four dimensional quintics, Invent. Math. **87** (1987) 303–329

[Reid94] M. Reid, Nonnormal del Pezzo surfaces, Publ. RIMS Kyoto Univ. **30** (1994) 695–727

[Sarkisov81] V. G. Sarkisov, Birational automorphisms of conic bundles, Math. USSR Izv. **17** (1981) 177–202

[Sarkisov82] V. G. Sarkisov, On the structure of conic bundles, Math. USSR Izv. **20** (1982) 355–390

[Segre43] B. Segre, A note on arithmetical properties of cubic surfaces, J. London Math. Soc. **18** (1943) 24–31

[Segre51] B. Segre, The rational solutions of homogeneous cubic equations in four variables, Notae Univ. Rosario **2** (1951) 1–68

[Shokurov83] V. Shokurov, Prym varieties: theory and applications, Izv. A. N. SSSR Ser. Mat. **47** (1983) 785–855 = Math. USSR. Izv. **23** (1984) 83–147

[Zariski58] O. Zariski, Introduction to the problem of minimal models in the theory of algebraic surfaces, Math. Soc. Japan, 1958

János Kollár,
Princeton University, Princeton, NJ 08544-1000
e-mail: kollar@math.princeton.edu

# Essentials of the method
# of maximal singularities

Aleksandr V. Pukhlikov

## 1 Historical introduction

### Noether's theorem

We start with a brief description of the principal events in the history of our subject. Around 1870, M. Noether [N] discovered that the group of birational automorphisms of the projective plane Bir $\mathbb{P}^2$, which is also known as the *Cremona group* Cr $\mathbb{P}^2$, is generated by its subgroup Aut $\mathbb{P}^2 = $ Aut $\mathbb{C}^3/\mathbb{C}^*$ together with any *standard quadratic Cremona transformation* $\tau$, that is, the map

$$\tau\colon (x_0 : x_1 : x_2) \mapsto (x_1 x_2 : x_0 x_2 : x_0 x_1)$$

in any system of homogeneous coordinates.

Noether's argument is as follows: take any Cremona transformation

$$\chi\colon \mathbb{P}^2 \dashrightarrow \mathbb{P}^2.$$

Then either $\chi$ is a projective isomorphism, or the proper inverse image of the linear system of lines of $\mathbb{P}^2$ is a linear system $|\chi|$ of curves of degree $n = n(\chi) \geq 2$ with assigned base points $a_1, \ldots, a_N$ (possibly including infinitely near points). Let $\nu_1, \ldots, \nu_N$ be their multiplicities with respect to the linear system $|\chi|$, and assume that $\nu_1 \geq \nu_2 \geq \cdots \geq \nu_N$. Then, because two lines intersect in one point, the *free intersection* of two curves of $|\chi|$ (that is, their intersection outside the base locus) equals 1. Hence

$$n^2 = \sum_{i=1}^{N} \nu_i^2 + 1.$$

Moreover, the curves in $|\chi|$ are rational and nonsingular outside the base locus, and so, computing their geometric genus in terms of their arithmetic genus, we get

$$(n-1)(n-2) = \sum_{i=1}^{N} \nu_i(\nu_i - 1).$$

73

It is easy to deduce from these two equalities that $N \geq 3$ and that the three maximal multiplicities satisfy *Noether's inequality*

$$\nu_1 + \nu_2 + \nu_3 > n.$$

If $a_1, a_2, a_3$ are actual points of the plane $\mathbb{P}^2$ (that is, not infinitely near points), we can then consider the composite

$$\chi \circ \tau \colon \mathbb{P}^2 \dashrightarrow \mathbb{P}^2,$$

where $\tau$ is the standard quadratic transformation constructed from these three points $(a_1, a_2, a_3)$.

Let us prove that $n(\chi \circ \tau) < n(\chi)$: indeed, the degree of a curve is its intersection number with a generic line $L$. But the intersection points of a curve $C \in |\chi \circ \tau|$ with $L$ correspond one-to-one with the points of *free intersection* of their images $\tau(C)$ and $\tau(L)$. Now $\tau(C) \in |\chi|$, whereas $\tau(L)$ is a conic passing through $a_1, a_2, a_3$; thus the intersection number equals

$$n(\chi \circ \tau) = 2n(\chi) - \nu_1 - \nu_2 - \nu_3 < n(\chi).$$

Proceeding in the same way, we can "untwist" the "maximal triples" until we arrive at the case $n(\chi) = 1$. This would prove Noether's theorem, except that Noether's argument does not work for a maximal triple involving infinitely near points. It took about 30 years to complete the proof.

## Fano and the quartic threefold

The second part of our story begins in the early years of the present century, with G. Fano's first attempts to extend two dimensional birational methods to threefolds [F1, F2]. He started by trying to describe the birational transformations of a quartic threefold $V_4 \subset \mathbb{P}^4$. In this, he really made the best possible choice of variety to study: right up to the present, the quartic threefold remains one of the main touchstones for higher dimensional birational constructions.

Following Noether's argument, Fano considers a birational transformation $\chi \colon V \dashrightarrow V'$ between two smooth quartic threefolds. Taking the proper inverse image $|\chi|$ of the linear system of hyperplane sections of $V' \subset \mathbb{P}^4$, he arrives at the following conclusion: either $|\chi|$ is cut out on $V$ by hyperplanes, and then it is a projective isomorphism; or it is cut out by hypersurfaces of degree $n(\chi) \geq 2$, in which case the base locus $|\chi|$ satisfies certain conditions that are similar to Noether's inequality. These conditions are now called the *Noether–Fano inequalities*. If $n(\chi) \geq 2$, Fano asserts that something like one of the following two cases happens:

1. *either* there is a point $x \in V$ such that $\mathrm{mult}_x |\chi| > 2n$,

2. *or* there is a curve $C \subset V$ such that $\mathrm{mult}_C |\chi| > n$.

We admit that Fano never asserted that these two cases are the *only* possibilities; thus, for example, he mentions the following additional case: a base point $x \in V$ and a base line $L \subset E$, where $E \subset \tilde{V}$ is the exceptional divisor of the blowup of $x$, satisfying the inequality

$$\mathrm{mult}_x |\chi| + \mathrm{mult}_L |\tilde{\chi}| > 3n.$$

But the general level of understanding and the technical weakness of his time prevented him from giving a rigorous and complete description of what happens when $n(\chi) \geq 2$.

Fano then asserts that neither of the above cases can happen. Since practically all of his arguments are absolutely invalid, it is really amazing that this conclusion is true. (Still more amazingly, this was invariably the case with Fano: wrong arguments often led him to true and deep conclusions.) For instance, to exclude the possibility of a curve $C$ with $\mathrm{mult}_C |\chi| > n$, he attempts to argue on the arithmetic genus of a general surface in $|\chi|$, apparently hoping to reproduce Noether's arguments in terms of the genus of a curve in $|\chi|$. However, Iskovskikh and Manin [IM] found out that in fact these arguments do not lead to any conclusion.

Having convinced himself that the case $n(\chi) \geq 2$ is impossible, Fano formulated one of his most impressive claims: any birational transformation between two nonsingular quartics in $\mathbb{P}^4$ is a projective isomorphism. In particular, the group of birational automorphisms $\mathrm{Bir}\, V = \mathrm{Aut}\, V$ is finite (trivial for sufficiently general $V$); therefore, $V$ is nonrational.

Fano did a lot of work in threefold birational geometry along these lines [F3]. He gave a description (however incomplete and unsubstantiated it may be) of birational transformations of cubic threefolds, complete intersections $V_{2\cdot 3}$ in $\mathbb{P}^5$, and many other varieties. Many of his results have not been completed to this day. However, because of the very style of Fano's work, his numerous mistakes and, generally speaking, incompatibility of his geometry with the universally adopted standards of mathematical arguments, his ideas and computations were abandoned for about twenty years.

## The work of Manin and Iskovskikh

In the late 1960s, Yu.I. Manin and V.A. Iskovskikh in Moscow started their pioneering program in threefold birational geometry, after a series of papers on two dimensional birational geometry. As a result, in 1970, they developed a method strong enough to prove Fano's claim on the quartic threefold [IM];

we refer to this as the *method of maximal singularities*. Using this method, Iskovskikh [I] subsequently proved several more of Fano's claims and corrected some of his mistakes. In the 1970s and 1980s, Iskovskikh and several of his students – A.A. Zagorskii, V.G. Sarkisov [S1, S2], S.L. Tregub [T1, T2], S.I. Khashin [Kh] and the author [P1]–[P5] – worked in this field, trying to describe birational maps of certain classes of algebraic varieties. Their work was often successful, but unfortunately not always: the method of maximal singularities was extended to a number of classes of varieties, including some of arbitrary dimension and some possibly singular, and including a large class of conic bundles. The well-known Sarkisov program [R1, C1] also has its origins in the framework of this field. At the same time, the method really only works for varieties of very small degree. We must admit that at present we have no good method of studying the birational geometry of higher dimensional Fano varieties and Fano fibrations in general.

Nevertheless, the results obtained using the method of maximal singularities cannot be proved at present in any other way. (See [K1], [K2] for an alternative approach in the spirit of characteristic $p$ tricks.)

## Acknowledgements

This paper is an extract from lectures given by the author during his stay at the University of Warwick in September–December 1995. Since [IP] has been published, it does not make sense to reproduce here all the details of the excluding/untwisting procedures. At the same time, [IP] was actually written in 1988. The real meaning of the "test class" construction has been clarified in more recent work, and some new methods of excluding maximal singularities have appeared [P5–P7]. The aim of the present paper is to give an easy introduction to the method of maximal singularities. We restrict ourselves to explaining only the crucial points. The principal and most difficult part of the method – that is, excluding infinitely near maximal singularities – is presented here in a new form, which is simple and easy; this version of the method has never been published before.

I would like to express my gratitude to Professor M. Reid who invited me to the University of Warwick and arranged my lecture course on birational geometry. I am thankful to all the staff of the Warwick Mathematics Institute for hospitality.

I would like to thank Professor V.A. Iskovskikh, who introduced me to the problem of the quintic four-fold in 1982 and thus determined the direction of my work in algebraic geometry. I am grateful to Professor Yu.I. Manin for constant and valuable support.

The author was financially supported by International Science Foundation, grant M90000, by ISF and Government of Russia, grant M90300, and by Russian Fundamental Research Fund, grant 93–011–1539.

# 2   Maximal singularities of birational maps

## The first step

Fix a projective $\mathbb{Q}$-factorial variety $V$ with at worst terminal singularities over the field $\mathbb{C}$ of complex numbers, and let $Y$ be a divisor on $V$ moving in a linear system $|Y|$ (for example, an ample divisor). We assume that we are given a birational map

$$\chi\colon V \dashrightarrow W.$$

Take the proper inverse image $|\chi| \subset |D|$ on $V$ of the linear system $|Y|$. We write $\operatorname{Bs}|\chi|$ for its base subscheme. This linear system $|\chi|$ and its base subscheme $\operatorname{Bs}|\chi|$ are our main objects of study.

**Examples 2.1** We list some of the main classes of varieties that have been studied more or less successfully using the method of maximal singularities over the last 25 years:

1. smooth quartics $V_4 \subset \mathbb{P}^4$;

2. complete intersections $V_{2\cdot 3} \subset \mathbb{P}^5$;

3. singular quartics $x \in V_4 \subset \mathbb{P}^4$;

4. smooth hypersurfaces $V_M \subset \mathbb{P}^M$;

5. double projective spaces $\sigma\colon V \to \mathbb{P}^n$ branched over a smooth hypersurface $Z_{2n} \subset \mathbb{P}^n$.

## Test pairs

**Definition 2.2** Let $W$ be a projective variety such that $\dim W = \dim V$, $\operatorname{codim} \operatorname{Sing} W \geq 2$, and $Y$ a divisor on $W$ that moves in a linear system $|Y|$. We say that $(W, Y)$ is a *test pair* if the following conditions hold:

(a) $|Y|$ is free from fixed components.

(b) Termination of adjunction: there exists $\alpha \in \mathbb{R}_+$ such that

$$|MY + NK_W| = \emptyset$$

for all $M, N \in \mathbb{Z}_+$ with $N/M > \alpha$.

The minimal $\alpha \in \mathbb{R}_+$ satisfying condition (b) is the *index* (or *threshold*) of the pair $(W, Y)$. We denote it by $\alpha(W, Y)$.

**Examples 2.3** We now give the main examples of test pairs, explaining briefly the applications for which we need them.

1. $\mathbb{P}^n$ with $Y =$ hyperplane: used to decide whether $V$ is rational. This pair has index $1/(n + 1)$.

2. A Fano fibration $\varphi: W \to S$ with $Y = \varphi^{-1}$(very ample divisor on $S$): used to decide whether there are structures of Fano fibrations on $V$; for instance, take a conic bundle or Del Pezzo fibration. The index here is obviously zero.

3. A Fano variety $V$ with $Y = -MK_V$: used to describe the group Bir $V$ and to give the birational classification within the same family of Fano varieties.

## The language of discrete valuations

We recall briefly the definitions and facts we need about discrete valuations; for more details see [P5, P6]. For $X$ a quasiprojective variety, we denote by $\mathcal{N}(X)$ the set of *geometric* discrete valuations

$$\nu: \mathbb{C}(X)^* \to \mathbb{Z},$$

having a centre on $X$. If $X$ is complete, then $\mathcal{N}(X)$ includes all the geometric discrete valuations. The centre of a discrete valuation $\nu \in \mathcal{N}(X)$ is denoted by $Z(X, \nu)$.

**Examples 2.4**    (1) Let $D \subset X$ be a prime divisor, $D \not\subset \operatorname{Sing} X$. Then $D$ determines a discrete valuation

$$\nu_D = \operatorname{ord}_D.$$

(2) Let $B \subset X$ be an irreducible subvariety, $B \not\subset \operatorname{Sing} X$. Then $B$ determines a discrete valuation:

$$\nu_B(f) = \operatorname{mult}_B(f)_0 - \operatorname{mult}_B(f)_\infty.$$

Note that if $\sigma_B: X(B) \to X$ is the blowup of $B$ and $E(B) = \sigma_B^{-1}(B)$ its exceptional divisor, then

$$\nu_B = \nu_{E(B)}$$

under the natural identification of $\mathbb{C}(X)$ and $\mathbb{C}(X(B))$.

**Definition 2.5** Let $\nu \in \mathcal{N}(X)$ be a discrete valuation. A *realization* of $\nu$ is a triple $(\widetilde{X}, \varphi, H)$, where $\varphi\colon \widetilde{X} \to X$ is a birational morphism and $H \not\subset \mathrm{Sing}\,\widetilde{X}$ a prime divisor such that $\nu = \nu_H$.

## Multiplicities

Let $\mathcal{J} \subset \mathcal{O}_X$ be a sheaf of ideals and $\nu \in \mathcal{N}(X)$.

**Definition 2.6** The multiplicity of $\mathcal{J}$ at a discrete valuation $\nu$ is given by

$$\nu(\mathcal{J}) = \mathrm{mult}_H\, \varphi^*\mathcal{J},$$

where $(\widetilde{X}, \varphi, H)$ is a realization of $\nu$.

Let $|\lambda| \subset |D|$ be a linear system of Cartier divisors and $\mathcal{L}_{|\lambda|} \subset \mathcal{O}_X(D)$ the subsheaf generated by the global sections in $|\lambda|$. Set

$$\mathcal{J}_{|\lambda|} = \mathcal{L}_{|\lambda|} \otimes \mathcal{O}_X(-D) \subset \mathcal{O}_X.$$

Obviously, $\mathcal{J}_{|\lambda|}$ is the ideal sheaf of the base subscheme $\mathrm{Bs}\,|\lambda|$.

**Definition 2.7** The multiplicity of $|\lambda|$ at a discrete valuation $\nu$ equals

$$\nu(|\lambda|) = \nu(\mathcal{J}_{|\lambda|}).$$

Now let $X$ be ($\mathbb{Q}$-)Gorenstein, and $\pi\colon X_1 \to X$ a resolution. Then

$$K_{X_1} = \pi^*K_X + \sum_i d_i E_i$$

where the $E_i \subset X_1$ are exceptional prime divisors. Consider a realization $(\widetilde{X}, \varphi, H)$ of $\nu \in \mathcal{N}(X_1) = \mathcal{N}(X)$. Then we get an inclusion

$$\varphi^*\omega_{X_1} \hookrightarrow \omega_{\widetilde{X}},$$

and the ideal sheaf

$$K(X_1, \varphi) = \varphi^*\omega_{X_1} \otimes \omega_{\widetilde{X}}^{-1} \hookrightarrow \mathcal{O}_{\widetilde{X}}$$

on $\widetilde{X}$.

**Definition 2.8** The *canonical multiplicity (discrepancy)* of $\nu$ is equal to $d_i$ if $\nu = \nu_{E_i}$, and is equal to

$$K(X, \nu) = \mathrm{mult}_H\, K(X_1, \varphi) + \sum_i d_i\nu(E_i)$$

otherwise.

**Example 2.9** Let $B \subset X$, $B \not\subset \operatorname{Sing} X$ be an irreducible subvariety of codimension $\geq 2$. Then

$$\nu_B(\mathcal{J}) = \operatorname{mult}_B \mathcal{J}, \quad \nu_B(|\lambda|) = \operatorname{mult}_B |\lambda|,$$
$$\text{and} \quad K(X, \nu_B) = \operatorname{codim} B - 1.$$

## Maximal singularities

We now return to our variety $V$ and birational map $\chi \colon V \dashrightarrow W$. Denote by $n(\chi)$ the index (threshold) of the pair $(V, D)$ (see Definition 2.2).

**Definition 2.10** A discrete valuation $\nu \in \mathcal{N}(V)$ is said to be a *maximal singularity* of $\chi$ if the following inequality holds:

$$\nu(|\chi|) > n(\chi)K(V, \nu).$$

**Theorem 2.11** *Either* $\alpha(V, D) \leq \alpha(W, Y)$, *or* $\chi$ *has a maximal singularity.*

**Proof** See [P5, P6]; this is actually so easy that it can be left as an exercise for the reader. The idea of the proof can be found in any paper concerned with these problems (for instance, [IM, I, P1, IP]). However, please bear in mind that the proof should not depend upon resolution of singularities.

**Example 2.12** Let $V$ be smooth with $\operatorname{Pic} \cong \mathbb{Z}K_V$, and assume that the anticanonical system $|-K_V|$ is free. Then

$$|\chi| \subset |-n(\chi)K_V|,$$

and for a birational automorphism $\chi \in \operatorname{Bir} V$ either $n(\chi) = 1$, or $\chi$ has a maximal singularity.

## Maximal cycles

Suppose that $V$ is nonsingular.

**Definition 2.13** An irreducible subvariety $B \not\subset \operatorname{Sing} V$ of codimension $\geq 2$ is said to be a *maximal cycle* if $\nu_B$ is a maximal singularity. Explicitly:

$$\operatorname{mult}_B |\chi| > n(\chi)(\operatorname{codim} B - 1).$$

A maximal singularity $\nu \in \mathcal{N}(V)$ is said to be *infinitely near* if it is not a maximal cycle. For singular $V$, these definitions should be modified slightly by adding some valuations sitting at the singularities (see [P3, P6]).

# 3   The untwisting scheme

Assume that $\alpha(v, D) > \alpha(W, Y)$. Then $\chi$ has a maximal singularity. The untwisting scheme is a strategy aiming to simplify $\chi$ according to its maximal singularities.

**Definition 3.1 (The basic conjecture)** We say that $V$ *satisfies the basic conjecture* if for any $\chi\colon V \dashrightarrow W$ for which the assumptions of Theorem 2.11 hold, we can replace "maximal singularity" by "maximal cycle": in other words, $\chi$ has a maximal cycle whenever

$$\alpha(v, D) > \alpha(W, Y).$$

**Remark 3.2** The point of Definition 2.13 is to distinguish the maximal cycles, which are "shallow" maximal singularities occurring at ground level on $V$, from the "deeper" infinitely near ones, which take several blowups to dig out. The point of the basic conjecture is that it many cases, it divides our study into treating the maximal cycles (which we exclude or untwist as appropriate), and the infinitely near maximal singularities, which we exclude in good cases by the uniform method of §§6–7.

## Excluding maximal cycles

Assume that $V$ satisfies the basic conjecture. The first thing we must do is describe all the subvarieties $B \subset V$ that can occur as maximal cycles, in other words, all $B$ such that

$$|D - \nu B|$$

is free from fixed components for some $D \in \operatorname{Pic} V$ and some

$$\nu > (\operatorname{codim} B - 1)\alpha(V, D).$$

## Untwisting maps

The second step of the scheme is to construct a birational automorphism $\tau_B \in \operatorname{Bir} V$ for each $B$ singled out at the previous step; $B$ should be a maximal cycle for $\tau_B$.

If $B$ is a maximal cycle for $\chi\colon V \dashrightarrow W$, take the composite

$$\chi \circ \tau_B\colon V \dashrightarrow W.$$

Then we must be able to prove that

$$n(\chi \circ \tau_B) < n(\chi).$$

Iterating, we come to a sequence of maximal cycles $B_1, \ldots, B_k$ such that

$$n(\chi \circ \tau_{B_1} \circ \cdots \circ \tau_{B_k}) \le \alpha(W, Y).$$

## Birationally rigid varieties

Informally speaking, a variety $V$ is birationally rigid if the untwisting scheme works on it.

**Definition 3.3** $V$ is said to be *birationally rigid* if for any test pair $(W, Y)$ (see Definition 2.2) and any map $\chi\colon V \dashrightarrow W$, there exists $\chi^* \in \operatorname{Bir} V$ such that

$$n(\chi \circ \chi^*) \leq \alpha(W, Y).$$

If, moreover, $\operatorname{Bir} V = \operatorname{Aut} V$, then $V$ is said to be *birationally superrigid*.

**Remark 3.4** When the untwisting scheme works, it not only proves that $V$ is birationally rigid, but also gives a natural set of generators of the group $\operatorname{Bir} V$ – namely, the maps $\tau_B$.

**Proposition 3.5** *Assume that $V$ is birationally rigid and $\operatorname{Pic} V \cong \mathbb{Z}$. Then $V$ does not admit any structure of Fano fibration; that is, $V$ has no structure as a conic bundle or a Del Pezzo fibration.*

**Proof**    Assume that there is a birational map

$$\chi\colon V \dashrightarrow W,$$

where $p\colon W \to S$ is a Fano fibration. Take $Y$ to be $p^{-1}(Q)$, where $Q \subset S$ is a very ample divisor. Then

$$n(\chi \circ \chi^*) = 0$$

for some $\chi^* \in \operatorname{Bir} V$, so that

$$|\chi \circ \chi^*| \subset |-n(\chi \circ \chi^*) K_V| = |0|.$$

This is a contradiction.    Q.E.D.

# 4    Excluding maximal cycles

The general idea of exclusion is very simple: we construct a sufficiently large family of curves or surfaces intersecting as often as possible (or containing) the cycle we are trying to exclude, while having "degree" as small as possible. Then we restrict our linear system $|\chi|$ to such a curve or surface and derive a contradiction using an intersection calculation.

We now illustrate the method of excluding maximal cycles in a specific case. Many other examples in the original papers [IM, I, P1–P6, IP] work by the same mechanism.

## Double spaces

Let $\pi\colon V \to \mathbb{P}^m \supset W_{2m}$ for $m \geq 3$ be a smooth double space of index 1, branched over a smooth hypersurface $W$ of degree $2m$. Let $|\chi| \subset |-nK_V|$ be a linear system free from fixed components.

**Theorem 4.1** $|\chi|$ *has no maximal cycles.*

**Corollary 4.2** *Modulo the basic conjecture (see Definition 3.1), $V$ is super-rigid.*

**Proof** This breaks up into two parts: we use separate arguments to exclude maximal points and maximal cycles of positive dimension.

First to exclude points: a point $x \in V$ obviously cannot be maximal: for take a plane $\overline{P} \ni \overline{x} = \pi(x)$ such that $P = \pi^{-1}(\overline{P})$ is a nonsingular surface and $|\chi|_P$ has no fixed curves. Then

$$D_1 \cdot D_2 = 2n^2 \quad \text{for any } D_1, D_2 \in |\chi|_P.$$

But $x \in V$ maximal says that $\text{mult}_x D_i > 2n$, a contradiction.    Q.E.D.

**Proposition 4.3** *For any curve $C \subset V$,*

$$\text{mult}_C |\chi| \leq n.$$

Our theorem is obviously an immediate consequence of this fact.

**Proof of the Proposition** We write $\overline{C} = \pi(C)$ and consider the following three cases:

(1) Easy case: $\pi\colon C \to \overline{C}$ is a double cover and $\overline{C} \not\subset W$.

(2) Moderately easy case: $\overline{C} \subset W$.

(3) Nontrivial case: $\pi\colon C \to \overline{C}$ is birational and $\overline{C} \not\subset W$.

**Case (1)** Take a generic line $\overline{L}$ intersecting $\overline{C}$, so that $L = \pi^{-1}(\overline{L})$ is a smooth curve. The restricted linear series

$$|\chi|_{\big|L}$$

has degree $2n$, but has as base points at least the two points $C \cap \pi^{-1}(\overline{L})$ with multiplicity $\text{mult}_C |\chi|$.

**Case (2)**  Take a generic point $x \in \mathbb{P}^m$ and consider the cone $Z(x)$ over $\overline{C}$ with vertex $x$. Then $Z(x) \cap W = \overline{C} \cup \overline{R}(x)$, where the residual curve $\overline{R}(x)$ intersects $\overline{C}$ at $\deg \overline{R}(x)$ distinct points (see [P5]). Let $R(x)$ be the curve $\pi^{-1}(\overline{R}(x))$; then $\pi \colon R(x) \to \overline{R}(x)$ is an isomorphism, and

$$\left. |\chi| \right|_{R(x)}$$

is a linear series of degree $n \times \deg \overline{R}(x)$ which has $\deg \overline{R}(x)$ base points of multiplicity $\text{mult}_C |\chi|$.

**Case (3)**  Again, take a generic point $x \in \mathbb{P}^m$ and consider the cone $Z(x)$ over $\overline{C}$ with the vertex $x$. Let

$$\varphi \colon X \to \mathbb{P}^m$$

be the blowup at $x$ with exceptional divisor $E$, so that the projection

$$\pi \colon X \to \mathbb{P}^{m-1} = E$$

is a regular map, making $X$ into a $\mathbb{P}^1$-bundle over $E$. Let

$$\alpha \colon Q \to \overline{C}$$

be the desingularization of $\overline{C}$, and

$$\overline{S} = Q \times_{\pi(\overline{C})} X,$$

which is a $\mathbb{P}^1$-bundle over $Q$. Obviously, $\text{Num} \, \overline{S} = A^1(S) = \mathbb{Z}f \oplus \mathbb{Z}e$, where $f$ is the class of a fiber and $e$ the class of the exceptional section coming from the vertex of the cone. Obviously, $f^2 = 0$, $f \cdot e = 1$, and $e^2 = -d$, where $d = \deg \overline{C} = \deg \pi(\overline{C})$. Let $h$ be the class of a hyperplane section; then $h = e + df$, so that $h^2 = d$.

Denote by $\widetilde{C}$ the inverse image of $\overline{C}$ on $\overline{S}$. Obviously, the class $\widetilde{c}$ of $\widetilde{C}$ equals $h$. For generic $x$, the set $\pi^{-1}(Z(x)) \cap \text{Bs} \, |\chi|$ contains at most two curves: $C$ itself and possibly the other component of $\pi^{-1}(\overline{C})$; moreover, the inverse image $\overline{W}$ of $W$ on $\overline{S}$ is a nonsingular curve.

Now let us take the surface $S = \overline{S} \times_{Z(x)} V$, the double cover of $\overline{S}$ with the smooth branch divisor $\overline{W}$. Denote the image of $C$ on $S$ by $C$ again, and the other component of $\pi^{-1}(\overline{C})$ on $S$ by $C^*$. The inverse image of the linear system $|\chi|$ on $S$ has at most two fixed components $C$, $C^*$ of multiplicities $\nu, \nu^*$ respectively. Therefore the system $|nh - \nu c - \nu^* c^*|$ on $S$ is free from fixed components, and we get the following inequalities:

$$\left( nh - \nu c - \nu^* c^* \right) \cdot c \geq 0 \quad \text{and} \quad \left( nh - \nu c - \nu^* c^* \right) \cdot c^* \geq 0.$$

It is easy to compute the multiplication table for the classes $h, c$ and $c^*$ on $S$. The only intersection number we need is

$$(c \cdot c^*)_S = \frac{1}{2}(\bar{c} \cdot \bar{w})_{\bar{S}} = md,$$

the others being obvious. Now we get the following system of linear inequalities:

$$(n - \nu^*) + (m - 1)(\nu - \nu^*) \geq 0 \quad \text{and} \quad (n - \nu) + (m - 1)(\nu^* - \nu) \geq 0.$$

If, for instance, $\nu \geq \nu^*$, then $\nu \leq n$ by the second inequality. By symmetry, we are done.   Q.E.D.

## What do we know about maximal cycles?

We list some of the known facts. First, maximal cycles do not exist in the following 3 cases:

1.  for a smooth hypersurface of degree $m$ in $\mathbb{P}^m$ with $m \geq 4$ [P5];

2.  for a smooth double space $V_2 \to \mathbb{P}^m \supset W_{2m}$ with $m \geq 3$: see [I] for $m = 3$, [P2] for $m \geq 4$, and see also [IP] (and also for a slightly singular $V_2$, [P6]);

3.  for a smooth double quadric $V_4 \to Q_2 \subset \mathbb{P}^{m+1}$, branched over the intersection $Q_2 \cap W_{2m-2}$ with $m \geq 4$: see [P2], and also [IP].

Next, in many cases, there are strong restrictions on the type of maximal singularities that can occur, for example:

4.  For a singular quartic $V_4 \subset \mathbb{P}^4$ having a unique ordinary double singular point $x$, there can be only 25 maximal cycles: the point $x$ itself, and the 24 lines on $V$ passing through $x$ (see [P3]). Moreover, a maximal cycle of a map $V \dashrightarrow W$ is always unique.

5.  For a double quadric $\pi \colon V_4 \to Q_2 \subset \mathbb{P}^4$, branched over $Q_2 \cap W_4$, there can be at most one maximal cycle – namely, a line $L \subset V$, $L \cdot K_V = -1$, with $\pi(L) \not\subset W_4$ (see [I, IP]).

6.  For a complete intersection $V = V_{2 \cdot 3} = Q_2 \cap Q_3 \subset \mathbb{P}^5$, a maximal cycle $B$ is a curve: either a line $L$, or a smooth conic $Y$ spanning a plane $\Pi(Y)$ contained in the quadric $Q_2$. Moreover, a map $V \dashrightarrow W$ can have at most two maximal curves, and if there are exactly two, they are lines $L_1$ and $L_2$ spanning a plane $\Pi(L_1 \cup L_2)$ contained in $Q_2$ (see [I, IP]).

# 5   Untwisting maximal cycles

We give what is probably the simplest example of an untwisting: the untwisting procedure of [P3] for the maximal singular point $x \in V_4 \subset \mathbb{P}^4$ on a quartic $V$ with an ordinary double point.

## Constructing the untwisting

Let $\pi \colon V \setminus \{x\} \to \mathbb{P}^3$ be the projection from $x$, so that $\deg \pi = 2$. Then the untwisting map $\tau \colon V \dashrightarrow V$ interchanges the points in the fibers of $\pi$.

Let $\sigma \colon V_0 \to V$ be the blowup of $x$, $E = \sigma^{-1}(x) \cong \mathbb{P}^1 \times \mathbb{P}^1$ its exceptional divisor, and $L_i$ for $i = 1, \dots, 24$ the proper inverse images of the lines on $V$ passing through $x$.

**Lemma 5.1** $\tau$ *extends to a biregular automorphism of*

$$V_0 \setminus \bigcup_{1 \leq i \leq 24} L_i,$$

*so that it has a well-defined action $\tau^*$ on $\operatorname{Pic} V_0 = \mathbb{Z}h \oplus \mathbb{Z}e$. This action is given by the following relations:*

$$\tau^* h = 3h - 4e \quad \text{and} \quad \tau^* e = 2h - 3e.$$

**Proof**   $\pi$ extends to a morphism $V_0 \to \mathbb{P}^3$ of degree 2. The covering involution $\tau$ is well defined away from the one dimensional fibers, which are exactly the 24 lines $L_i$. Thus $\tau$ is an automorphism of the complement of a set of codimension 2 of $V_0$, so that it has a well-defined action $\tau^*$ on $\operatorname{Pic} V_0$.

Obviously, for any plane $\Pi \subset \mathbb{P}^3$ its proper inverse image $\pi^{-1}(\Pi)$ represents an invariant class, so that

$$\tau^*(h - e) = h - e.$$

Furthermore, $\pi(E)$ is a quadric in $\mathbb{P}^3$ and $\pi(H)$ a quartic in $\mathbb{P}^3$, where $H \subset V$ is a hyperplane section disjoint from $E$. Thus

$$e + \tau^* e = 2(h - e) \quad \text{and} \quad h + \tau^* h = 4(h - e). \quad \text{Q.E.D.}$$

## Untwisting

Let $\chi\colon V \dashrightarrow W$ be our birational map. We define the number $\nu_x(\chi) \in \mathbb{Z}_+$ as follows: the class of the proper inverse image of the linear system $|\chi|$ on $V_0$ is

$$n(\chi)h - \nu_x(\chi)e.$$

The condition that the singular point $x$ is a maximal cycle for $|\chi|$ means that

$$\nu_x(\chi) > n(\chi).$$

Now consider the composite $\chi \circ \tau\colon V \dashrightarrow W$.

**Lemma 5.2**   *(i)* $n(\chi \circ \tau) = 3n(\chi) - 2\nu_x(\chi)$.

*(ii)* $\nu_x(\chi \circ \tau) = 4n(\chi) - 3\nu_x(\chi)$.

**Proof**   Since $\tau$ is an automorphism in codimension 1, we can write

$$n(\chi \circ \tau)h - \nu_x(\chi \circ \tau)e = \tau^*\Big(n(\chi)h - \nu_x(\chi)e\Big).$$

Applying the formulas obtained in the preceding Lemma 5.1 gives the result.   Q.E.D.

Now if $x$ is a maximal point for $\chi$, then $\nu_x(\chi) > n(\chi)$, so that

$$n(\chi \circ \tau) < n(\chi) \quad \text{and} \quad \nu_x(\chi \circ \tau) < n(\chi \circ \tau).$$

We say that the maximal cycle $x$ is *untwisted*, because passing from $\chi$ to $\chi \circ \tau$ decreases the degree, and $x$ is no longer a maximal cycle of $\chi \circ \tau$.

# 6   Infinitely near maximal singularities. I

Our exclusion arguments so far have only dealt with maximal cycles (see Definition 2.13), so that we have only obtained results modulo the basic conjecture (see Definition 3.1 and Remark 3.2). This section develops the techniques required to exclude infinitely near maximal singularities.

## Resolution

Let $X$ be any quasiprojective variety and $\nu \in \mathcal{N}(X)$ a discrete valuation. Suppose that $\nu$ has centre $B = Z(X, \nu)$ on $X$ of $\operatorname{codim} B \geq 2$, and that $B \not\subset \operatorname{Sing} X$.

**Proposition 6.1** *Either $\nu = \nu_B$, or after making the blowup $\sigma_B\colon X(B) \to X$ with exceptional divisor $E(B) = \sigma_B^{-1}(B)$, we find that $\nu \in \mathcal{N}(X(B))$ has centre $Z(X(B), \nu) \subset E(B)$ an irreducible subvariety of codimension $\geq 2$ in $X(B)$, and*

$$\sigma_B(Z(X(B), \nu)) = B.$$

**Proof** Easy.

Now set $X_0 = X$, and consider the sequence of blowups

$$\varphi_{i,i-1}\colon X_i \to X_{i-1} \quad \text{for } i \geq 1,$$

where $\varphi_{i,i-1}$ blows up the subvariety $B_{i-1} = Z(X_{i-1}, \nu) \subset X_{i-1}$ of co-dimension $\geq 2$, and $E_i = \varphi_{i,i-1}^{-1}(B_{i-1}) \subset X_i$. Note that although the $X_i$ are possibly singular, $B_i \not\subset \operatorname{Sing} X_i$ for all $i$. We set

$$\varphi_{i,j} = \varphi_{j+1,j} \circ \cdots \circ \varphi_{i,i-1}\colon X_i \to X_j, \quad \text{for } i > j,$$

and $\varphi_{i,i} = \operatorname{id}_{X_i}$. Note that $\varphi_{i,j}(B_i) = B_j$ for $i \geq j$.

We denote the proper inverse image on $X_i$ of a subvariety or cycle $(\dots)$ by a superscript $i$: $(\dots)^i$.

**Proposition 6.2** *This sequence is finite: in other words, for some $k \in \mathbb{Z}_+$, the triple $(X_k, \varphi_{k,0}, E_k)$ is a realization of $\nu$, that is, $\nu = \nu_{E_k}$ (compare Definition 2.5.*

**Proof** This is easy. See [P6], or prove it for yourself.

**Definition 6.3** The sequence $\{\varphi_{i,i-1}\}_{i=1,\dots,k}$ is called the *resolution* of the discrete valuation $\nu$ (with respect to the model $X$).

## The graph structure

**Definition 6.4** For $\mu, \nu \in \mathcal{N}(X)$, we write

$$\nu \underset{X}{\geq} \mu$$

if $\nu$ is infinitely near to $\mu$, that is, if $\mu = \nu_{E_l}$ for some $l \leq k$, and

$$(X_l, \varphi_{l,0}, E_l)$$

is a realization of $\mu$; in other words, the resolution of $\mu$ occurs as an initial segment of the resolution of $\nu$.

We introduce an oriented graph structure on $\mathcal{N}(X)$, drawing an arrow

$$\nu \xrightarrow{X} \mu,$$

if $\nu \underset{X}{\geq} \mu$ and $B_{k-1} \subset E_l^{k-1}$.

Denote by $P(\nu, \mu)$ the set of all paths from $\nu$ to $\mu$ in $\mathcal{N}(X)$, which is nonempty if and only if $\nu \underset{X}{\geq} \mu$. Set

$$p(\nu, \mu) = |P(\nu, \mu)| \quad \text{if } \nu \neq \mu,$$

and $p(\nu, \nu) = 1$. We define $\mathcal{N}(X, \nu)$ to be the subgraph of $\mathcal{N}(X)$ with the set of vertices $\underset{X}{\leq} \nu$.

## Intersections, degrees and multiplicities

Let $B \subset X$ with $B \not\subset \operatorname{Sing} X$ be an irreducible subvariety of codimension $\geq 2$; as usual, let $\sigma_B \colon X(B) \to X$ be its blowup, and $E(B) = \sigma_B^{-1}(B)$ the exceptional divisor. Let

$$Z = \sum m_i Z_i, \quad \text{with} \quad Z_i \subset E(B)$$

be a $k$-cycle for some $k \geq \dim B$. We define the *degree* of $Z$ as

$$\deg Z = \sum_i m_i \deg \left( Z_i \cap \sigma_B^{-1}(b) \right),$$

where $b \in B$ is a generic point, and the degree on the right-hand side is the ordinary degree in the projective space $\sigma_B^{-1}(b) \cong \mathbb{P}^{\operatorname{codim} B - 1}$. Note that $\deg Z_i = 0$ if and only if $\sigma_B(Z_i)$ is a proper closed subset of $B$.

Our computations in what follows are based on the following statement.

**Lemma 6.5** *Let $D$ and $Q$ be distinct prime Weil divisors on $X$, and write $D^B$ and $Q^B$ for their proper inverse images on $X(B)$. We write $\bullet$ for the codimension $2$ cycle of the scheme theoretic intersection.*

*(i) Assume that* $\operatorname{codim} B \geq 3$. *Then*

$$D^B \bullet Q^B = (D \bullet Q)^B + Z,$$

*where* $\operatorname{Supp} Z \subset E(B)$, *and*

$$\operatorname{mult}_B(D \bullet Q) = (\operatorname{mult}_B D)(\operatorname{mult}_B Q) + \deg Z.$$

*(ii) Assume that* $\operatorname{codim} B = 2$. *Then*

$$D^B \bullet Q^B = Z + Z_1,$$

*where* $\operatorname{Supp} Z \subset E(B)$, $\operatorname{Supp} \sigma_B(Z_1)$ *does not contain* $B$, *and*

$$D \bullet Q = \left[ (\operatorname{mult}_B D)(\operatorname{mult}_B Q) + \deg Z \right] B + (\sigma_B)_* Z_1.$$

**Proof**  Let $b \in B$ be a generic point, $S \ni b$ a germ of a nonsingular surface in general position with $B$, $S^B$ its proper inverse image on $X(B)$. We get an elementary two dimensional problem: to compute the intersection number of two different irreducible curves at a smooth point on a surface in terms of its blowup. This is easy.   Q.E.D.

## Multiplicities in terms of the resolution

We divide the resolution $\varphi_{i,i-1} \colon X_i \to X_{i-1}$ into the *lower part* with $i = 1, \ldots, l \le k$, for which codim $B_{i-1} \ge 3$, and the *upper part*, $i = l+1, \ldots, k$, for which codim $B_{i-1} = 2$. It may occur that $l = k$ and the upper part is empty.

Let $|\lambda|$ be a linear system on $X$ with no fixed components, $|\lambda|^j$ its proper inverse image on $X_j$. Set

$$\nu_j = \text{mult}_{B_{j-1}} |\lambda|^{j-1}.$$

Obviously,

$$\nu_{E_j}(|\lambda|) = \sum_{i=1}^{j} p(\nu_{E_j}, \nu_{E_i}) \nu_i$$

and

$$K(X, \nu_{E_j}) = \sum_{i=1}^{j} p(\nu_{E_j}, \nu_{E_i})(\text{codim}\, B_{i-1} - 1).$$

For simplicity of notation we write $i \to j$ instead of

$$\nu_{E_i} \xrightarrow{\ X\ } \nu_{E_j}$$

in the graph of Definition 6.4.

Now everything is ready for the main step of the theory.

# 7   Infinitely near maximal singularities. II. The main computation

We now prove the crucial inequalities which enable us to exclude infinitely near maximal singularities in cases of low degree.

## Counting multiplicities

Let $D_1, D_2 \in |\lambda|$ be generic divisors. We define a sequence of codimension 2 cycles on the blowups $X_i$, setting

$$
\begin{aligned}
D_1 \bullet D_2 &= Z_0, \\
D_1^1 \bullet D_2^2 &= Z_0^1 + Z_1, \\
&\;\;\vdots \\
D_1^i \bullet D_2^i &= (D_1^{i-1} \bullet D_2^{i-1})^i + Z_i, \\
&\;\;\vdots
\end{aligned}
$$

where $Z_i \subset E_i$. Thus for any $i \leq l$ we get

$$
D_1^i \bullet D_2^i = Z_0^i + Z_1^i + \cdots + Z_{i-1}^i + Z_i.
$$

For any $j$ with $i < j \leq l$, set

$$
m_{i,j} = \mathrm{mult}_{B_{j-1}}(Z_i^{j-1});
$$

here the multiplicity of an irreducible cycle along a smaller cycle is understood in the usual sense; for an arbitrary cycle we extend the multiplicity by linearity.

## The crucial point

**Lemma 7.1** *If $m_{i,j} > 0$, then $i \to j$.*

**Proof** If $m_{i,j} > 0$, then some component of $Z_i^{j-1}$ contains $B_{j-1}$. But $Z_i^{j-1} \subset E_i^{j-1}$. Q.E.D.

## Degree and multiplicity

Set $d_i = \deg Z_i$.

**Lemma 7.2** *For any $i \geq 1$ and $j \leq l$ we have*

$$
m_{i,j} \leq d_i.
$$

**Proof** Each $B_a$ is nonsingular at its generic point. But since $\varphi_{a,b} \colon B_a \to B_b$ is surjective, we can count multiplicities at generic points. Now the multiplicities are nonincreasing with respect to blowup of a nonsingular subvariety, so we are reduced to the obvious case of a hypersurface in a projective space. Q.E.D.

## The computation

We get the following system of equalities:

$$\left.\begin{aligned}
\nu_1^2 + d_1 &= m_{0,1}, \\
\nu_2^2 + d_2 &= m_{0,2} + m_{1,2}, \\
&\ \ \vdots \\
\nu_i^2 + d_i &= m_{0,i} + \cdots + m_{i-1,i}, \\
&\ \ \vdots \\
\nu_l^2 + d_l &= m_{0,l} + \cdots + m_{l-1,l}.
\end{aligned}\right\} \qquad (*)$$

Now

$$d_l \geq \sum_{i=l+1}^{k} \nu_i^2 \deg(\varphi_{i-1,l})_* B_{i-1} \geq \sum_{i=l+1}^{k} \nu_i^2.$$

**Definition 7.3** A function $a: \{1, \ldots, l\} \to \mathbb{R}_+$ is *compatible* with the graph structure if

$$a(i) \geq \sum_{j \to i} a(j)$$

for any $i = 1, \ldots, l$.

**Examples 7.4** $a(i) = p(l, i)$, $a(i) = p(K, i)$.

**Theorem 7.5** *Let $a(\cdot)$ be any compatible function. Then*

$$\sum_{i=1}^{l} a(i) m_{0,i} \geq \sum_{i=1}^{l} a(i) \nu_i^2 + a(l) \sum_{i=l+1}^{k} \nu_i^2.$$

**Proof** Multiply the $i$th equality in $(*)$ by $a(i)$ and add them all together: on the right-hand side, for any $i \geq 1$, we get the expression

$$\sum_{\substack{j \geq i+1}} a(j) m_{i,j} = \sum_{\substack{j \geq i+1 \\ m_{i,j} \neq 0}} a(j) m_{i,j} \leq d_i \sum_{j \to i} a(j) \leq a(i) d_i.$$

On the left-hand side, for any $i \geq 1$, we get

$$a(i) d_i.$$

So we can throw away all the $m_{i,*}$ from the right-hand side for $i \geq 1$, and all the $d_i$ from the left-hand side for $i \geq 1$, replacing $=$ by $\leq$.   Q.E.D.

**Corollary 7.6** *Set* $m = m_{0,1} = \mathrm{mult}_{B_0}(D_1 \bullet D_2), D_i \in |\chi|.$ *Then*

$$m \sum_{i=1}^{l} a(i) \geq \sum_{i=1}^{l} a(i)\nu_i^2 + a(l) \sum_{i=l+1}^{k} \nu_i^2.$$

## Applications

**Corollary 7.7** *Set* $r_i = p(K,i).$ *Then*

$$m \sum_{i=1}^{l} r_i \geq \sum_{i=1}^{k} r_i \nu_i^2.$$

**Proof** For $i \geq l+1$ obviously $r_i \leq r_l$.   Q.E.D.

**Corollary 7.8 (Iskovskikh and Manin [IM])** *Suppose that* $\dim V = 3,$ *and let* $\nu \in \mathcal{N}(V)$ *be a maximal singularity such that* $Z(V,\nu) = x$ *is a smooth point,* $m = \mathrm{mult}_x C,$ *where the curve* $C = (D_1 \bullet D_2)$ *is the intersection of two generic divisors in* $|\chi|,$ $n = n(\chi)$ *and assume that* $|-K_V|$ *is free. Then*

$$m \left( \sum_{i=1}^{l} r_i \right) \left( \sum_{i=1}^{k} r_i \right) > n^2 \left( 2 \sum_{i=1}^{l} r_i + \sum_{i=l+1}^{k} r_i \right)^2.$$

*In particular,* $m > 4n^2.$

**Proof** This follows immediately from the fact that $\nu$ is a maximal singularity and from the preceding Corollary 7.7. We prove the final statement. Denoting

$$\sum_{i=1}^{l} r_i, \quad \sum_{i=l+1}^{k} r_i$$

by $\Sigma_0, \Sigma_1,$ respectively, we get

$$4\Sigma_0(\Sigma_0 + \Sigma_1) \leq (\Sigma_1 + 2\Sigma_0)^2,$$

which is exactly what we want.

**Corollary 7.9 (Iskovskikh and Manin [IM])** *The basic conjecture (see Definition 3.1) holds for a smooth quartic* $V \subset \mathbb{P}^4.$

**Proof**   Obviously, $m \leq 4n^2$. This contradicts the previous corollary.

Since it is easy to show that $|\chi|$ has no maximal cycles on $V_4$ ([IM] or [P5]), we get:

**Corollary 7.10 (Iskovskikh and Manin [IM])** *A smooth quartic three-fold $V \subset \mathbb{P}^4$ is a birationally superrigid variety.*

# 8   Sarkisov's theorem on conic bundles

We give an extremely short version of the proof of Sarkisov's theorem [S1, S2]. The idea of the proof is essentially the same as in these well-known papers of Sarkisov. At the same time, our general viewpoint of working in codimension 1 makes the arguments brief and very clear.

## Statement of the theorem

Let $S$ be a smooth projective variety of dimension $\dim S \geq 2$, and let $\mathcal{E}$ be a locally free sheaf of rank 3 over $S$. Let

$$X \subset \mathbb{P}(\mathcal{E}) \xrightarrow{\pi} S$$

be a standard conic bundle, that is, a smooth hypersurface with

$$\operatorname{Pic} X = \mathbb{Z}K_X \oplus \pi^* \operatorname{Pic} S.$$

Denote by $C \subset S$ the discriminant divisor. Recall that $C$ has at most normal crossings, the fiber over any point outside $C$ is a smooth conic, the fiber over the generic point of any component of $C$ is a pair of distinct lines, and the inverse image of any component of $C$ on $X$ is irreducible.

Let $\tau \colon V \to F$ be another conic bundle of the same dimension (not necessarily smooth).

**Theorem 8.1** *If $|4K_S + C| \neq \emptyset$, then any birational map*

$$\chi \colon X \dashrightarrow V$$

*takes fibers into fibers, that is, there exists a map $\overline{\chi} \colon S \dashrightarrow F$ such that*

$$\tau \circ \chi = \overline{\chi} \circ \pi.$$

## Start of the proof

We write
$$\mathcal{F} = \{C_u \mid u \in U\}$$
for the proper inverse image of the family of conics $\tau^{-1}(q)$ for $q \in F$, and by
$$\overline{\mathcal{F}} = \{\overline{C}_u = \pi(C_u) \mid u \in U\}$$
its image on the base $S$. When we perform a birational operation with respect to these families, we sometimes replace the parametrizing set $U$ by some dense open subset; for brevity, we omit mention of this change, and just bear in mind that we use the same symbol $U$, meaning it to be as small as necessary.

Let $\sigma \colon S^* \to S$ be a birational morphism such that:

(1) $S^*$ is projective and nonsingular in codimension 1;

(2) the proper inverse image
$$\mathcal{F}^* = \{L_u \mid u \in U\}$$
of the family $\overline{\mathcal{F}}$ on $S^*$ is free in the following sense: for any cycle $Z \subset S^*$ of codimension $\geq 2$ a general curve $L_u$ does not meet $Z$.

The existence of such a morphism $\sigma$ can be proved by quite elementary methods, without using Hironaka's results (see [P6]). Set
$$\mathbb{P}^* = \mathbb{P}(\sigma^* \mathcal{E}) \quad \text{and} \quad X^* = X \times_S S^* \subset \mathbb{P}^*.$$
Then $X^*$ is a singular conic bundle over $S^*$. For ease of notation, the natural morphisms of $X^*$ to $S^*$ and $X$ will be denoted by $\pi, \sigma$ respectively, and the map $\chi \circ \sigma$ just by $\chi$.

**Proposition 8.2** *There exist: a closed subset $Y \subset S^*$ of codimension $\geq 2$, a nonsingular conic bundle*
$$\pi \colon W \to S^* \setminus Y$$
*with nonsingular discriminant divisor*
$$C^* \subset S^* \setminus Y$$
*and*
$$\operatorname{Pic} W \cong \mathbb{Z}K_W \oplus \pi^* \operatorname{Pic} S^*,$$
*and a fiberwise map*
$$\lambda \colon X^* \dashrightarrow W,$$
*such that $\pi \circ \lambda = \pi$. Moreover,*
$$|4K_{S^*} + C^*| \neq \emptyset.$$

**Proof**   We obtain $W$ by fiberwise restructuring of $X^*$ over the prime divisors $T \subset S^*$ such that $\operatorname{codim} \sigma(T) \geq 2$. If $t \in \mathbb{C}(S^*)$ is a local equation of $T$ on $S^*$, then at the generic point of $T$ the variety $X^*$ is given by one of the two following types of equations:

$$\text{Case 1:} \quad x^2 + t^k a y^2 + t^l b z^2, \quad k \leq l$$
$$\text{Case 2:} \quad x^2 + y^2 + t^k a z^2,$$

where $(x : y : z)$ are homogeneous coordinates on $\mathbb{P}^2$, and $a, b$ are regular and nonvanishing at a generic point of $T$. In Case 1 for $k \geq 2$, the variety $X^*$ has a whole divisor of singular points, that is, $\pi^{-1}(T)$. Blow it up $[k/2]$ times. Now in either case, the singularity of our variety over $T$ is of type either $A_n$ or $D_n$. Blowing up the singularities, *covering* $T$, and contracting afterwards $-1$-components in fibers, we get the proposition. The last statement is easily obtained by computing the discrepancy of $\nu_T$ on $S$.   Q.E.D.

Denote $\chi \circ \lambda^{-1}$ again by $\chi \colon W \dashrightarrow V$.

Let $Z \subset W \times V$ be the (closed) graph of $\chi$, and let $\varphi$ and $\psi$ be the projections (birational morphisms) onto $W$ and $V$ respectively. Obviously, $Z$ is projective over $W$.

**Proposition 8.3**   *For any closed set $Y^* \supset Y$ of codimension $\geq 2$ there exists an open set $U \subset F$ such that*

$$\psi^{-1} \tau^{-1}(U) \subset \varphi^{-1} \pi^{-1}(S^* \setminus Y^*)$$

*and $\psi^{-1} \tau^{-1}(U)$ is projective over $\tau^{-1}(U) \subset V$.*

**Proof**   This follows immediately from the fact that the family of curves $\mathcal{F}^*$ is free on $S^*$.   Q.E.D.

## The test surface construction

Now let $|H^*|$ be any linear system which is the inverse image of a very ample linear system on $F$, and $|\chi|$ its proper inverse image on $W$. Write

$$|\chi| \subset |-\mu K_W + \pi^* A|$$

for some $\mu \in \mathbb{Z}_+$ and $A \in \operatorname{Pic} S^*$. If $\mu = 0$, we get the statement of the Theorem. So we assume that $\mu \geq 1$. Let us show that this is impossible.

In the notation of the preceding Proposition 8.3, set $Q = \psi^{-1} \tau^{-1}(U)$. Obviously, we may assume that

$$\psi \colon Q \to \tau^{-1}(U) \subset V$$

is an isomorphism. For a generic conic $R_u$ with $u \in U$, we have

$$H^* \cdot R_u = 0, \quad \text{and} \quad K_V \cdot R_u = -2.$$

So the same is true on $Q$. Hence for some prime divisors $T_1, \ldots, T_m \subset Q$, we get

$$\left( -\mu\varphi^* K_W + \varphi^*\pi^* A - \sum_{i=1}^{m} a_i T_i \right) \cdot \psi^{-1}(R_u) = 0$$

and

$$\left( \varphi^* K_W + \sum_{i=1}^{m} d_i T_i \right) \cdot \psi^{-1}(R_u) = -2.$$

Making the set $U$ smaller if necessary, we may assume that

$$T_i \cdot \psi^{-1}(R_u) \geq 1 \quad \text{for all } i.$$

Thus the cycles

$$\pi \circ \varphi(T_i)$$

have codimension 1 in $S^*$ and the $T_i$ can be realized by the successions of blowups

$$\varphi_{j,j-1}^{(i)} : \quad X_j^{(i)} \longrightarrow X_{j-1}^{(i)}$$
$$\cup \qquad\qquad \cup$$
$$E_j^{(i)} \longrightarrow B_{j-1}^{(i)},$$

where $B_0^{(i)} = \varphi(T_i)$, $B_{j+1}^{(i)}$ covers $B_j^{(i)}$, $E_{K(i)}^{(i)} = T_i$. Since $|\chi|$ has no fixed components, $\deg(B_{j+1}^{(i)} \to B_j^{(i)}) = 1$ and the corresponding graph of discrete valuations is a chain. Taking the union of these blowups (that is, throwing away some more cycles of codimension 2 from $S^*$), we get on $Q$ that

$$|\widetilde{\chi}| \subset \left| -\mu\varphi^* K_W + \varphi^*\pi^* A - \sum_{i,j} \nu_{i,j} E_j^{(i)} \right|,$$

whereas the canonical divisor on $Q$ equals

$$\varphi^* K_W + \sum_{i,j} E_j^{(i)}.$$

Consequently, in view of $\mu \geq 1$, the divisor

$$\varphi^* \pi^* A - \sum_{i,j} (\nu_{i,j} - \mu) E_j^{(i)}$$

intersects $\psi^{-1}(R_u)$ negatively. Of course, we may assume that

$$\nu_{i,K(i)} \geq \mu + 1 \quad \text{for all } i = 1, \ldots, m.$$

Now consider the surface $\Lambda_u = \pi^{-1}(\pi \circ \varphi(\psi^{-1}(R_u))$ (the *test surface*, see [P5, P6]) and its proper inverse image $\Lambda_u^*$ on $Q$. These surfaces are projective and, since $\mathcal{F}^*$ is free, we get

$$D^2 \cdot \Lambda^* \geq 0,$$

where $D$ is the class of $\psi^{-1}(|H^*|)$. On the other hand, setting $L = \psi^{-1}(R_u)$, and $\overline{L} = \pi(L)$, we can write $D^2 \cdot \Lambda^*$ as

$$4\mu A \cdot \overline{L} - \mu^2 (4K_{S^*} + C^*) \cdot \overline{L} - \sum_{i,j} \nu_{i,j}^2 E_j^{(i)} \cdot L$$

(since for a generic $u \in U$ the curve $\psi^{-1}(R_u)$ intersects all the $T_i$ transversally). At the same time, according to the above remark,

$$A \cdot \overline{L} < \sum_{i,j} (\nu_{i,j} - \mu) E_j^{(i)} \cdot L,$$

so that

$$4\mu A \cdot \overline{L} < \sum_{i,j} 4\mu(\nu_{i,j} - \mu) E_j^{(i)} \cdot L$$

$$\leq \sum_{i,j} \nu_{i,j}^2 E_j^{(i)} \cdot L.$$

Since the intersection

$$(4K_{S^*} + C^*) \cdot \overline{L}$$

is obviously nonnegative, we get a contradiction:

$$D^2 \cdot \Lambda^* < 0. \quad \text{Q.E.D.}$$

# References

[C1] Corti A., Factoring birational maps of threefolds after Sarkisov, J. Alg. Geom. **4** (1995) 223–254

[F1] Fano G., Sopra alcune varieta algebriche a tre dimensioni aventi tutti i generi nulli. Atti Acc. Torino. **43** (1908) 973–977

[F2] Fano G., Osservazioni sopra alcune varieta non razionali aventi tutti i generi nulli. Atti Acc. Torino. **50** (1915) 1067–1072

[F3] Fano G., Nuove ricerche sulle varieta algebriche a tre dimensioni a curve-sezioni canoniche. Comm. Rend. Ac. Sci. **11** (1947) 635–720

[IM] Iskovskikh V. A. and Manin Yu. I., Three-dimensional quartics and counterexamples to the Lüroth problem. Mat. Sb. **86** (128) (1971) 140–166 = Math. USSR Sbornik **15**:1 (1971) 141–166

[I] Iskovskikh V. A., Birational automorphisms of three-dimensional algebraic varieties. J. Soviet Math. **13** (1980) 815–868

[IP] Iskovskikh V. A. and Pukhlikov A. V., Birational automorphisms of multi-dimensional algebraic manifolds. Algebraic geometry, 1. J. Math. Sci. **82**:4 (1996) 3528–3613

[K1] Kollár J., Nonrational hypersurfaces, Jour. AMS **8** (1995) 241–249

[K2] Kollár J., Nonrational covers of $\mathbb{P}^m \times \mathbb{P}^m$, this volume, 51–71

[Kh] Khashin S. I., Birational automorphisms of a double Veronese cone of dimension three, Vestnik Moskov. Univ. Ser. I Mat. Mekh. (1984) no. 1, 13–16 = Moscow Univ. Math. Bull. **1** (1984) 13–16

[N] Noether M., Ueber Flächen welche Schaaren rationaler Curven besitzen. Math. Ann. **3** (1871) 161–227

[P1] Pukhlikov A. V., Birational isomorphisms of four-dimensional quintics. Invent. Math. **87** (1987) 303–329

[P2] Pukhlikov A. V., Birational automorphisms of a double space and a double quadric, Izv. Akad. Nauk SSSR Ser. Mat. **52** (1988) 229–239 = Math. USSR Izv. **32** (1989) 233–243

[P3] Pukhlikov A. V., Birational automorphisms of a three-dimensional quartic with the simplest singularity, Mat. Sb. **135** (1988) 472–496 = Math. USSR Sbornik **63** (1989) 457–482

[P4] Pukhlikov A. V., Maximal singularities on the Fano variety $V_6^3$, Vestnik Moskov. Univ. Ser. I Mat. Mekh. (1989) no. 2, 47–50 = Moscow Univ. Math. Bull. **44** (1989) 70–75

[P5] Pukhlikov A. V., A note on the theorem of V. A. Iskovskikh and Yu. I. Manin on the three-dimensional quartic. Proc. Steklov Math. Inst. **208** (1995) 278–289

[P6] Pukhlikov A. V., Birational automorphisms of double spaces with singularities. Algebraic geometry, **2** J. Math. Sci. (New York) **85** (1997) 2128–2141

[P7] Pukhlikov A. V., Birational automorphisms of Fano hypersurfaces. Invent. Math. **134** (1998) 401–426

[R1] Reid M., Birational geometry of 3-folds according to Sarkisov, Notes from a lecture at Johns Hopkins, Apr 1991.

[S1] Sarkisov V.G., Birational automorphisms of conic bundles. Izv. Akad. Nauk SSSR Ser. Mat. **44** (1980) 918–945 = Math. USSR Izv. **17** (1981)

[S2] Sarkisov V.G., On the structure of conic bundles. Izv. Akad. Nauk SSSR Ser. Mat. **46** (1982) 371–408 = Math. USSR Izv. **20**:2 (1982) 354–390

[T1] Tregub S.L., Birational automorphisms of a three-dimensional cubic, Uspekhi Mat. Nauk **39** (1984) 159–160 = Russian Math. Surveys. **39**:1 (1984) 159–160

[T2] Tregub S.L., Construction of a birational isomorphism of a three-dimensional cubic and a Fano variety of the first kind with $g = 8$, connected with a rational normal curve of degree 4, Vestnik Moskov. Univ. Ser. I Mat. Mekh. (1985) no. 6, 99–101 = Moscow Univ. Math. Bull. **40**:6 (1985) 78–80

Alexandr V. Pukhlikov,
Number Theory Section, Steklov Mathematics Institute,
Gubkina, 8,
117966 Moscow, Russia
e-mail: dost@dost.mccme.rssi.ru and pukh@mi.ras.ru

# Working with
# weighted complete intersections

## A. R. Iano-Fletcher

## Contents

1 Introduction                                                    102

2 Acknowledgments                                                 103

3 Notation                                                        103

4 Preamble                                                        105

5 Definitions and theorems on weighted projective spaces         105

6 Definitions and theorems on weighted complete intersections 108

7 Cohomology of weighted complete intersections                  114

8 Quasismoothness                                                 116

9 Cyclic singularities and counting points                       123

10 Determination of singularities on weighted complete
   intersections                                                 128

11 Preamble                                                       132

12 Weighted curve hypersurfaces                                   133

13 Weighted surface complete intersections                        136

14 Weighted 3-fold complete intersections                         145

15 Canonically embedded weighted 3-folds                          150

16 Q-Fano 3-folds                                                 154

17 The plurigenus formula                                    162

18 The Reid table method                                     163

References                                                   171

# 1    Introduction

This article contains the following:

(I) A presentation of the basic definitions, theorems and techniques of weighted complete intersections, along with many examples. This information was collected from a variety of sources (mainly [WPS]) but also includes some original results.

(II) Lists of various types of weighted complete intersections of dimensions 1, 2 and 3 with isolated canonical cyclic quotient singularities.

Weighted complete intersections occur naturally in many disguises. Enriques' famous example of a surface of general type for which $\varphi_{4K_S}$ is not birational can be expressed as the weighted complete intersection $S_{10}$ in $\mathbb{P}(1,1,2,5)$.

For certain classes of variety $V$ of general type (e.g., minimal surfaces of general type) the canonical maps $\varphi_{nK_V} : V \to \tilde{V}$ are birational onto the canonical model $\tilde{V}$ for large enough $n$. Define the canonical ring $R_V$ by

$$R_V = \bigoplus_{n \geq 0} \mathrm{H}^0(V, nK_V).$$

The ring $R_V$ is known to be finitely generated in these cases, although not necessarily generated in degree 1. So $\tilde{V} \simeq \mathrm{Spec}\, R_V$ is a subvariety of some weighted projective space.

Weighted complete intersections are similar to complete intersections in ordinary projective space $\mathbb{P}^n$ but are usually singular and hence have some pathologies. However they are still very easy to visualise and to work with; their basic invariants are calculated using combinatorics. So they form a large quagmire of *good* examples. This article sets out to familiarise the reader with weighted complete intersections and to give certain combinatoric conditions for their important properties. Some of these are already known (see [Da], [Di], [Du], [WPS], etc.), but some are new. This constitutes Part I.

In Part II we present various lists of weighted complete intersections of dimension 1, 2 and 3; all with at worst isolated canonical cyclic quotient singularities. The canonical 3-fold weighted complete intersections are interesting since they are all canonical models (see [R1], [R2], [R4], Section 2.5) and hence are of interest for classification purposes as well as in their own right. These were all calculated using a set of combinatoric conditions and a computer. We also give a complete list of the 95 families of weighted hypersurface K3 surfaces (see [R1], Section 4.5) found by Reid in 1979 after a long hand calculation. We also calculate the corresponding singularities.

Another method originally used by Reid to produce examples of K3 surfaces is to be found in Section 18. It is used to produce canonically and anticanonically embedded canonical 3-folds. From the Poincaré series of the graded ring corresponding to a weighted complete intersection, the degrees of the generators and the relations can be determined. This technique uses repeated differencing to evaluate the power series. Using the Riemann–Roch formula for canonical 3-folds (see Section 17) a Poincaré series can be produced from a list (or *record*) of invariants, which we hope will correspond to either a canonically or an anticanonically embedded canonical 3-fold. Clearly there will be a large number of rejected records and hence this is very hit-and-miss. However in practice it works very well.

This article started life as the third chapter of my Ph.D. thesis [F2] and grew.

## 2 Acknowledgments

I would like to thank Miles Reid for all his help and A. Dimca and A. Parusiński for many useful conversations. My thanks to Maria Iano and Duncan Dicks for reading through previous versions and suggesting changes. I would also thank all those at the Mathematics Institute, University of Warwick, the mathematics department of the University of Leicester, and the Max-Planck-Institut für Mathematik, Bonn. I am grateful to Prof. Hirzebruch and the institute for the invitation and their kind hospitality during 1987 and 1988.

## 3 Notation

All varieties will be assumed to be quasiprojective over an algebraically closed field $\mathbb{K}$ of characteristic zero. Let $V$ be such a variety, of dimension $m$.

- $\mathbb{K}^*$ is the multiplicative group of nonzero elements of $\mathbb{K}$.

- $\mathbb{Z}$, $\mathbb{Q}$ are the rings of integers and rational numbers respectively.

- $\mathbb{Z}_r$ is the Abelian group $\{0, 1, \ldots, r-1\}$ under addition modulo $r$.

- $\mathbb{Z}_r^*$ is the group of units of $\mathbb{Z}_r$ under multiplication modulo $r$.

- $\{a, \ldots, \widehat{b}, \ldots, c\}$ is a list with the element $b$ omitted.

- $\mathbb{A}^m$ is affine $m$-space.

- $\mathbb{P}^m$ is projective $m$-space.

- $\mathbb{P}(a_0, \ldots, a_m)$ is used to denote weighted projective space with weighting $a_0, \ldots, a_m$. When there is no ambiguity this is denoted simply by $\mathbb{P}$.

- $V^0$ is the nonsingular locus of $V$.

- $\mathcal{O}_V$ is the sheaf of regular functions on $V$.

- $\Omega_V^1 = \Omega_{V/\mathbb{K}}^1$ is the sheaf of regular 1-forms on $V^0$.

- $\Omega_V^n = \Lambda^n \Omega_{V/\mathbb{K}}^1$ is the sheaf of regular $n$-forms on $V^0$.

- $\omega_V = \Omega_V^m$ is the sheaf of regular canonical differentials on $V^0$.

- $K_V$ is the canonical divisor corresponding to $\omega_V = \mathcal{O}_V(K_V)$.

- Let $\mathcal{L}$ be a coherent sheaf on $V$. Then

   (a) $\mathrm{h}^i(\mathcal{L}) = \mathrm{h}^i(V, \mathcal{L}) = \dim \mathrm{H}^i(V, \mathcal{L})$,
   (b) $\chi(\mathcal{L}) = \sum_i (-1)^i \mathrm{h}^i(\mathcal{L})$
   (c) and $\varphi_{\mathcal{L}}$ is the rational map corresponding to the sheaf $\mathcal{L}$.

- Let $D$ be a Cartier divisor on $V$. Then

   (a) $\mathrm{h}^i(D) = \mathrm{h}^i(\mathcal{O}_V(D))$,
   (b) $\chi(D) = \sum_i (-1)^i \mathrm{h}^i(\mathcal{O}_V(D))$.
   (c) and $\varphi_D$ is the rational map corresponding to the sheaf $\mathcal{O}_V(D)$.

- In particular $\varphi_{nK_V}$ is called the $n$th canonical map.

- $p_g(V) = \mathrm{h}^0(\omega_V)$ is the geometric genus of $V$.

- $P_n(V) = \mathrm{h}^0(\omega_V^{\otimes n})$ is the $n$th plurigenus of $V$. For negative $n$ these are referred to as the anti-plurigenera.

The words smooth and nonsingular are used interchangeably.

# Part I. Weighted complete intersections

## 4  Preamble

In this chapter we give a brief summary of the facts about weighted complete intersections, along with many examples. We also prove necessary and sufficient conditions for a weighted hypersurface $X_d$ in $\mathbb{P}(a_0, \ldots, a_n)$ to be quasismooth and well formed.

Sections 5 and 6 recap the main definitions and theorems about weighted projective spaces and weighted complete intersections. Section 7 sets out various facts about the cohomology of weighted complete intersections. Section 8 contains necessary and sufficient conditions for quasismoothness in the hypersurface and codimension 2 cases. Information about cyclic quotient canonical singularities in dimensions 1, 2 and 3 is to be found in Section 9, along with two technical lemmas used to count points of intersection along singular strata of $\mathbb{P}$. Examples of how to calculate the singularities of various weighted complete intersections are included in Section 10.

## 5  Definitions and theorems on weighted projective spaces

We start by reviewing some definitions and notation concerned with weighted complete intersections.

**5.1 Definition**  Let $a_0, \ldots, a_n$ be positive integers; define $S = S(a_0, \ldots, a_n)$ to be the graded polynomial ring $\mathbb{K}[x_0, \ldots, x_n]$, graded by $\deg x_i = a_i$. The *weighted projective space* $\mathbb{P}(a_0, \ldots, a_n)$ is defined by

$$\mathbb{P}(a_0, \ldots, a_n) = \operatorname{Proj} S.$$

**5.2 Note**  Let $x_0, \ldots, x_n$ be affine coordinates on $\mathbb{A}^{n+1}$ and let the group $\mathbb{K}^*$ act via:

$$\lambda(x_0, \ldots, x_n) = (\lambda^{a_0} x_0, \ldots, \lambda^{a_n} x_n).$$

Then $\mathbb{P}(a_0, \ldots, a_n)$ is the quotient $(\mathbb{A}^{n+1} \setminus \{0\})/\mathbb{K}^*$. Under this group action $x_0, \ldots, x_n$ are the *homogeneous coordinates* on $\mathbb{P}(a_0, \ldots, a_n)$. Clearly $\mathbb{P}(a_0, \ldots, a_n)$ is a rational $n$-dimensional projective variety.

**5.3 Affine coordinate pieces**  Let $\{x_0, \ldots, x_n\}$ be the homogeneous coordinates on $\mathbb{P}(a_0, \ldots, a_n)$. The affine piece $x_i \neq 0$ is isomorphic to $\mathbb{A}^n / \mathbb{Z}_{a_i}$. Let $\varepsilon$ be a primitive $a_i$th root of unity. The group acts via:

$$z_j \mapsto \varepsilon^{a_j} z_j$$

for all $j \neq i$, on the coordinates $\{z_0, \ldots, \widehat{z_i}, \ldots, z_n\}$ of $\mathbb{A}^n$; here $z_j$ is thought of as $x_j/x_i^{a_j/a_i}$. Compare this with the case of $\mathbb{P}^n$ where the affine coordinates on $x_i \neq 0$ are $z_j = x_j/x_i$.

**5.4 Examples**    (i) $\mathbb{P}^n = \mathbb{P}(1, \ldots, 1)$.

(ii)  Consider $\mathbb{P}(1, 1, 2)$ with homogeneous coordinates $u$, $v$ and $w$. The affine piece $w = 1$ is $\mathbb{A}^2/\mathbb{Z}_2$ with group action

$$u \mapsto -u, \quad v \mapsto -v.$$

The coordinate ring $R$ is given by:

$$R = \mathbb{K}[u, v]^{\mathbb{Z}_2} = \mathbb{K}[u^2, v^2, uv]$$
$$= \mathbb{K}[x, y, z]/(xy - z^2).$$

Thus $\mathbb{P}(1, 1, 2)$ is the projective completion of the ordinary quadratic cone $xy = z^2$ in $\mathbb{A}^3$.

**5.5 Lemma** *For all positive integers $q$ we have*

$$\operatorname{Proj} S(a_0, \ldots, a_n) \simeq \operatorname{Proj} S(qa_0, \ldots, qa_n).$$

**Proof**    This follows from the fact that the 2 graded rings are isomorphic.    $\square$

From [EGA], Proposition 2.4.7 (see also [Hart], Exercise II.5.13) we have:

**5.6 Lemma** *Let $S$ be a graded ring; define the truncation $S^{(q)} = \bigoplus_{m \geq 0} S_{qm}$ to be the graded subring whose $m$th graded part is $S_{qm}$. Then there exists a canonical isomorphism $\operatorname{Proj} S^{(q)} \simeq \operatorname{Proj} S$.*

This is called the $q$th *Veronese embedding*, and is used in the proof of the following:

**5.7 Lemma** *Let $a_0, \ldots, a_n$ be positive integers with no common factor. If $q = \operatorname{hcf}(a_1, \ldots, a_n)$ then*

$$\operatorname{Proj} S(a_0, \ldots, a_n) \simeq \operatorname{Proj} S(a_0, a_1/q, \ldots, a_n/q).$$

**Proof**    Define $S' = \bigoplus_{m \geq 0} S_{qm}$ with the same grading as $S$. So $S' \simeq S^{(q)}$. By the previous lemma we have $\operatorname{Proj} S' \simeq \operatorname{Proj} S$.

Suppose that $x_0^{p_0} \cdots x_n^{p_n}$ is a monomial of degree $mq$ for any $m$. Then $p_0 a_0 + \cdots + p_n a_n = qm$, and so $q \mid p_0 a_0$. As the $\{a_i\}$ have no common factor, $q \mid p_0$. Hence $x_0$ only appears in $S'$ as $x_0^q$. Thus $S' = \mathbb{K}[x_0^q, x_1, \ldots, x_n]$, which is isomorphic to $S(qa_0, a_1, \ldots, a_n)$. Therefore

$$\operatorname{Proj} S(a_0, \ldots, a_n) \simeq \operatorname{Proj} S' \simeq \operatorname{Proj} S(a_0, a_1/q, \ldots, a_n/q). \quad \square$$

**5.8 Quasireflections** Let $G$ be a finite group acting on a variety $X$. A *quasireflection* is any element of $G$ whose fixed locus is a hyperplane. No singularities are produced by the action of any group generated by quasireflections.

The cancellation which occurs in Lemma 5.7 is nothing other than the elimination of quasireflections from the actions of each $\mathbb{Z}_{a_i}$ on the corresponding affine coordinate piece.

This lemma leads to the following corollary from [WPS], 1.3.1 (see also [De], Proposition 1.3):

**5.9 Corollary** $\mathbb{P}(a_0, \ldots, a_n) \simeq \mathbb{P}(b_0, \ldots, b_n)$ *for some* $\{b_0, \ldots, b_n\}$ *such that*

$$\mathrm{hcf}(b_0, \ldots, \widehat{b_i}, \ldots, b_n) = 1 \quad \text{for each } i.$$

**Proof** By Lemma 5.5 we can cancel any common factor of the $\{a_i\}$. By renumbering as necessary and by repeated applications of Lemma 5.7 we can reduce $\mathbb{P}(a_0, \ldots, a_n)$ to the case $\mathbb{P}(b_0, \ldots, b_n)$. A maximum of $n + 1$ applications of Lemma 5.7 are required. $\square$

**5.10 Examples** (i) $\mathbb{P}(a, b) \simeq \mathbb{P}^1$ for all $a$ and $b$.

(ii) $\mathbb{P}(2, 3, 3) \simeq \mathbb{P}(2, 1, 1)$.

(iii) Let $f = x^5 + y^3 + z^2 \in \mathbb{K}[x, y, z]$ with weights 6, 10 and 15 respectively. Define $X : (f = 0) \subset \mathbb{P} = \mathbb{P}(6, 10, 15)$. By the previous lemma $\mathbb{P} \simeq \mathbb{P}^2$.

$$\mathbb{P}(6, 10, 15) \simeq \mathbb{P}(6, 2, 3) \simeq \mathbb{P}(3, 1, 3) \simeq \mathbb{P}(1, 1, 1).$$

The monomials transform as:

$$(x^5, y^3, z^2) \mapsto (x, y^3, z^2) \mapsto (x, y^3, z) \mapsto (x, y, z).$$

Thus $X \subset \mathbb{P} \simeq (x + y + z = 0) \subset \mathbb{P}^2 = \mathbb{P}^1 \subset \mathbb{P}^2$. Of course the coordinate rings of the affine cones (see 6.1) over $X \subset \mathbb{P}$ and $\mathbb{P}^1 \subset \mathbb{P}^2$ are not isomorphic.

In view of Corollary 5.9 we make the following:

**5.11 Definition** The expression $\mathbb{P}(a_0, \ldots, a_n)$ is *well formed* if

$$\mathrm{hcf}(a_0, \ldots, \widehat{a_i}, \ldots, a_n) = 1 \quad \text{for each } i.$$

**5.12  The quotient map**  Let $T = \mathbb{K}[y_0, \ldots, y_n]$, where the $\{y_i\}$ all have weight 1, and so $\mathbb{P}^n \simeq \operatorname{Proj} T$. Consider the inclusion map $S \hookrightarrow T$ given by:

$$x_i \mapsto y_i^{a_i} \quad \text{for all } i.$$

This induces a quotient map $\sigma \colon \mathbb{P}^n \to \mathbb{P}$. In terms of the coordinates $\{Y_i\}$ on $\mathbb{P}^n$

$$[Y_0, \ldots, Y_n] \mapsto [Y_0^{a_0}, \ldots, Y_n^{a_n}].$$

The map $\mathbb{P}^n \to \mathbb{P}$ is a ramified Galois covering with Galois group $\bigoplus_i \mathbb{Z}_{a_i}$.

**5.13  Definition**  Let $r > 0$ and $a_1, \ldots, a_n$ be integers and let $x_1, \ldots, x_n$ be coordinates on $\mathbb{A}^n$. Suppose that $\mathbb{Z}_r$ acts on $\mathbb{A}^n$ via:

$$x_i \mapsto \varepsilon^{a_i} x_i \quad \text{for all } i,$$

where $\varepsilon$ is a fixed primitive $r$th root of unity. A singularity $Q \in X$ is a *quotient singularity* of type $\frac{1}{r}(a_1, \ldots, a_n)$ if $(X, Q)$ is isomorphic to an analytic neighbourhood of $(\mathbb{A}^n, 0)/\mathbb{Z}_r$.

**5.14  Notation**  Write $P_i \in \mathbb{P}$ for the point $[0, \ldots, 0, 1, 0, \ldots, 0]$, where the 1 is in the $i$th position. We will call $P_i$ a vertex, the 1-dimensional toric stratum $P_i P_j$ an edge, etc. The fundamental simplex (that is, the union of all the coordinate hyperplanes $P_0 \ldots \widehat{P_i} \ldots P_n$) will be denoted by $\Delta$.

**5.15  The singular locus $\mathbb{P}_{\text{sing}}$ of $\mathbb{P}$**  Define $h_{i,j,\ldots} = \operatorname{hcf}(a_i, a_j, \ldots)$. The vertex $P_i$ is a singularity of type $\frac{1}{a_i}(a_0, \ldots, \widehat{a_i}, \ldots, a_n)$; this singularity is not necessarily isolated. Each generic point $P$ of the edge $P_i P_j$ has an analytic neighbourhood $P \in U$ which is analytically isomorphic to $(0, Q) \in \mathbb{A}^1 \times Y$, where $Q \in Y$ is a singularity of type $\frac{1}{h_{i,j}}(a_0, \ldots, \widehat{a_i}, \ldots, \widehat{a_j}, \ldots, a_n)$. Similar results hold for higher dimensional toric strata. The singularities only occur on the fundamental simplex $\Delta$.

Notice that $\operatorname{codim}_{\mathbb{P}}(\mathbb{P}_{\text{sing}}) \geq 2$.

# 6  Definitions and theorems on weighted complete intersections

The first few definitions come from [WPS].

**6.1  Definition**  Let $X$ be a closed subvariety of a weighted projective space $\mathbb{P}$ and $p \colon \mathbb{A}^{n+1} \setminus \{0\} \to \mathbb{P}$ the canonical projection. The *punctured affine cone* $C_X^*$ over $X$ is given by $C_X^* = p^{-1}(X)$, and the *affine cone* $C_X$ over $X$ is the completion of $C_X^*$ in $\mathbb{A}^{n+1}$.

Notice that $\mathbb{K}^*$ acts on $C_X^*$ to give $X = C_X^*/\mathbb{K}^*$.

**6.2 Lemma** $C_X^*$ *has no isolated singularities.*

**Proof** If $P \in C_X^*$ is singular then every point on the same fibre of the $\mathbb{K}^*$-action will be singular. $\square$

**6.3 Definition** $X$ in $\mathbb{P}(a_0, \ldots, a_n)$ is *quasismooth* of dimension $m$ if its affine cone $C_X$ is smooth of dimension $m + 1$ outside its vertex $\underline{0}$.

When $X \subset \mathbb{P}$ is quasismooth, the singularities of $X$ are due to the $\mathbb{K}^*$-action and hence are cyclic quotient singularities. Notice that this definition is not equivalent to the smoothness of the inverse image $\sigma^{-1}(X)$ under the Galois cover of Section 5.12 (e.g., $X_8$ in $\mathbb{P}(2,3,5)$).

Another important fact ([WPS], Theorem 3.1.6) is that a quasismooth subvariety $X$ of $\mathbb{P}$ is a $V$-manifold (that is, a complex space locally isomorphic to the quotient of a complex manifold by a finite group of holomorphic automorphisms). This is used later to define the canonical sheaf of $X$, which is usually singular.

**6.4 Definition** Let $I$ be a homogeneous ideal of the graded ring $S$ and define $X_I$ to be:

$$X_I = \operatorname{Proj} S/I \subset \mathbb{P}.$$

Suppose furthermore that $I$ is generated by a regular sequence $\{f_i\}$ of homogeneous elements of $S$. $X_I \subset \mathbb{P}$ is called a *weighted complete intersection* of multidegree $\{d_i = \deg f_i\}$. In this case, we denote by $X_{d_1,\ldots,d_c}$ in $\mathbb{P} = \mathbb{P}(a_0, \ldots, a_n)$ a sufficiently general element of the family of all weighted complete intersections of multidegree $\{d_i\}$.

$X_{d_1,\ldots,d_c}$ in $\mathbb{P}(a_0, \ldots, a_n)$ is of dimension $n - c$. We will usually write $C_{d_1,\ldots,d_c}$ in $\mathbb{P}(a_0, \ldots, a_{c+1})$ for a dimension 1 complete intersection and $S_{d_1,\ldots,d_c}$ in $\mathbb{P}(a_0, \ldots, a_{c+2})$ for a surface.

**6.5 Definition** $X_d$ in $\mathbb{P}(a_0, \ldots, a_n)$ will be said to be a *linear cone* if $d = a_i$ for some $i$ (that is, the defining equation $f$ can be written as $f = x_i + g$). Clearly in this case $X_d$ in $\mathbb{P}(a_0, \ldots, a_n)$ is isomorphic to $\mathbb{P}(a_0, \ldots, \widehat{a_i}, \ldots, a_n)$.

**6.6 Examples** (i) $X_{46}$ in $\mathbb{P}(4,5,6,7,23)$ is a general element in the family of all degree 46 hypersurfaces in $\mathbb{P}(4,5,6,7,23)$.

(ii) $X_8$ in $\mathbb{P}(1,1,1,1,4)$ is a double cover of $\mathbb{P}^3$ branched along a smooth octic surface.

**6.7  The coefficient convention**  When a general polynomial of a given weighted homogeneous degree occurs in a calculation then it is usually written without the nonzero coefficients. For example the defining polynomial for $X_2$ in $\mathbb{P}(1,1,1)$ is:

$$f = c_0 x^2 + c_1 xy + c_2 xz + c_3 y^2 + c_4 yz + c_5 z^2$$

and will be simply written as:

$$f = x^2 + xy + xz + y^2 + yz + z^2.$$

**6.8  The canonical sheaf $\omega_X$**  All weighted complete intersections (and weighed projective spaces) are $V$-manifolds (that is, locally quotients of $\mathbb{A}^n$ by a finite group action) and so the dualising sheaf $\omega_X$ is given by

$$\omega_X \simeq i_* \omega_{X^0},$$

where $i\colon X^0 \hookrightarrow X$ is the inclusion of the smooth part $X^0$ into $X$. This sheaf is a divisorial sheaf (see [R1], Appendix to Section 1, Theorem 7) and can be written as

$$\omega_X \simeq \mathcal{O}_X(K_X),$$

where $K_X$ is a $\mathbb{Q}$-Cartier divisor (that is, $rK_X$ is a Cartier divisor for some nonzero integer $r$). In fact $K_X|_{X^0}$ is Cartier. For the general definition of the canonical sheaf for varieties with at worst canonical singularities see [R4], Section 1.4.

We now introduce an important concept which was not mentioned (and possibly missed) by Dolgachev in [WPS].

**6.9  Definition**  A subvariety $X \subset \mathbb{P}$ of codimension $c$ is *well formed* if the expression for $\mathbb{P}$ is well formed (see Definition 5.11) and $X$ contains no codimension $c + 1$ singular stratum of $\mathbb{P}$.

This means that any codimension 1 stratum of $X$ is either nonsingular on $\mathbb{P}$, or an intersection $X \cap S$, where $S$ is a codimension 1 stratum of $\mathbb{P}$, that is, $\operatorname{codim}_X(X \cap \mathbb{P}_{\text{sing}}) \geq 2$.

**6.10  Well-formedness for hypersurfaces in $\mathbb{P}(a_0, \ldots, a_n)$**  The hypersurface $X_d$ in $\mathbb{P}(a_0, \ldots, a_n)$ is well formed if and only if

(i)  $\operatorname{hcf}(a_0, \ldots, \widehat{a_i}, \ldots, \widehat{a_j}, \ldots, a_n) \mid d$

(ii)  $\operatorname{hcf}(a_0, \ldots, \widehat{a_i}, \ldots, a_n) = 1$

for all distinct $i$, $j$.

**6.11 Well-formedness in codimension 2** The codimension 2 weighted complete intersection $X_{d_1,d_2}$ in $\mathbb{P}(a_0,\ldots,a_n)$ is well formed if and only if

(i) for all distinct $i$, $j$ and $k$, with $h = \mathrm{hcf}(a_0,\ldots,\widehat{a_i},\ldots,\widehat{a_j},\ldots,\widehat{a_k},\ldots,a_n)$, either $h \mid d_1$ or $h \mid d_2$,

(ii) for all distinct $i$ and $j$, with $h = \mathrm{hcf}(a_0,\ldots,\widehat{a_i},\ldots,\widehat{a_j},\ldots,a_n)$, then $h \mid d_1$ and $h \mid d_2$,

(iii) $\mathrm{hcf}(a_0,\ldots,\widehat{a_i},\ldots,a_n) = 1$ for all $i$.

**6.12 Well-formedness in higher codimensions** The above conditions can be generalised to higher codimensions. $X_{d_1,\ldots,d_c}$ in $\mathbb{P}(a_0,\ldots,a_n)$ is well formed if and only if

(i) $\mathbb{P}(a_0,\ldots,a_n)$ is well formed

(ii) for all $\mu = 1,\ldots,c$ the highest common factor of any $(n-1-c+\mu)$ of the $\{a_i\}$ must divide at least $\mu$ of the $\{d_j\}$.

**6.13 Note** Dimca also defines well-formedness (see [Di]) under a different name. He gives the following equivalent set of arithmetic conditions in the quasismooth case. Define:

$$m(h) = |\{i : h \mid a_i\}|, \quad k(h) = |\{i : h \mid d_i\}| \quad \text{and}$$
$$q(h) = \dim X + 1 - m(h) + k(h)$$

for all $h \in \mathbb{Z}$. Then the quasismooth weighted complete intersection $X_{d_1,\ldots,d_c}$ in $\mathbb{P}(a_0,\ldots,a_n)$ is well formed if and only if $q(p) \geq 2$ for all primes $p$. This follows from a theorem essentially due to Hamm (see [Di], Proposition 2).

In fact a weighted complete intersection (not necessarily quasismooth) is well formed if and only if $q(h) \geq 2$ for all integers $h \geq 2$. This is easy to show from the conditions in Section 6.12.

**6.14 The adjunction formula** If $X_{d_1,\ldots,d_c}$ in $\mathbb{P}(a_0,\ldots,a_n)$ is well formed and quasismooth then $\omega_X \simeq \mathcal{O}_X(\sum d_i - \sum a_i)$ (see [WPS], Theorem 3.3.4). We define the *amplitude* to be this difference of sums, and usually denoted it by $\alpha$.

**6.15 Note** The adjunction formula does not hold if the weighted complete intersection is not well formed. We give two examples in dimensions 1 and 2 respectively.

(i) Consider the curve $C_7$ in $\mathbb{P}(1,2,3)$. Let $D \subset \mathbb{P}^2$ be the curve $\sigma^{-1}(C)$ where $\sigma\colon \mathbb{P}^2 \to \mathbb{P}$ is the quotient map (see Section 5.12) Then the curve $D$ is nonsingular of degree 7 and so is of genus 15. By the Hurwitz Theorem (see [Hart], Corollary IV.2.4) we calculate that $g(C) = 1$ and so $\omega_C \simeq \mathcal{O}_C$. On the other hand, since the amplitude is 1, this contradicts the adjunction formula.

(ii) An example in dimension 2 is the surface $S_9$ in $\mathbb{P}(1,2,2,3)$. A quick calculation shows that this surface is both quasismooth and nonsingular. If it is well formed then the amplitude $\alpha = 1$ and so $K_S^2 = \frac{3}{4}$. This contradicts the fact that $K_S^2 \in \mathbb{Z}$ whenever $S$ is nonsingular. In fact $S_9$ in $\mathbb{P}$ is a smooth K3 surface.

**6.16  Well-formedness in dimensions greater than 2**    However, it turns out that well-formedness only needs to be checked in dimensions 1 and 2. We have the following generalisation of a proposition due to Dimca (see [Di], Proposition 6).

**6.17 Theorem** *Let* $X = X_{d_1,\ldots,d_c}$ *in* $\mathbb{P}(a_0,\ldots,a_n)$ *be a quasismooth weighted complete intersection of dimension greater than 2. Then*

*either X is well formed*

*or X is the intersection of a linear cone with other hypersurfaces (that is, $a_i = d_\lambda$ for some i and $\lambda$).*

**6.18 Note**    (i) In case (ii) the weighted complete intersection is isomorphic to an intersection of lower codimension, that is, $X_{d_1,\ldots,\widehat{d_\lambda},\ldots,d_c}$ in $\mathbb{P}(a_0,\ldots,\widehat{a_i},\ldots,a_n)$ or possibly a weighted projective space.

(ii) Cases (i) and (ii) are not mutually exclusive. Consider the hypersurface $X_2$ in $\mathbb{P}(1,1,1,1,2)$ given by

$$f = z + \sum_{i,j} x_i x_j.$$

This is both a linear cone and well formed, and is, of course, isomorphic to $\mathbb{P}^3$.

We need a preliminary result.

**6.19 Lemma** *Let Z be the affine variety of all points P which satisfy the determinantal condition:*

$$\operatorname{rank} \begin{pmatrix} g_1^1(P) & \cdots & g_1^m(P) \\ \vdots & & \vdots \\ g_c^1(P) & \cdots & g_c^m(P) \end{pmatrix} \leq k,$$

where $\{g_i^j\}$ are general weighted homogeneous nonzero polynomials. If $Z$ is nonempty then codim $Z \leq (m - k)(c - k)$.

This is an elementary fact (see [ACGH], p. 83).

**6.20  Proof of Theorem 6.17**  Let $X = (f_1, \ldots, f_c) \subset \mathbb{P} = \mathbb{P}(a_0, \ldots, a_n)$. Suppose that $\mathbb{P}$ is well formed and assume that $X$ is quasismooth with $\dim X \geq 3$ but not well formed. So there is a singular stratum $\widetilde{\Pi}$ of $\mathbb{P}$ such that $\mathrm{codim}_X(\widetilde{\Pi} \cap X) \leq 1$.

If $\mathrm{codim}_X(\widetilde{\Pi} \cap X) = 0$ then $X \subset \widetilde{\Pi}$ and so $X$ is contained in some coordinate hyperplane. Thus some of the defining polynomials are of the form $f_\lambda = x_i$ for some $\lambda$ and $i$. So $X$ is the intersection of at least one linear cone with other hypersurfaces.

So assume that $\mathrm{codim}_X(\widetilde{\Pi} \cap X) = 1$. By reordering we can assume that

$$\widetilde{\Pi} = (x_k = \cdots = x_n = 0) \subset \mathbb{P}$$

for some $k$. Let $\Pi = p^{-1}\widetilde{\Pi} \subset \mathbb{A}^{n+1} \setminus \{0\}$, where $p : \mathbb{A}^{n+1} \setminus \{0\} \to \mathbb{P}$ is the natural projection. Since $\mathrm{codim}_X \widetilde{\Pi} = 1$ then $k = \dim \Pi = n - c$. As $\Pi$ is a fixed component of $C_X$ then we can write the $\{f_\lambda\}$ as:

$$f_\lambda = \sum_{i=k}^{n} x_i g_\lambda^i(x_0, \ldots, x_{k-1}) + \left\{ \begin{matrix} \text{higher order terms} \\ \text{in } x_k, \ldots, x_n \end{matrix} \right\}$$

for all $\lambda = 1, \ldots, c$.

Define $M_P$ to be the matrix

$$M_P = \begin{pmatrix} \partial f_1/\partial x_0(P) & \cdots & \partial f_1/\partial x_n(P) \\ \vdots & & \vdots \\ \partial f_c/\partial x_0(P) & \cdots & \partial f_c/\partial x_n(P) \end{pmatrix}.$$

Singular points on $C_X$ occur whenever rank $M_P < c$. Consider this matrix restricted to $\Pi$:

$$M_{P \in \Pi} = \begin{pmatrix} 0, \ldots, 0 & g_1^k(P) & \cdots & g_c^k(P) \\ \vdots & \vdots & & \vdots \\ 0, \ldots, 0 & g_1^n(P) & \cdots & g_c^n(P) \end{pmatrix}.$$

So $P \in \Pi \cap C_X$ is singular whenever $\mathrm{rank}(g_i^j) \leq c - 1$. Let $Z$ be just this set.

If $Z$ is empty then, in particular, $\underline{0} \notin Z$. As the entries of $M_P$ are all weighted homogeneous polynomials, they must all be of degree 0. Thus, using the coefficient convention 6.7,

$$f_\lambda = \sum x_i + \left\{ \begin{matrix} \text{higher order terms} \\ \text{in } x_k, \ldots, x_n \end{matrix} \right\}$$

for all $\lambda = 1, \ldots, c$. So $X$ is the intersection of a linear cone with other hypersurfaces.

So assume that $Z$ is nonempty. By the previous lemma, codim $Z \le n - k - c + 2$. Remembering that $k = n - c$ we have

$$\dim Z \ge k - (n - k - c + 2) = n - c - 2 = \dim X - 2 \ge 1.$$

Thus $Z \setminus \{0\}$ is nonempty and thus $C_X$ is not smooth away from the origin, a contradiction. $\square$

# 7    Cohomology of weighted complete intersections

From [WPS], Section 3.4.3 we have:

**7.1 Lemma** *Let* $X = (f_1, \ldots, f_c) \subset \mathbb{P}(a_0, \ldots, a_n)$ *be a well-formed quasi-smooth weighted projective complete intersection. Let* $A$ *be the graded ring* $S(a_0, \ldots, a_n)/(f_1, \ldots, f_c)$ *and* $A_n$ *be the* $n$*th graded part of* $A$. *Then*

$$H^i(X, \mathcal{O}_X(n)) \simeq \begin{cases} A_n & \text{if } i = 0 \\ 0 & \text{if } i = 1, \ldots, \dim X - 1 \\ A_{-n-\alpha} & \text{if } i = \dim X \end{cases}$$

*for all* $n \in \mathbb{Z}$.

In particular if $S$ is a well-formed quasismooth weighted projective complete intersection of dimension 2 then the following are equivalent:

(i) $S$ is a K3 surface.

(ii) $\omega_S \simeq \mathcal{O}_S$.

(iii) the amplitude $\alpha = \sum_\lambda d_\lambda - \sum_i a_i = 0$.

For hypersurfaces we have the following result due to Steenbrink [S]:

**7.2 Theorem** *Let* $X$ *be the weighted hypersurface* $X_d$ *in* $\mathbb{P}(a_0, \ldots, a_n)$ *with defining equation* $f$ *and* $\alpha = d - \sum a_i$. *Then the Hodge structure is given by:*

$$h^{i,j}(X) = \begin{cases} 0 & \text{if } i + j \neq n - 1 \text{ and } i \neq j \\ 1 & \text{if } i + j \neq n - 1 \text{ and } i = j \\ \dim_K \left( \frac{S(a_0, \ldots, a_n)}{\theta_f} \right)_{jd + \alpha} & \text{if } i + j = n - 1 \text{ and } i \neq j \\ \dim_K \left( \frac{S(a_0, \ldots, a_n)}{\theta_f} \right)_{jd + \alpha} + 1 & \text{if } i + j = n - 1 \text{ and } i = j \end{cases}$$

*where* $\theta_f = (\partial f / \partial x_i)_{i=0, \ldots, n}$ *is the Jacobian ideal of* $f$.

**Proof**  This follows from [WPS], Section 4 and duality.  □

**7.3 Note**  The above formula satisfies the duality relations $h^{i,j} = h^{j,i} = h^{n-1-i,n-1-j}$ for all $i$ and $j$ because

$$\dim_{\mathbb{K}} \left( \frac{S(a_0, \ldots, a_n)}{\theta_f} \right)_{jd+\alpha} = \dim_{\mathbb{K}} \left( \frac{S(a_0, \ldots, a_n)}{\theta_f} \right)_{(n-1-j)d+\alpha}$$

**7.4  The Euler number**  The Euler number $e(V)$ of a variety $V$ is defined by

$$e(V) = \sum_{i,j} (-1)^{i+j} h^{i,j}(V).$$

For a smooth curve $C$ we have $e(C) = -\deg K_C = 2 - 2g$. For a surface S, with at worst Du Val singularities of types $\{Q_{n_i}\}_i$ where $Q = A, B$ or $E$, we have Noether's formula:

$$12\chi(\mathcal{O}_S) = K_S^2 + e(S) + \sum_i n_i.$$

In particular the case of a K3 surface $S$ with Du Val singularities of types $\{Q_{n_i}\}_i$ gives that $h^{1,1}(S) = 20 - \sum_i n_i$ and so $e(S) = 24 - \sum_i n_i$.

When $X$ is a well-formed quasismooth weighted hypersurface of dimension 3 most of the Hodge numbers cancel or are zero and so

$$e(X) = 2(1 - h^{1,2}(X)).$$

**7.5 Examples**  (i) The hypersurface $S_3$ in $\mathbb{P}(1,1,1,2)$ has Euler number 5. There are two ways to check this.

(a) It is easy to see that this surface has exactly one singularity, of type $\frac{1}{2}(1,1)$ (that is, of Du Val type $A_1$). Also the amplitude is $-2$ and $K_S^2 = (-2)^2 \cdot \frac{3}{2} = 6$. By Noether's formula we have $e(S_3) = 5$.

(b) Alternatively, the Hodge numbers are simple to calculate. Let $w$, $x$, $y$ and $z$ be generators of weights 1, 1, 1 and 2 respectively in $S(1,1,1,2)$. Then

$$h^{1,1} = \dim \left( \frac{\mathbb{K}[w,x,y,z]}{(w^2, x^2, y^2, w+x+y)} \right)_1 = 2.$$

Thus the Hodge structure is:

| $h^{i,j}$ | $i=0$ | $i=1$ | $i=2$ |
|---|---|---|---|
| $j=0$ | 1 | 0 | 0 |
| $j=1$ | 0 | 3 | 0 |
| $j=2$ | 0 | 0 | 1 |

(c) Thus $e(S_3) = 1 + 3 + 1 = 5$.

(ii) The hypersurface $X_{10}$ in $\mathbb{P}(1,1,1,2,5)$ has the following Hodge structure.

| $h^{i,j}$ | $i=0$ | $i=1$ | $i=2$ | $i=3$ |
|---|---|---|---|---|
| $j=0$ | 1 | 0 | 0 | 1 |
| $j=1$ | 0 | 1 | 145 | 0 |
| $j=2$ | 0 | 145 | 1 | 0 |
| $j=3$ | 1 | 0 | 0 | 1 |

Let $v$, $w$, $x$, $y$ and $z$ be generators of weights 1, 1, 1, 2 and 5 respectively in $S(1,1,1,2,5)$. The only hard Hodge number is

$$h^{1,2}(X) = \dim_{\mathbb{K}} \left( \frac{\mathbb{K}[v,w,x,y,z]}{(v^9, w^9, x^9, y^4, z)} \right)_{20} = 145.$$

This gives an Euler number of $-288$.

# 8    Quasismoothness

In this section we prove conditions for quasismoothness for hypersurfaces and codimension 2 weighted complete intersections.

First we consider the problem of a hypersurface.

**8.1 Theorem** *The general hypersurface $X_d$ in $\mathbb{P} = \mathbb{P}(a_0, \ldots, a_n)$ of degree $d$, where $n \geq 1$ is quasismooth if and only if*

*either (1) there exists a variable $x_i$ for some $i$ of weight $d$ (that is, $X$ is a linear cone)*

*or (2) for every nonempty subset $I = \{i_0, \ldots, i_{k-1}\}$ of $\{0, \ldots, n\}$,*

  *either (a) there exists a monomial $x_I^M = x_{i_0}^{m_0} \cdots x_{i_{k-1}}^{m_{k-1}}$ of degree $d$,*

  *or (b) for $\mu = 1, \ldots, k$, there exist monomials*

$$x_I^{M_\mu} x_{e_\mu} = x_{i_0}^{m_{0,\mu}} \cdots x_{i_{k-1}}^{m_{k-1,\mu}} x_{e_\mu}$$

  *of degree $d$, where $\{e_\mu\}$ are $k$ distinct elements.*

**8.2 Note** If $X_d$ is a linear cone then $f$ can be written as $f = x_i + g$ for some $x_i$ and $X_d$ is clearly quasismooth. So we need only consider the case where $f$ is not linear in any of the variables (that is, $\deg x_i = a_i \neq d$ for all $i$).

**Proof**  Assume that $X_d$ in $\mathbb{P}$ is not a linear cone. Let $F$ be the linear system of all homogeneous polynomials of degree $d$ with respect to the weights $a_i$. Let $f \in F$ be a sufficiently general polynomial. Define $X_d : (f = 0) \subset \mathbb{P}$.

$$
\begin{array}{ccc}
C_X^* & \xrightarrow{\;i\;} & \mathbb{A}^{n+1} \setminus \{0\} \\
\downarrow & & \downarrow \\
X_d & \xrightarrow{\;i\;} & \mathbb{P}
\end{array}
$$

Note that the point $\underline{0}$ is a base point and is usually singular; as this point does not lie in $C_X^*$ this does not affect quasismoothness. By Bertini's Theorem (see [Hart], Remark III.10.9.2) the only singularities of the general $C_X^*$ lie on the base locus of the linear system $F$. Any component of the base locus is just a coordinate $k$-plane for some $k = 0, \ldots, n$. So the general hypersurface $X_d$ is quasismooth if and only if the general hypersurface $C_X^*$ is nonsingular at each point of its intersection with every coordinate $k$-plane contained in the base locus.

Let $\Pi$ be a coordinate $k$-plane for some $k = 1, \ldots, n$. By renumbering, assume that $\Pi$ is given by $x_k = \cdots = x_n = 0$, corresponding to the subset $I = \{0, \ldots, k-1\}$. Let $\Pi^0 \subset \Pi$ be the open toric stratum where $x_0, \ldots, x_{k-1}$ are nonzero. Expand $f$ in terms of the coordinates $x_k, \ldots, x_n$:

$$
f = h(x_0, \ldots, x_{k-1}) + \sum_{i=k}^{n} x_i g_i(x_0, \ldots, x_{k-1}) + \left\{ \begin{array}{c} \text{higher order terms} \\ \text{in } x_k, \ldots, x_n \end{array} \right\}.
$$

Assume that one of conditions $(a)$ and $(b)$ hold for $I$. If $(a)$ holds (that is, $h$ is nonzero) then $\Pi$ is not part of the base locus, and so by Bertini's Theorem $\Pi^0$ contains no singular points. Geometrically this means that $C_X^*$ intersects $\Pi^0$ transversally and so $\Pi^0$ is normal to the hypersurface at the points of intersection.

Assume that only $(b)$ holds. So $h \equiv 0$ and $\Pi \subset C_X^*$. By $(b)$ there are at least $k$ of the $g_i$ which are nonzero. Singular points occur exactly on the locus $Z = \bigcap_i (g_i = 0) \subset \Pi^0$, which is an intersection of at least $k$ free linear systems on $\Pi^0$. Thus $\dim Z \leq 0$. As $Z$ is a quasicone, it is at worst the origin (compare Lemma 6.2). Therefore $C_X^*$ is nonsingular along $\Pi^0$.

As one of these two conditions holds for every nonempty subset $I$, $C_X^*$ is nonsingular.

Conversely assume that conditions $(a)$ and $(b)$ do not hold for all $I$. Let $I$ be a subset for which these two conditions fail. Without loss of generality assume that $I = \{0, \ldots, k-1\}$. Let $\Pi$ be the corresponding coordinate $k$-plane $x_k = \cdots = x_n = 0$. As $(a)$ and $(b)$ do not hold

$$
f = \sum_{i=k}^{n} x_i g_i(x_0, \ldots, x_{k-1}) + \left\{ \begin{array}{c} \text{higher order terms} \\ \text{in } x_k, \ldots, x_n \end{array} \right\}
$$

and at most $k - 1$ of the $g_i$ are nonzero.

As above, singular points occur exactly on the intersection

$$Z = \bigcap_{i \geq k}(g_i = 0) \cap \Pi.$$

Since there are at most $k-1$ of the $g_i$ which are nonzero, $\dim Z \geq k-(k-1) = 1$. Thus $Z$ is nonempty and so $C_X^*$ is singular on $\Pi$.

Therefore conditions $(a)$ and $(b)$ are both sufficient and necessary for quasismoothness when $X_d$ in not a linear cone. $\square$

**8.3 Note** (i) The only quasismooth cones are the linear cones. Suppose a variable $x_i$ does not occur in the defining equation $f$. So $C_X \simeq C_{X'} \times \mathbb{A}^1$ where $X' : (f = 0) \subset \mathbb{P}(a_0, \ldots, \widehat{a_i}, \ldots, a_n)$. Suppose that $C_{X'}$ has a singularity at the origin. Thus $C_{X'} \times \mathbb{A}^1$ has a line of singularities along $\underline{0} \times \mathbb{A}^1$; a contradiction. So $C_{X'}$ is nonsingular at the origin and so $f$ must be linear in a variable; this is the linear cone case.

(ii) Without loss of generality we can assume in $(b)$ that $e_\mu \in \{0, \ldots, n\} - I$, since otherwise this is condition $(a)$.

(iii) For $2|I| \geq n + 1$ condition $(b)$ implies condition $(a)$, since there are simply not enough variables $x_i$.

(iv) Condition $(b)$, with $|I| = 1$, of the theorem gives that for all $i = 0, \ldots, n$ there must exist a monomial $x_i^n x_{e_i}$, for some $e_i$, of degree $d$. This is equivalent to requiring that $C_X^*$ is smooth along the coordinate axes (that is, $X_d$ is quasismooth at the vertices) and is in practice the most substantial case. Weighted hyperspaces (and polynomials) which satisfy this condition will be said to be *semi-quasismooth*.

(v) $C_X$ contains no coordinate stratum of dimension $\geq (n + 1)/2$ except possibly in the linear cone case.

So we have the following corollaries for curves, surfaces and 3-folds.

**8.4 Corollary** *The curve $C_d$ in $\mathbb{P}(a_0, a_1, a_2)$, where $d > a_i$, is quasismooth if and only if the following hold for all $i$:*

(1) *there exists a monomial $x_i^n x_{e_i}$, for some $e_i$, of degree $d$.*

(2) *there exists a monomial of degree $d$ which does not involve $x_i$.*

**Proof** Since $d > a_i$ for all $i$, $X_d$ is not a linear cone. Conditions (1) and (2) come from considering the conditions of the above theorem for $|I| = 1$ and $|I| = 2$ respectively. □

The proofs of the following corollaries are similar to the above.

**8.5 Corollary** *The surface $S_d$ in $\mathbb{P}(a_0, \dots, a_3)$, where $d > a_i$, is quasismooth if and only if the following hold:*

(1) *for all $i$ there exists a monomial $x_i^n x_{e_i}$ for some $e_i$ of degree $d$.*

(2) *for all distinct $i$, $j$*

*either there exists a monomial $x_i^m x_j^n$ of degree $d$,*

*or there exist monomials $x_i^{n_1} x_j^{m_1} x_{e_1}$ and $x_i^{n_2} x_j^{m_2} x_{e_2}$ of degree $d$ such that $e_1$ and $e_2$ are distinct.*

(3) *there exists a monomial of degree $d$ which does not involve $x_i$.*

**8.6 Corollary** *The 3-fold $X_d$ in $\mathbb{P}(a_0, \dots, a_4)$, where $d > a_i$, is quasismooth if and only if the following hold:*

(1) *for all $i$ there exists a monomial $x_i^n x_{e_i}$ of degree $d$.*

(2) *for all distinct $i$, $j$*

*either there exists a monomial $x_i^m x_j^n$ of degree $d$,*

*or there exist monomials $x_i^{n_1} x_j^{m_1} x_{e_1}$ and $x_i^{n_2} x_j^{m_2} x_{e_2}$ of degree $d$ such that $e_1$ and $e_2$ are distinct.*

(3) *there exists a monomial of degree $d$ which does not involve either $x_i$ or $x_j$.*

In the codimension 2 case we have:

**8.7 Theorem** *Suppose the general codimension 2 weighted complete inter-section $X_{d_1,d_2}$ in $\mathbb{P} = \mathbb{P}(a_0, \dots, a_n)$, where $n \geq 2$, of multidegree $\{d_1, d_2\}$ is not the intersection of a linear cone with another hypersurface. $X_{d_1,d_2}$ in $\mathbb{P}$ is quasismooth if and only if for each nonempty subset $I = \{i_0, \dots, i_k - 1\}$ of $\{0, \dots, n\}$ one of the following holds:*

(a) *there exists a monomial $x_I^{M_1}$ of degree $d_1$ and there exists a monomial $x_I^{M_2}$ of degree $d_2$*

(b) *there exists a monomial $x_I^M$ of degree $d_1$, and for $\mu = 1, \dots, k-1$ there exist monomials $x_I^{M_{m^u}} x_{e_{m^u}}$ of degree $d_2$, where $\{e_\mu\}$ are $k-1$ distinct elements.*

(c) there exists a monomial $x_I^M$ of degree $d_2$, and for $\mu = 1, \ldots, k-1$ there exist monomials $x_I^{Mm^\mu} x_{e_m u}$ of degree $d_1$, where $\{e_\mu\}$ are $k-1$ distinct elements.

(d) for $\mu = 1, \ldots, k$, there exist monomials $x_I^{M_\mu^1} x_{e_\mu^1}$ of degree $d_1$, and $x_I^{M_\mu^2} x_{e_\mu^2}$ of degree $d_1$, such that $\{e_\mu^1\}$ are $k$ distinct elements, $\{e_\mu^2\}$ are $k$ distinct elements and $\{e_\mu^1, e_\mu^2\}$ contains at least $k+1$ distinct elements.

**Proof** Let $F_1$ and $F_2$ be linear systems of all homogeneous polynomials of degrees $d_1$ and $d_2$ respectively with respect to the weights $a_0, \ldots, a_n$. Let $f_1 \in F_1$ and $f_2 \in F_2$ be sufficiently general polynomials. Define

$$X = X_{d_1,d_2} : (f_1 = f_2 = 0) \subset \mathbb{P}.$$

We have the following commutative diagram:

$$
\begin{array}{ccc}
C_X^* & \xrightarrow{\ i\ } & \mathbb{A}^{n+1} \setminus \{0\} \\
\downarrow & & \downarrow \\
X & \xrightarrow{\ i\ } & \mathbb{P}
\end{array}
$$

The only singularities that can occur in the general member of the family occur on the coordinate strata. So as in the proof of quasismoothness for hypersurfaces, $X$ is quasismooth if and only if $C_X^*$ is smooth along all the coordinate strata.

Assume that one of conditions $(a)$, $(b)$, $(c)$ or $(d)$ holds for each nonempty subset $I$. Let $\Pi$ be a coordinate $k$-plane for some $k$. By renumbering, we can assume that $\Pi$ is given by $x_k = \cdots = x_n = 0$, corresponding to the subset $I = \{0, \ldots, k-1\}$. As before let $\Pi^0$ be the open toric strata where $x_0, \ldots, x_{k-1}$ are all nonzero. Expand both $f_1$ and $f_2$ in terms of the coordinates $x_k, \ldots, x_n$:

$$f_\lambda = h_\lambda(x_0, \ldots, x_{k-1}) + \sum_{i=k}^n x_i g_\lambda^i(x_0, \ldots, x_{k-1}) + \left\{ \begin{array}{l} \text{higher order terms} \\ \text{in } x_k, \ldots, x_n \end{array} \right\}$$

for $\lambda = 1, 2$.

Suppose $(a)$ holds. So $h_1$ and $h_2$ are nonzero on $\Pi^0$. If either $h_1$ or $h_2$ involves only one monomial then $\Pi^0 \cap C_X^*$ is empty. This includes the case when $k = 1$. So without loss of generality assume that $h_1$ and $h_2$ each involve at least 2 monomials and hence $k \geq 2$. $\Pi^0$ is not part of the base locus of $F_1$ or $F_2$. By Bertini's Theorem $(f_1 = 0)$ and $(f_2 = 0)$ are nonsingular on $\Pi^0$. Since $(h_1 = 0)$ and $(h_2 = 0)$ are free linear systems on $\Pi^0$, $(h_1 = 0)$ and $(h_2 = 0)$ intersect transversally. Thus, at each point of $(h_1 = h_2 = 0) \cap \Pi^0$, there exist two distinct normals. Therefore $C_X^*$ is nonsingular along $\Pi^0$.

Suppose (b) holds. So $h_1$ is nonzero and there are at least $k - 1$ of the $\{g_1^i\}$ which are nonzero. So $\Pi^0$ is not part of the base locus for $F_1$, and so by Bertini's Theorem we have that $(f_1 = 0)$ is nonsingular on $\Pi^0$. Singular points occur exactly on the locus

$$Z = (h_1 = 0) \bigcap_i (g_2^i = 0) \subset \Pi^0,$$

which is an intersection of at least $k - 1$ free linear systems on $(h_1 = 0) \cap \Pi^0$. Thus $\dim Z \leq 0$ and hence is at worst the origin. Therefore $C_X^*$ is nonsingular along $\Pi^0$.

The case where condition (c) holds is similar to the case for condition (b). Suppose that only condition (d) holds. We have

$$f_\lambda = \sum_{i=k}^n x_i g_\lambda^i(x_0, \ldots, x_{k-1}) + \left\{ \begin{matrix} \text{higher order terms} \\ \text{in } x_k, \ldots, x_n \end{matrix} \right\}$$

for $\lambda = 1, 2$. The normal directions, perpendicular to the plane $\Pi$, to the hypersurfaces are $(g_1^k, \ldots, g_1^n)$ and $(g_2^k, \ldots, g_2^n)$. Define the matrix $M_P$ by

$$M_P = \begin{pmatrix} g_1^k(P) & \cdots & g_1^n(P) \\ g_2^k(P) & \cdots & g_2^n(P) \end{pmatrix}.$$

Singular points occur exactly on the locus $Z = \{P : \text{rank } M_P \leq 1\}$. As there are at least $k$ monomials of the form $x_I^M x_e$ of degree $d_\lambda$, at least $k$ of the $\{g_\lambda^i\}$ are nonzero. As these are free on $\Pi^0$, each row of the matrix $M_P$ is nonzero for each $P \in \Pi^0$. Furthermore this matrix for any $P \in Z$ has at least $k + 1$ nonzero columns, since there are at least $k + 1$ distinct elements in $\{e_\mu^1, e_\mu^2\}$. By renumbering we can assume that the first $k + 1$ columns of $M^P$ are not identically zero on $\Pi^0$.

Fix $P \in \Pi^0$. Without loss of generality we can assume that $g_1^k(P) \neq 0$. If $g_2^k(P) = 0$ then $g_2^i(P) \neq 0$ for some $i > k$, and so $M^P$ has rank 2. In this case $P \in C_X^*$ is nonsingular. Suppose that $g_2^k(P) \neq 0$. Define $a = g_1^k(P)$, $b = g_2^k(P)$ and

$$Z_P = \bigcap_{i > k} (a g_2^i(Q) - b g_1^i(Q) = 0) \subset \Pi^0.$$

Notice that $P \in Z_P$ if and only if rank $M_P \leq 1$, which is equivalent to $P \in C_X^*$ being singular. Since $Z_P$ is the intersection of $k$ free linear systems on $\Pi^0$, $\dim Z_P \leq 0$ and so $Z_P$ is at worst the origin. In particular $P \notin Z_P$ and hence $P \in C_X^*$ is nonsingular. Therefore $C_X^*$ is nonsingular along $\Pi^0$.

As one of these four conditions holds for every nonempty subset $I$, $C_X^*$ is nonsingular.

Conversely assume that none of the conditions $(a)$, $(b)$, $(c)$ or $(d)$ hold for some nonempty subset $I$. Without loss of generality we can assume that $I = \{0, \dots, k-1\}$. Let $\Pi$ be the corresponding coordinate plane $x_k = \dots = x_n = 0$. There are three cases:

(i) $\Pi \not\subset C_{X_{d_1}}$. So $h_1$ is nonzero and there are at most $k-2$ of the $\{g_2^i\}$ which are nonzero. The singular points are exactly the locus $Z = (h_1 = 0) \cap_i (g_2^i = 0)$. However

$$\dim Z \geq k - (k-2) - 1 = 1,$$

and so $Z$ contains more than the origin. Thus $C_X^*$ is singular along $\Pi$.

(ii) $\Pi \not\subset C_{X_{d_2}}$. Similarly in this case $C_X^*$ is singular along $\Pi$.

(iii) $\Pi \subset C_{X_{d_1}} \cap C_{X_{d_2}}$. In this case both $h_1$ and $h_2$ are identically zero. So

$$f_\lambda = \sum_{i=k}^{n} x_i g_\lambda^i(x_0, \dots, x_{k-1}) + \left\{ \begin{array}{l} \text{higher order terms} \\ \text{in } x_k, \dots, x_n \end{array} \right\}$$

for $\lambda = 1, 2$. As condition $(d)$ does not hold, one of two cases occurs:

either for some $\lambda$ there are at most $k-1$ of the $\{g_\lambda^i\}$ which are nonzero. Thus the intersection $Z_\lambda = \cap_i (g_\lambda^i = 0)$ has dimension at least 1 and so these $\{g_\lambda^i\}$ have a common solution. Therefore the matrix

$$M_P = \begin{pmatrix} g_1^k(P) & \cdots & g_1^n(P) \\ g_2^k(P) & \cdots & g_2^n(P) \end{pmatrix}$$

has rank less than 2 for some $P \in Z_\lambda$ and hence $C_X^*$ is singular along $\Pi$.

or there are at most $k$ distinct elements in $\{e_\mu^1, e_\mu^2\}$. Thus there are at most $k$ nonzero columns in the matrix $M_P$. Let $Z = \{P : \operatorname{rank} M_P \leq 1\}$. Therefore

$$\dim Z \geq k - (k-1) = 1,$$

and so contains more than just the origin. Therefore $C_X^*$ is singular along $\Pi$.

So if one of these four conditions are not satisfied for every subset $I$ then $C_X^*$ is singular. $\square$

**8.8 Corollary** *Suppose $X_{d_1, d_2}$ in $\mathbb{P}$ is quasismooth and is not the intersection of a linear cone with another hypersurface. We have the following:*

(i) *Every variable $x_i$ occurs in at least one of the defining equations.*

(ii) *All but at most one variable are in both equations.*

(iii) *If $x_i$ does not appear in one defining equation then there exists a monomial $x_i^m$ occurring in the other equation.*

**Proof**

(i) This follows from the previous theorem with $|I| = 1$.

(ii) Suppose, after renumbering, that $x_0$ and $x_1$ are not involved in $f_1$. Then none of the conditions can hold for $I = \{0, 1\}$, a contradiction.

(iii) Suppose that $x_i$ does not appear in $f_1$. Conditions $(a)$, $(b)$ and $(d)$ cannot hold and so there must be a monomial $x_i^m$ of degree $d_2$. Geometrically if one of the hypersurfaces is singular along a coordinate axis, because the equation $f_i$ does not involve that variable, then the other hypersurface cannot pass through that axis. □

# 9   Cyclic singularities and counting points

In this section we give combinatorial conditions for cyclic quotient singularities to be isolated and canonical (see [R4], Definition 1.1 for the definitions of canonical and terminal singularities). The last two lemmas of this section are used to count the number of intersections along 1 and 2 dimensional strata. We also give an alternative proof of the first of these lemmas in terms of the Minkowski mixed volume of integral polyhedra.

**9.1 Lemma** *A canonical curve point is smooth.*

This is clear since canonical singularities are normal. For dimension 2 we have:

**9.2 Lemma** *The following are equivalent:*

(i) *$Q$ in $S$ is a cyclic quotient canonical surface singularity.*

(ii) *$Q$ is of type $\frac{1}{r}(a, -a)$ for some index $r$ and $a$ coprime to $r$.*

(iii) *$Q$ is of type $\frac{1}{r}(1, -1)$ for some index $r$.*

The above singularities are Du Val singularities of type $A_{r-1}$.

For 3-folds we have the following due to White, Morrison, Stevens, Danilov and Frumkin:

**9.3 Lemma** *The following are equivalent:*

(i) *$S$ is an isolated cyclic quotient terminal 3-fold singularity.*

(ii) *$S$ is of type $\frac{1}{r}(b_0, b_1, b_2)$, for some positive integers $r$, $b_0$, $b_1$, $b_2$, with $r \geq 2$, $r$ and $b_i$ coprime and $r \mid b_i + b_j$ for a pair of distinct $i$, $j$.*

(iii) *$S$ is of the form $\frac{1}{r}(1, -1, b)$ for some $r \geq 2$ and $b$ coprime to $r$.*

The following two lemmas are very useful for calculating the number and arrangement of singularities on a complete intersection.

**9.4 Lemma** *Let $x$ and $y$ be of weight $a_0$ and $a_1$ respectively, such that $\mathrm{hcf}(a_0, a_1) = 1$. Suppose $f(x, y)$ is a homogeneous polynomial of degree $d$, semi-quasismooth (see Note 8.3, iv) and sufficiently general. Let $P_0 = [1, 0]$ and $P_1 = [0, 1]$. Then $X_d : (f = 0)$ in $\mathbb{P}(a_0, a_1)$ is a finite set and:*

(i) *$P_i$ is in $X_d$ if and only if $a_i \nmid d$ for $i = 0, 1$,*

(ii) *there are exactly $\lfloor \frac{d}{a_0 a_1} \rfloor$ other points in $X_d$.*

**Proof**    Notice that $x^{a_1}/y^{a_0}$ is an invariant of the group action of $\mathbb{K}^*$ on $\mathbb{A}^2 - \underline{0}$ which defines $\mathbb{P}(a_0, a_1)$. There are four cases:

(i) $a_0 \mid d$ and $a_1 \mid d$. Then $f$ is of the form

$$f = x^{d/a_0} + \cdots + y^{d/a_1},$$

written using the coefficient convention (see Section 6.7). So

$$\frac{f}{y^{d/a_1}} = \left( \frac{x_1^a}{y_0^a} \right)^{d/a_0 a_1} + \cdots + 1,$$

which has exactly $\frac{d}{a_0 a_1}$ roots.

(ii) $a_0 \nmid d$ and $a_1 \mid d$. Since $X_d$ is semi-quasismooth, $f$ is of the form

$$f = y(x^{(d-a_1)/a_0} + \cdots + y^{(d-a_1)/a_1}).$$

The solution $y = 0$ gives the point $P_0$.

$$\frac{f}{y^{d/a_1}} = \left( \frac{x_1^a}{y_0^a} \right)^{(d-a_1)/a_0 a_1} + \cdots + 1.$$

This has exactly $n = \frac{d-a_1}{a_0 a_1}$ roots. So $d = n a_0 a_1 + a_1$. As $a_0 \nmid d$ then $a_0 > 1$, and so $a_1 < a_0 a_1$. Thus $n = \lfloor \frac{d}{a_0 a_1} \rfloor$.

(iii) $a_0 \mid d$ and $a_1 \nmid d$. Similar to (ii).

(iv) $a_0 \nmid d$ and $a_1 \nmid d$.

$$f = xy(x^{(d-a_0-a_1)/a_0} + \cdots + y^{(d-a_0-a_1)/a_1})$$

So the two vertices $P_0$ and $P_1$ are solutions. Also

$$\frac{f}{xy^{d/a_1}} = \left(\frac{x_1^a}{y_0^a}\right)^{(d-a_0-a_1)/a_0\,a_1} + \cdots + 1,$$

which has exactly $n = \frac{d-a_0-a_1}{a_0 a_1}$ roots on $\mathbb{P} \setminus \{P_0, P_1\}$. So $d = na_0a_1 + (a_0 + a_1)$. As $a_0 \nmid d$ and $a_1 \nmid d$ then $a_0, a_1 \geq 2$ and not both equal to 2. Thus

$$a_0\, a_1 = (a_0 - 1)(a_1 - 1) - 1 + a_0 + a_1 a_0 + a_1.$$

Therefore $n = \lfloor \frac{d}{a_0 a_1} \rfloor$. $\square$

**9.5 Lemma** *Suppose that* $x_0$, $x_1$ *and* $x_2$ *have weights* $a_0$, $a_1$ *and* $a_2$, *where* $\mathrm{hcf}(a_0, a_1, a_2) = 1$. *Let* $f$ *and* $g$ *be sufficiently general semi-quasismooth homogeneous polynomials in* $\mathbb{K}[x_0, x_1, x_2]$ *of degrees* $d$ *and* $e$ *respectively. Suppose that* $X_{d,e} : (f = 0, g = 0)$ *in* $\mathbb{P}(a_0, a_1, a_2)$ *is a finite set. Let*

- $n_{i,j}$ *be the number of points of* $X_{d,e}$ *along the edge* $P_i P_j$,

- $h_{i,j} = \mathrm{hcf}(a_i, a_j)$,

- $n_i$ *be the number of points at the vertex* $P_i$ *(that is,* $n_i = 0, 1$*),*

- $N$ *be the number of points in* $\mathbb{P} \setminus \Delta$.

*Then:*

$$\frac{de}{a_0\, a_1\, a_2} = \sum_i \frac{n_i}{a_i} + \sum_{i>j} \frac{n_{i,j}}{h_{i,j}} + N.$$

**9.6 Note**    (i) $X_{d,e}$ in $\mathbb{P}$ is not automatically finite (consider for example $X_{5,9}$ in $\mathbb{P}(1, 2, 4)$).

(ii) Similar results hold for higher codimensions and involve induction on the dimension.

(iii) Notice that Lemma 9.4 can be deduced from the above (consider $X_{d,1}$ in $\mathbb{P}(a_0, a_1, 1)$).

(iv) This also has connections with the Minkowski mixed volumes of Newton polyhedra (see after the proof).

**Proof**  Let $\sigma\colon \mathbb{P}^2 \to \mathbb{P}$ be the quotient map defined in Section 5.12. Let $F = \sigma^* f$ and $G = \sigma^* g$. Since $X_{d,e}$ is finite, $V(F)$ and $V(G)$ have no common components. By Bezout's theorem $Y = V(F,G)$ in $\mathbb{P}^2$ consists of exactly $de$ points counted with multiplicity.

The restriction of $\sigma$ to $\mathbb{P}^2 \setminus \Delta$ is $a_0 a_1 a_2$-to-1, onto $\mathbb{P} \setminus \Delta$. As there are $N$ points on $\mathbb{P} \setminus \Delta$ this accounts for $a_0 a_1 a_2 N$ points on $\mathbb{P}^2 \setminus \Delta$.

The restriction of $\sigma$ to the line $Q_i Q_j$ is $a_i a_j / h_{i,j}$-to-1, onto $P_i P_j$. Without loss of generality assume that $h_{i,j} \mid d$ but that $h_{i,j} \nmid e$. Let $k$ be such that $\{i,j,k\} = \{0,1,2\}$. Notice that $x_k \mid g$, or else there would exist a monomial $x_i^a x_j^b$ of degree $e$, contradicting $h_{i,j} \nmid e$. Then $f$ and $g$ are of the form:

$$f = x_i^m x_j + x_j^m x_i + \cdots$$
$$g = x_k(x_i^{n'} + x_j^{m'} + \cdots).$$

Thus $F$ and $G$ are of the form:

$$F = X_i^{ma_i} X_j^{a_j} + X_j^{na_j} X_i^{a_i} + \cdots$$
$$G = X_k^{a_k}(X_i^{n'a_i} + X_j^{m'a_j} + \cdots).$$

We localise $F$ and $G$ by setting $X_i = 1$, to give the corresponding affine equations $\overline{F}$ and $\overline{G}$. Let $[X_i, X_j, X_k] = [1, \xi, 0]$ be a point of intersection along the line $Q_i Q_j$. The multiplicity $\mu$ of the intersection is given by:

$$\begin{aligned}
\mu &= \operatorname{mult}(F, G, [1, \xi, 0]) \\
&= \operatorname{mult}(\overline{F}, \overline{G}, (\xi, 0)) \\
&= \operatorname{mult}(X_i^{a_i} + X_i^{ma_i} + \cdots, X_k^{a_k}, (\xi, 0)) \\
&= \operatorname{mult}(X_i' + \cdots, X_k^{a_k}, (0, 0)) \\
&= a_k,
\end{aligned}$$

where $X_i' = X_i - \xi$. So this line contributes $(n_{i,j} a_k) a_i a_j / h_{i,j}$ points (counted with multiplicity) to Bezout's theorem.

Consider the vertex $Q_i$. If $P_i$ is contained in $X$ then $a_i \nmid d$ and $a_i \nmid e$. As $X$ is semi-quasismooth, $a_i \mid d - a_j$ and $a_i \mid e - a_k$ for distinct $i$, $j$, and $k$. So $f$ and $g$ are of the form:

$$f = x_i^n x_j + \cdots$$
$$g = x_i^m x_k + \cdots$$

Thus:

$$F = X_i^{na_i} X_j + \cdots$$
$$G = X_i^{ma_i} X_k + \cdots$$

The intersection multiplicity $\mu$ at $Q_i$ is:

$$\mu = \text{mult}(F, G, Q_i).$$

Localising at $X_i = 1$ gives:

$$\begin{aligned}
\mu &= \text{mult}(\overline{F}, \overline{G}, (0,0)) \\
&= \text{mult}(X_j^{a_j} + \cdots, X_k^{a_k} + \cdots, (0,0)) \\
&= a_j a_k.
\end{aligned}$$

Clearly $X_j^{a_j}$ and $X_k^{a_k}$ are the smallest degree monomials in $\overline{F}$ and $\overline{G}$. So this gives a contribution of $a_j a_k n_i$.

Combining the above gives:

$$de = \sum_{\text{distinct } i,j,k} n_i a_j a_k + \sum_{i>j,\ k\neq i,j} \frac{n_{i,j} a_i a_j a_k}{h_{i,j}} + N a_0 a_1 a_2,$$

which rearranges to give the formula in the lemma. $\quad\square$

An alternative proof of the above two lemmas is via Newton polyhedra and the Minkowski mixed volume (see both [Be] and [Ku]).

**9.7 Definition** An *integral polyhedron* $S$ is a polyhedron in $\mathbb{R}^n$ with vertices in $\mathbb{Z}^n$. The $n$-dimensional volume of $S$ will be denoted by $V_n(S)$, where the volume of the unit parallelepiped is 1.

**9.8 Definition** For each $m = (m_1, \ldots, m_n) \in \mathbb{Z}^n$ define

$$x^m = x_1^{m_1} \cdots x_n^{m_n}.$$

Let $f \in \mathbb{K}[x_1, x_1^{-1}, \ldots, x_n, x_n^{-1}]$ be a Laurent polynomial. Then

$$f = \sum_{m \in \mathbb{Z}^n} c_m x^m,$$

where all but a finite number of the $\{c_m\}$ are zero. The *Newton polyhedron* Newton($f$) of $f$ is the convex hull of $\{m \in \mathbb{Z}^n : c_m \neq 0\}$, and is an integral polyhedron.

**9.9 Definition** Let $\mathcal{S} = \{S_i : i = 1, \ldots, n\}$ be a set of integral polyhedra. The *Minkowski mixed volume* $V(\mathcal{S})$ of $\mathcal{S}$ is given by:

$$V(\mathcal{S}) = (-1)^{n-1} \sum V_n(S_i) + (-1)^{n-2} \sum_{i>j} V_n(S_i + S_j) +$$

$$\cdots + V_n(S_1 + \cdots + S_n)$$

where $S_i + S_j = \{s_i + s_j : s_i \in S_i, s_j \in S_j\}$.

This is the classical formula, up to a multiple of $n!$.

Let $T^n$ be the $n$-dimensional torus $(\mathbb{K}^*)^n$. This corresponds to the open toric stratum in $\mathbb{P}$. Let $\mathcal{F}$ be a system of $n$ sufficiently general Laurent polynomials $\{f_i : T^n \to \mathbb{K}\}$ with corresponding Newton polyhedra $\mathcal{S} = \{S_i\}$. The roots of these $n$ polynomials in $T^n$ are isolated. Let $L(\mathcal{F})$ be the number of such roots, counted with multiplicity. Then [Be], Theorem A gives:

$$L(\mathcal{F}) = V(\mathcal{S}).$$

**9.10   Alternative proof of Lemma 9.4**   Let $T^1$ be the torus $x_0 x_1 \neq 0$ in $\mathbb{P} = \mathbb{P}(a_0, a_1)$. Suppose that $a_0, a_1 \mid d$. Then $f = x_0^{d/a_0} + \cdots + x_1^{d/a_1}$. So

$$N_f = \text{Newton}(f) = [(d/a_0, 0), (0, d/a_1)],$$

where $[P, Q]$ denotes the line segment in $\mathbb{Z}^2$ from $P$ to $Q$. So $V_1(N_f) + 1$ is the number of integral points on $N_f$, that is, the number of solutions to

$$\{(\alpha, \beta) \in \mathbb{Z}^2 : \alpha \geq 0, \beta \geq 0, \alpha a_0 + \beta a_1 = d\}.$$

For a solution $(\alpha, \beta)$ we have $\alpha = (d - \beta a_1)/a_0 \in \mathbb{Z}$, that is, $d \equiv \beta a_1$ mod $a_1$. As $a_0$ and $a_1$ are coprime, then $a_1$ is invertible modulo $a_0$, with inverse $s$. So $\beta \equiv ds$ mod $a_0$, that is, $\beta = ds + na_0$ for some $n$. Also $0 \leq \beta \leq d/a_1$. So

$$-\frac{ds}{a_0} \leq n \leq \frac{d}{a_0 a_1} - \frac{ds}{a_0}.$$

There are $\frac{d}{a_0 a_1} + 1$ such solutions. Thus $f$ has $\frac{d}{a_0 a_1}$ roots on the torus $T^1$ in $\mathbb{P}$.
Similarly when $a_0 \nmid d$, etc.   $\square$

Lemma 9.5 can be proved using analogous methods.

# 10   Determination of singularities on weighted complete intersections

In this section we shall determine the singularities of three weighted complete intersections, presenting the calculations in detail. These examples are a good introduction to the theorems giving arithmetic conditions for weighted complete intersections to have at worst isolated canonical singularities.

**10.1   The surface $S = S_{36}$ in $\mathbb{P}(7, 8, 9, 12)$**   We shall see that this surface has four singularities, one each of type $A_2$, $A_3$, $A_6$ and $A_7$. The Euler number of such a K3 surface is 6, the lowest Euler number found in any of the lists of weighted complete intersection K3 surfaces.

Let $w$, $x$, $y$ and $z$ be the homogeneous coordinates on $\mathbb{P} = \mathbb{P}(7,8,9,12)$ of weights 7, 8, 9 and 12 respectively. Let $f$ be a general polynomial of homogeneous degree 36. Using the coefficient convention (see Section 6.7) we have:

$$f = w^4 x + x^3 z + y^4 + z^3 + \text{others}.$$

So $S$ is well formed and, by Theorem 8.1, is quasismooth. Hence its singularities arise only due to the singularities of $\mathbb{P}$ and occur only on the edges and vertices of $\mathbb{P}$. Consider the vertices.

(1) $P_0$: $f$ contains no monomial of the form $w^n$ for any $n$ and so $P_0 \in S$. Consider the affine piece $(w = 1)$. The point $P_0 \in S$ looks like:

$$(\widetilde{f} = f(1,x,y,z) = x + \cdots = 0) \subset \mathbb{A}^3/\varepsilon$$

where $\varepsilon$ is a primitive 7th root of unity and acts on the coordinates of $\mathbb{A}^3$ via:

$$x \mapsto \varepsilon^8 x = \varepsilon x$$
$$y \mapsto \varepsilon^9 y = \varepsilon^2 y$$
$$z \mapsto \varepsilon^{12} z = \varepsilon^5 z.$$

Notice that $\partial f / \partial x = w^4 + \cdots$ is nonzero at $P_0$. By the Inverse Function Theorem $y$ and $z$ are local coordinates around $P_0 \in S$. This gives a singularity of type $\frac{1}{7}(2,5)$, which is Du Val of type $A_6$.

(2) $P_1$: Again $f$ contains no monomial of type $x^n$ and so $P_1 \in S$. As above, this gives a Du Val singularity of type $A_7$.

(3) $P_2$, $P_3$: Since $f$ contains the monomials $y^4$ and $z^3$ then $P_2$, $P_3 \notin S$.

There are only two singular edges in $\mathbb{P}$, $P_1 P_3$ which is analytically isomorphic to $\mathbb{K}^* \times \frac{1}{4}(3,1)$ and $P_2 P_3$ which is $\mathbb{K}^* \times \frac{1}{3}(2,1)$.

(4) $P_1 P_3$: Since $f_{|P_1 P_2} = x^3 z + z^3 = z(x^3 + z^2)$ then $S$ does not contain the edge $P_1 P_3$. As $x \neq 0$ and $z \neq 0$ on the edge $P_1 P_3$ then the affine piece $(z = 1)$ contains all of the intersection points. Since $(\partial f / \partial x)_{|z=1} = x^2 + \cdots$ is nonzero then $w$ and $y$ are local coordinates on $S$ at each of the points of $S \cap P_1 P_3$. This is clear geometrically since $S$ is a general element of all degree 36 hypersurfaces and so it must cross this line transversally. Thus each point is a singularity locally analytically isomorphic to $\mathbb{A}^2/\varepsilon$ where the coordinates of $\mathbb{A}^2$ are $w$ and $y$ and $\varepsilon$ is a 4th root acting via:

$$w \mapsto \varepsilon^7 w = \varepsilon^3 w$$
$$y \mapsto \varepsilon^9 y = \varepsilon y.$$

This gives a Du Val singularity of type $A_3$.

(5) We must now count the number of intersection points on this edge. Each point of the intersection is given by the equation $x^3 + z^2 = 0$ in $\mathbb{P}(8,12)$. This is just $X_{24}$ in $\mathbb{P}(8,12)$, that is, $X_6$ in $\mathbb{P}(2,3)$. Either from first principles or from Lemma 9.4 we can see that this is exactly one point.

(6) $P_2 P_3$: As above, there is exactly one Du Val singularity, which is of type $A_2$, along this edge.

**10.2   The 3-fold $X = X_{46}$ in $\mathbb{P}(4,5,6,7,23)$**   The canonical 3-fold hypersurface $X_{46}$ in $\mathbb{P}(4,5,6,7,23)$ has 3 singularities of type $\frac{1}{2}(1,1,1)$, 1 of type $\frac{1}{4}(3,1,1)$, 1 of type $\frac{1}{5}(4,1,2)$, 1 of type $\frac{1}{6}(5,1,1)$, and 1 of type $\frac{1}{7}(6,1,3)$.

The singularities are checked as follows. Let $v$, $w$, $x$, $y$ and $z$ be the homogeneous coordinates of $\mathbb{P} = \mathbb{P}(4,5,6,7,23)$ of weights 4, 5, 6, 7 and 23 respectively. Let $f$ be a general polynomial of homogeneous degree 46. Then, using the coefficient convention, we write $f$ in the form:

$$f = v^{10}x + w^8 x + x^7 v + y^6 v + z^2 + \text{others}.$$

This is well formed and quasismooth (see Theorem 8.1). So the singularities of the hypersurface occur only on the edges and at the vertices of $\mathbb{P}$. Consider the vertices in reverse order:

(1) $P_4$: Since $f$ contains the monomial $z^2$ with nonzero coefficient, $f(P_4) \neq 0$ and so $P_4 \notin X_{46}$.

(2) $P_3$: There is no monomial of the form $y^n$ for any $n$ in $f$, and so $P_3 \in X_{46}$. Consider the affine piece $(y = 1)$. $P_3 \in X_{46}$ looks like:

$$(\tilde{f} = f(v, w, x, 1, z) = v + \cdots = 0) \subset \mathbb{A}^4/\varepsilon,$$

where $\varepsilon$ is a primitive 7th root of unity and acts as:

$$(v, w, x, z) \mapsto (\varepsilon^4 v, \varepsilon^5 w, \varepsilon^6 x, \varepsilon^{23} z).$$

Notice that $\partial f/\partial v = y^6 + \cdots$ is nonzero at $P_3$. By the Inverse Function Theorem $w$, $x$ and $z$ are local coordinates on $X_{46}$ around $P_3 \in X_{46}$. Thus the singularity here is of type $\frac{1}{7}(5,6,23)$. This is equivalent to $\frac{1}{7}(6,1,3)$, which is terminal.

(3) $P_2$: Again there is no monomial of the form $x^n$ for any $n$ in $f$, and so $P_2 \in X_{46}$. Consider the affine piece $(x = 1)$. $P_2 \in X_{46}$ looks like:

$$(\tilde{f} = f(v, w, 1, y, z) = v + \cdots = 0) \subset \mathbb{A}^4/\varepsilon,$$

where $\varepsilon$ is a primitive 6th root of unity and acts as:

$$(v, w, y, z) \mapsto (\varepsilon^4 v, \varepsilon^5 w, \varepsilon^7 y, \varepsilon^{23} z).$$

Notice that $\partial f / \partial v = x^7 + \cdots$ is nonzero at $P_3$. By the Inverse Function Theorem, $w$, $y$ and $z$ are local coordinates on $X_{46}$ around $P_2 \in X_{46}$. Thus the singularity here is of type $\frac{1}{6}(5, 7, 23)$. This is equivalent to $\frac{1}{6}(5, 1, 1)$, which is terminal.

(4) $P_1$: $P_1 \in X_{46}$ is locally $f = x + \cdots = 0$ and gives a terminal singularity of type $\frac{1}{5}(4, 1, 2)$.

(5) $P_0$: $P_0 \in X_{46}$ is locally $f = x + \cdots = 0$ and gives a terminal singularity of type $\frac{1}{4}(3, 1, 1)$.

Consider the edges of $\mathbb{P}$. An edge $P_i P_j$ is singular if and only if $h = \mathrm{hcf}(a_i, a_j) \neq 1$. In which case it is analytically equivalent to $\mathbb{K}^* \times \frac{1}{h}(a_0, \ldots, \widehat{a_i}, \ldots, \widehat{a_j}, \ldots, a_4)$. So only the edge $P_0 P_2$ is singular and looks like $\mathbb{K}^* \times \frac{1}{2}(1, 1, 1)$. Since $2 = \mathrm{hcf}(4, 6) \mid 46$, the hypersurface does not contain this line. Lemma 9.4 is used on $X_{46}$ in $\mathbb{P}(4, 6)$, after cancelling the common factor, to give three points of intersection. Alternatively,

$$f_{|P_0 P_2} = u x g_{36}(u, x) = u x g_3(u^3, x^2),$$

where $g_{36}$ and $g_3$ are polynomials of degree 36 and 3 respectively. There are exactly three solutions to $g_3 = 0$, and so there are three points of intersection. So $X_{46}$ crosses $P_0 P_2$ transversally and hence there are three singularities, each of type $\frac{1}{2}(1, 1, 1)$, along $P_0 P_2$.

**10.3  The 3-fold $X_{12,14}$ in $\mathbb{P}(2, 3, 4, 5, 6, 7)$**  The family of codimension two complete intersections $X_{12,14}$ in $\mathbb{P}(2, 3, 4, 5, 6, 7)$ is an anticanonically embedded Fano 3-fold having only the following isolated terminal singularities: 1 of type $\frac{1}{5}(4, 1, 2)$, 2 of type $\frac{1}{3}(2, 1, 1)$ and 7 of type $\frac{1}{2}(1, 1, 1)$.

The singularities are checked as follows. Let $u$, $v$, $w$, $x$, $y$ and $z$ be the homogeneous coordinates of weights 2, 3, 4, 5, 6 and 7 respectively. Let $f$, $g$ be homogeneous polynomials of degrees 12 and 14 respectively. Then $X = (f = g = 0) \subset \mathbb{P} = \mathbb{P}(2, 3, 4, 5, 6, 7)$.

Consider the vertices of the weighted projective space $\mathbb{P}$. Since $5 \nmid 12$ and $5 \nmid 14$, $P_3 \in X$. So

$$f = x^2 u + \cdots$$
$$g = x^2 w + \cdots$$

Thus $\{v, y, z\}$ are local coordinates around $P_3$, which is therefore a singularity of type $\frac{1}{5}(3, 6, 7)$, that is, $\frac{1}{5}(4, 1, 2)$. There are no other vertices contained in $X$.

Consider the 1-dimensional loci of $\mathbb{P}$.

(1) $P_0P_2$: $h = \text{hcf}(2,4) = 2$ and

$$f = u^6 + w^3 + \cdots$$
$$g = u^7 + w^2y + \cdots$$

Then the local coordinates are $\{v, x, z\}$ and the singularities are of type $\frac{1}{2}(1,1,1)$. There are three such intersection points (by Lemma 9.4 applied to $X_6$ in $\mathbb{P}(1,2)$).

(2) $P_0P_4$: Likewise $h = \text{hcf}(2,6) = 2$ and

$$f = u^6 + y^2 + \cdots$$
$$g = u^7 + u^5w + y^2u + \cdots$$

$(f = 0)$ in $\mathbb{P}(1,3)$ is two points by Lemma 9.4. So there are two singularities, each of type $\frac{1}{2}(1,1,1)$, along $P_0P_4$.

(3) $P_2P_4$: There is exactly one singularity on this line, of type $\frac{1}{2}(1,1,1)$.

(4) $P_1P_4$: This time $h = \text{hcf}(3,6) = 3$ and

$$f = v^4 + y^2 + \cdots$$
$$g = v^4u + y^2u + \cdots$$

So there are two of type $\frac{1}{3}(1,-1,1)$ on $P_1P_4$.

Consider the only singular 2-dimensional locus, $P_0P_2P_4$, of $\mathbb{P}$ where $h = \text{hcf}(2,4,6) = 2$. By Lemma 9.5, there are seven intersection points (some of which have already been counted), all of type $\frac{1}{2}(1,1,1)$.

# Part II. Lists of various weighted complete intersections

## 11   Preamble

The aim of this chapter is to produce lists of hypersurface and codimension 2 weighted complete intersections of dimension at most 3 with at worst isolated canonical singularities. We present various theorems giving combinatoric conditions on the weights and degrees of such intersections. From these conditions we can produce lists of intersections (along with their corresponding singularities). In most cases a computer was used for its speed and inability to become bored.

Sections 12 and 13 treat the cases of dimension 1 and 2 respectively and give corresponding lists. Section 14 deals with the 3-fold case (both hypersurfaces and codimension 2) and Sections 15 and 16 deal with the particular cases of canonical 3-folds and $\mathbb{Q}$-Fano 3-folds respectively. Section 18 gives an alternative method for producing canonically and anticanonically embedded 3-fold complete intersections using the Poincaré series of a ring.

# 12    Weighted curve hypersurfaces

**12.1 Theorem** *A weighted curve complete intersection is smooth if and only if it is quasismooth.*

**Proof**    Any 1-dimensional cyclic quotient singularity is of type $\frac{1}{r}(a)$ for some coprime $r$ and $a$. Let $x$ be the coordinate on $\mathbb{A}^1$. The group $\mathbb{Z}_r$ acts via:

$$x \mapsto \varepsilon^a x,$$

where $\varepsilon$ is a primitive $r$th root of unity. So

$$\mathbb{A}^1/\mathbb{Z}_r \simeq \operatorname{Spec} \mathbb{K}[x]^{\mathbb{Z}_r} \simeq \operatorname{Spec} \mathbb{K}[x^r] \simeq \operatorname{Spec} \mathbb{K}[x] \simeq \mathbb{A}^1.$$

So this is nonsingular. Notice that this group action is just a quasireflection (see Section 5.8).    $\square$

From [OW], Corollary 3.5 we have a formula for the genus of dimension 1 hypersurfaces.

**12.2 Theorem** *Let $C_d$ in $\mathbb{P}(a_0, a_1, a_2)$ be a nonsingular curve. Then the genus $g$ is given by:*

$$g = \frac{1}{2} \left( \frac{d^2}{a_0 a_1 a_2} - d \sum_{i>j} \frac{\operatorname{hcf}(a_i, a_j)}{a_i a_j} + \sum_{i=0}^{2} \frac{\operatorname{hcf}(d, a_i)}{a_i} - 1 \right).$$

**12.3 Theorem** *A weighted curve $C_d$ in $\mathbb{P}(a_0, a_1, a_2)$ is well formed, not a linear cone and quasismooth if and only if for each $i$ the following three conditions hold:*

*(1) $a_i < d$,*

*(2) $a_i \mid d$,*

*(3) $\operatorname{hcf}(a_i, a_j) = 1$ for all distinct $i$, $j$.*

**Proof**  $C$ is well formed if and only if $a_i \mid d$ for all $i$ and $\mathrm{hcf}(a_i, a_j) = 1$ for all distinct $i$, $j$ (see Section 6.10). These are conditions (2) and (3).

Suppose $C$ is not a linear cone and quasismooth. Then conditions (1) holds. Also $a_i \mid d - a_e$ for some $e$. But this is already satisfied by condition (2).

The converse follows immediately from conditions (1), (2) and (3).  $\square$

**12.4  Smooth weighted curve hypersurfaces with amplitude $\alpha = d - \sum a_i = 0$**  We list the only smooth weighted curves of codimension 1 with $\alpha = 0$ satisfying the above conditions.

| Curve | $D$ |
|-------|-----|
| $C_3 \subset \mathbb{P}(1,1,1)$ | $3P$ |
| $C_4 \subset \mathbb{P}(1,1,2)$ | $2P$ |
| $C_6 \subset \mathbb{P}(1,2,3)$ | $P$ |

All are elliptic curves (that is, $g = 1$ and $\omega \simeq \mathcal{O}_C$) and are given by $\mathrm{Proj}\, R_C$ where $R_C$ is:

$$R_C = \bigoplus_{n \geq 0} \mathrm{H}^0(\mathcal{O}_C(nD)),$$

and $D$ is given in the above table.

**12.5  The calculation**  The above curves are the only ones satisfying the conditions of Theorem 12.3. This is demonstrated as follows.

Order the $\{a_i\}$ by $a_0 \leq a_1 \leq a_2$. conditions (2) and (3) of Theorem 12.3 give $a_0 a_1 a_2 \mid d$. Let $d = \lambda a_2$. As $\alpha = 0$ then $3a_2 \geq a_0 + a_1 + a_2 = d = \lambda a_2$. So $\lambda \leq 3$ (that is, $\lambda = 2, 3$).

(i)  $\lambda = 2$. So $a_0 a_1 \mid 2$. Either $(a_0, a_1) = (1, 1)$ (that is, $C_4$ in $\mathbb{P}(1,1,2)$) or $(a_0, a_1) = (1, 2)$ (that is, $C_6$ in $\mathbb{P}(1,2,3)$).

(ii)  $\lambda = 3$. So $a_0 a_1 \mid 3$. Either $(a_0, a_1) = (1, 1)$ (that is, $C_3$ in $\mathbb{P}(1,1,1)$) or $(a_0, a_1) = (1, 3)$ in which case $a_2 = 2 < a_1$, a contradiction.

**12.6  The ring $R_C$**  Consider an elliptic curve $C$ and the divisor $D = 2P$, where $P$ is any point on $C$. By Riemann–Roch,

$$h^0(nD) - h^1(nD) = \deg(nD) + (1 - g).$$

As $D > K \equiv 0$, then $h^1(nD) = 0$ for all $n \geq 1$. Also $g = 1$ and so

$$h^0(nD) = \deg(nD) = 2n.$$

Thus $h^0(D) = 2$ and $h^0(2D) = 4$. Let $x_0, x_1$ be a basis for $H^0(D)$. Then $x_0^2$, $x_0 x_1$ and $x_1^2$ are linearly independent elements of $H^0(2D)$. As $h^0(2D) = 4$ then there exists an extra element $y$ of degree 4.

Consider the map:

$$\varphi_n \colon H^0(D) \otimes H^0((n-1)D) \to H^0(nD).$$

Notice that $x_0$ and $x_1$ have no common base points. By the base-point-free pencil trick (see [ACGH], p. 126),

$$\operatorname{Ker} \varphi_n \simeq H^0((n-1)D - D) = H^0((n-2)D),$$

which has dimension $2(n-2)$. Also $H^0(D) \otimes H^0((n-1)D)$ has dimension $2 \times 2(n-1)$. So $\dim \operatorname{Im} \varphi_n = 2n$, and hence $\varphi_n$ is onto for all $n \geq 2$. This means that $H^0(nD)$ is generated by $H^0(D)$ and $H^0((n-1)D)$.

We thus have the following table of bases for the $H^0(nD)$.

| $n$ | $h^0(nD)$ | monomials |
|---|---|---|
| 1 | 2 | $x_0, x_1$ |
| 2 | 4 | $x_0^2, x_0 x_1, x_1^2, y$ |
| 3 | 6 | $x_0^3, x_0^2 x_1, x_0 x_1^2, x_1^3, x_0 y, x_1 y$ |
| 4 | 8 | $x_0^4, x_0^3 x_1, x_0^2 x_1^2, x_0 x_1^3, x_1^4, x_0^2 y, x_0 x_1 y, x_1^2 y, y^2$ |

Notice that $H^0(4D)$ has dimension 8, but there are 9 monomials. Since $\varphi_4$ is onto then the first eight in the list are linear independent. So there must be a relation of the form:

$$f = y^2 + y h_2(x_0, x_1) - g_4(x_0, x_1),$$

where $h_2$ and $g_4$ are homogeneous polynomials of degrees 4 and 2 respectively.

The number $N_n$ of monomials in $H^0(nD)$ is given by:

$$N_n = 1 + n \left\lfloor \frac{n}{2} \right\rfloor \left\lfloor n + \frac{1}{2} \right\rfloor.$$

Suppose that $f$ was the only relation, then the dimension of the module generated by the monomials of degree $n$ is $N_n - 1 \cdot N_{n-4} = 2n$, which is the same as $h^0(nD)$.

Thus the ring $R$ is $\mathbb{K}[x_0, x_1, y]/(f)$, where $x_i$ has weight 1 and $y$ has weight 2, that is, the curve is $C_4$ in $\mathbb{P}(1,1,2)$. This technique should be compared to that in [M], Lecture 1, p. 17–21] and to Weierstrass normal form.

**12.7  Smooth weighted curve hypersurfaces with amplitude $\alpha = d - \sum a_i = 1$**  There are only two such curves which satisfy the conditions of Theorem 12.3:

| curve | genus | $\omega_C$ |
|-------|-------|-----------|
| $C_4 \subset \mathbb{P}(1,1,1)$ | 3 | $\mathcal{O}_C(1)$ |
| $C_6 \subset \mathbb{P}(1,1,3)$ | 2 | $\mathcal{O}_C(1)$ |

These were calculated in a similar way to those of Section 12.5 and the genera by the formula in Theorem 12.2.

# 13   Weighted surface complete intersections

In this section we give necessary and sufficient conditions for surface weighted complete intersections of codimension 1 and 2 to be quasismooth, well formed, and have at worst canonical singularities. We also include lists of such intersections.

**13.1 Theorem**  *Let $S_d$ in $\mathbb{P} = \mathbb{P}(a_0, a_1, a_2, a_3)$ be a general hypersurface of degree $d$ and let $\alpha = d - \sum a_i$. $S_d$ is quasismooth, well formed with at worst canonical quotient singularities and is not a linear cone if and only if all the following hold:*

*(1) For all $i$,*

    *(i) $d > a_i$.*

    *(ii) there exists $e$ such that $a_i \mid d - a_e$ (that is, there exists a monomial $x_i^n x_e$ of degree $d$).*

    *(iii) there exists a monomial of degree $d$ which does not involve $x_i$.*

    *(iv) if $a_i \nmid d$, then $a_i \mid \alpha$.*

*(2) For all distinct $i$, $j$, with $h = \mathrm{hcf}(a_i, a_j)$, then*

    *(i) $h \mid d$.*

    *(ii) $h \mid \alpha$.*

    *(iii) one of the following holds:*

      *either there exists a monomial $x_i^m x_j^n$ of degree $d$,*

      *or there exist monomials $x_i^{n_1} x_j^{m_1} x_{e_1}$ and $x_i^{n_2} x_j^{m_2} x_{e_2}$ of degree $d$ such that $e_1$ and $e_2$ are distinct.*

    *for all distinct $i$, $j$, $k$, $\mathrm{hcf}(a_i, a_j, a_k) = 1$.*

**13.2 Note**  Since the hypersurface is well formed then $\omega_S = \mathcal{O}_S(\alpha)$.

**Proof**  Let $f$ be a general homogeneous polynomial of degree $d$ in variables $x_0, \ldots, x_3$; define $S_d : (f = 0) \subset \mathbb{P}$.

$S_d$ is quasismooth and not a linear cone if and only if conditions $(1i)$, $(1ii)$, $(1iii)$ and $(2iii)$ hold (see Corollary 8.5).

Suppose furthermore that conditions $(1iv)$, $(2i)$, $(2ii)$ and $(3)$ hold. As $S_d$ is quasismooth the only singularities are due to the $\mathbb{K}^*$-action and hence are cyclic quotient singularities on the fundamental simplex $\Delta \subset \mathbb{P}$. By condition $(3)$ only vertices and edges need be checked.

Consider $P_i \in S_d$. By renumbering we can assume that $i = 0$. So $a_0 \nmid d$. Condition $(1ii)$ gives that there exists an $e \neq 0$ such that $a_0 \mid d - a_e$. Without loss of generality we can assume that $e = 1$. So $f$ is of the form $f = x_0^n x_1 + \cdots$. Thus $\partial f / \partial x_1$ is nonzero at $P_0$. By the Inverse Function Theorem $x_2$ and $x_3$ are local coordinates. So $P_0 \in S_d$ is of type $\frac{1}{a_0}(a_2, a_3)$. However $d = a_0 + \cdots + a_3 + \alpha$ and so $a_0 \mid a_2 + a_3 + \alpha$. By condition $(1iv)$, $a_0 \mid a_2 + a_3$. Let $h = \mathrm{hcf}(a_0, a_2)$. So $h \mid a_3$ and hence, by condition $(3)$, $h = 1$. Therefore $P_0 \in S_d$ is a canonical singularity.

Consider the edge $P_i P_j$. Again by renumbering assume that $i = 0$ and $j = 1$. $f$ restricted to $P_0 P_1$ is:

$$f = \sum x_0^n x_1^m,$$

where the sum is taken over the set $\{(n, m) : na_0 + ma_1 = d\}$. If $a_0 \nmid d$ then $a_0 \mid d - a_e$ for some $e \neq 0$. If $e \neq 1$ then $h = \mathrm{hcf}(a_0, a_1) \mid a_e$ and by condition $(4)$ $h = 1$. Then $P_0 P_1$ is nonsingular. So assume that either $a_0 \mid d$ or $a_0 \mid d - a_1$. Hence $f$ is not identically zero on $P_0 P_1$, and so $S_d \cap P_0 P_1$ is finite. Each point in this intersection is of type $\frac{1}{h}(a_2, a_3)$. Since $d = a_0 + \cdots + a_3 + \alpha$ and $h \mid \alpha$ then $h \mid a_2 + a_3$. Also $\mathrm{hcf}(h, a_2) = 1$. Thus each point is canonical.

Therefore $S_d$ in $\mathbb{P}$ has at worst canonical singularities.

Conversely assume that $S_d$ is quasismooth, well-formed, not a linear cone and has at worst only canonical singularities. Suppose $a_i \nmid d$. By renumbering we can assume that $i = 0$. So $P_0 \in S_d$ and $a_0 \mid d - a_e$ for some $e$. Without loss of generality assume that $e = 1$. As above the singularity at $P_0 \in S_d$ is of type $\frac{1}{a_0}(a_2, a_3)$. Since this is canonical we have $a_0 \mid a_2 + a_3$ and so $a_0 \mid \alpha$. This is condition $(1iv)$.

Suppose $h = \mathrm{hcf}(a_i, a_j)$. By renumbering assume that $i = 0$ and $j = 1$. As $S_d$ is well formed then $h \mid d$, which is condition $(2i)$. So $P_0 P_1 \cap S_d$ is a finite intersection, where each point is of type $\frac{1}{h}(a_2, a_3)$. This is canonical and so $h \mid \alpha$. This is condition $(2ii)$.

Suppose $h = \mathrm{hcf}(a_i, a_j, a_k)$. Without loss of generality assume that $i = 0$, $j = 1$ and $k = 2$. Let $h' = \mathrm{hcf}(a_0, a_1)$. So $h' \mid d$. Hence the line $P_0 P_1$ contains singularities of type $\frac{1}{h'}(a_2, a_3)$. As these are canonical $h = \mathrm{hcf}(h', a_2) = 1$. This is condition $(3)$.  $\square$

**13.3  Reid's 95 codimension 1 K3 surfaces**  In 1979, Reid produced the
list of all families of codimension 1 weighted K3 surfaces; 95 in all (see [R1],
Section 4.5). The full list follows along with their respective singularities.

| | Weighted K3 | Singularities |
|---|---|---|
| No. 1 | $X_4 \subset \mathbb{P}(1,1,1,1)$ | |
| No. 2 | $X_5 \subset \mathbb{P}(1,1,1,2)$ | $A_1$ |
| No. 3 | $X_6 \subset \mathbb{P}(1,1,1,3)$ | |
| No. 4 | $X_6 \subset \mathbb{P}(1,1,2,2)$ | $3 \times A_1$ |
| No. 5 | $X_7 \subset \mathbb{P}(1,1,2,3)$ | $A_1, A_2$ |
| No. 6 | $X_8 \subset \mathbb{P}(1,1,2,4)$ | $2 \times A_1$ |
| No. 7 | $X_8 \subset \mathbb{P}(1,2,2,3)$ | $4 \times A_1, A_2$ |
| No. 8 | $X_9 \subset \mathbb{P}(1,1,3,4)$ | $A_3$ |
| No. 9 | $X_9 \subset \mathbb{P}(1,2,3,3)$ | $A_1, 3 \times A_2$ |
| No. 10 | $X_{10} \subset \mathbb{P}(1,1,3,5)$ | $A_2$ |
| No. 11 | $X_{10} \subset \mathbb{P}(1,2,2,5)$ | $5 \times A_1$ |
| No. 12 | $X_{10} \subset \mathbb{P}(1,2,3,4)$ | $2 \times A_1, A_2, A_3$ |
| No. 13 | $X_{11} \subset \mathbb{P}(1,2,3,5)$ | $A_1, A_2, A_4$ |
| No. 14 | $X_{12} \subset \mathbb{P}(1,1,4,6)$ | $A_1$ |
| No. 15 | $X_{12} \subset \mathbb{P}(1,2,3,6)$ | $2 \times A_1, 2 \times A_2$ |
| No. 16 | $X_{12} \subset \mathbb{P}(1,2,4,5)$ | $3 \times A_1, A_4$ |
| No. 17 | $X_{12} \subset \mathbb{P}(1,3,4,4)$ | $3 \times A_3$ |
| No. 18 | $X_{12} \subset \mathbb{P}(2,2,3,5)$ | $6 \times A_1, A_4$ |
| No. 19 | $X_{12} \subset \mathbb{P}(2,3,3,4)$ | $3 \times A_1, 4 \times A_2$ |
| No. 20 | $X_{13} \subset \mathbb{P}(1,3,4,5)$ | $A_2, A_3, A_4$ |
| No. 21 | $X_{14} \subset \mathbb{P}(1,2,4,7)$ | $3 \times A_1, A_3$ |
| No. 22 | $X_{14} \subset \mathbb{P}(2,2,3,7)$ | $7 \times A_1, A_2$ |
| No. 23 | $X_{14} \subset \mathbb{P}(2,3,4,5)$ | $3 \times A_1, A_2, A_3, A_4$ |
| No. 24 | $X_{15} \subset \mathbb{P}(1,2,5,7)$ | $A_1, A_6$ |
| No. 25 | $X_{15} \subset \mathbb{P}(1,3,4,7)$ | $A_3, A_6$ |
| No. 26 | $X_{15} \subset \mathbb{P}(1,3,5,6)$ | $2 \times A_2, A_5$ |
| No. 27 | $X_{15} \subset \mathbb{P}(2,3,5,5)$ | $A_1, 3 \times A_4$ |
| No. 28 | $X_{15} \subset \mathbb{P}(3,3,4,5)$ | $5 \times A_2, A_3$ |
| No. 29 | $X_{16} \subset \mathbb{P}(1,2,5,8)$ | $2 \times A_1, A_4$ |
| No. 30 | $X_{16} \subset \mathbb{P}(1,3,4,8)$ | $A_2, 2 \times A_3$ |
| No. 31 | $X_{16} \subset \mathbb{P}(1,4,5,6)$ | $A_1, A_4, A_5$ |
| No. 32 | $X_{16} \subset \mathbb{P}(2,3,4,7)$ | $4 \times A_1, A_2, A_6$ |
| No. 33 | $X_{17} \subset \mathbb{P}(2,3,5,7)$ | $A_1, A_2, A_4, A_6$ |
| No. 34 | $X_{18} \subset \mathbb{P}(1,2,6,9)$ | $3 \times A_1, A_2$ |
| No. 35 | $X_{18} \subset \mathbb{P}(1,3,5,9)$ | $2 \times A_2, A_4$ |
| No. 36 | $X_{18} \subset \mathbb{P}(1,4,6,7)$ | $A_3, A_1, A_6$ |
| No. 37 | $X_{18} \subset \mathbb{P}(2,3,4,9)$ | $4 \times A_1, 2 \times A_2, A_3$ |

| No. 38 | $X_{18} \subset \mathbb{P}(2,3,5,8)$ | $2 \times A_1, A_4, A_7$ |
|--------|----|----|
| No. 39 | $X_{18} \subset \mathbb{P}(3,4,5,6)$ | $3 \times A_2, A_3, A_1, A_4$ |
| No. 40 | $X_{19} \subset \mathbb{P}(3,4,5,7)$ | $A_2, A_3, A_4, A_6$ |
| No. 41 | $X_{20} \subset \mathbb{P}(1,4,5,10)$ | $A_1, 2 \times A_4$ |
| No. 42 | $X_{20} \subset \mathbb{P}(2,3,5,10)$ | $2 \times A_1, A_2, 2 \times A_4$ |
| No. 43 | $X_{20} \subset \mathbb{P}(2,4,5,9)$ | $5 \times A_1, A_8$ |
| No. 44 | $X_{20} \subset \mathbb{P}(2,5,6,7)$ | $3 \times A_1, A_5, A_6$ |
| No. 45 | $X_{20} \subset \mathbb{P}(3,4,5,8)$ | $A_2, 2 \times A_3, A_7$ |
| No. 46 | $X_{21} \subset \mathbb{P}(1,3,7,10)$ | $A_9$ |
| No. 47 | $X_{21} \subset \mathbb{P}(1,5,7,8)$ | $A_4, A_7$ |
| No. 48 | $X_{21} \subset \mathbb{P}(2,3,7,9)$ | $A_1, 2 \times A_2, A_8$ |
| No. 49 | $X_{21} \subset \mathbb{P}(3,5,6,7)$ | $3 \times A_2, A_4, A_5$ |
| No. 50 | $X_{22} \subset \mathbb{P}(1,3,7,11)$ | $A_2, A_6$ |
| No. 51 | $X_{22} \subset \mathbb{P}(1,4,6,11)$ | $A_3, A_1, A_5$ |
| No. 52 | $X_{22} \subset \mathbb{P}(2,4,5,11)$ | $5 \times A_1, A_3, A_4$ |
| No. 53 | $X_{24} \subset \mathbb{P}(1,3,8,12)$ | $2 \times A_2, A_3$ |
| No. 54 | $X_{24} \subset \mathbb{P}(1,6,8,9)$ | $A_1, A_2, A_8$ |
| No. 55 | $X_{24} \subset \mathbb{P}(2,3,7,12)$ | $2 \times A_1, 2 \times A_2, A_6$ |
| No. 56 | $X_{24} \subset \mathbb{P}(2,3,8,11)$ | $3 \times A_1, A_{10}$ |
| No. 57 | $X_{24} \subset \mathbb{P}(3,4,5,12)$ | $2 \times A_2, 2 \times A_3, A_4$ |
| No. 58 | $X_{24} \subset \mathbb{P}(3,4,7,10)$ | $A_1, A_6, A_9$ |
| No. 59 | $X_{24} \subset \mathbb{P}(3,6,7,8)$ | $4 \times A_2, A_1, A_6$ |
| No. 60 | $X_{24} \subset \mathbb{P}(4,5,6,9)$ | $2 \times A_1, A_4, A_2, A_8$ |
| No. 61 | $X_{25} \subset \mathbb{P}(4,5,7,9)$ | $A_3, A_6, A_8$ |
| No. 62 | $X_{26} \subset \mathbb{P}(1,5,7,13)$ | $A_4, A_6$ |
| No. 63 | $X_{26} \subset \mathbb{P}(2,3,8,13)$ | $3 \times A_1, A_2, A_7$ |
| No. 64 | $X_{26} \subset \mathbb{P}(2,5,6,13)$ | $4 \times A_1, A_4, A_5$ |
| No. 65 | $X_{27} \subset \mathbb{P}(2,5,9,11)$ | $A_1, A_4, A_{10}$ |
| No. 66 | $X_{27} \subset \mathbb{P}(5,6,7,9)$ | $A_4, A_5, A_2, A_6$ |
| No. 67 | $X_{28} \subset \mathbb{P}(1,4,9,14)$ | $A_1, A_8$ |
| No. 68 | $X_{28} \subset \mathbb{P}(3,4,7,14)$ | $A_2, A_1, 2 \times A_6$ |
| No. 69 | $X_{28} \subset \mathbb{P}(4,6,7,11)$ | $2 \times A_1, A_5, A_{10}$ |
| No. 70 | $X_{30} \subset \mathbb{P}(1,4,10,15)$ | $A_3, A_1, A_4$ |
| No. 71 | $X_{30} \subset \mathbb{P}(1,6,8,15)$ | $A_1, A_2, A_7$ |
| No. 72 | $X_{30} \subset \mathbb{P}(2,3,10,15)$ | $3 \times A_1, 2 \times A_2, A_4$ |
| No. 73 | $X_{30} \subset \mathbb{P}(2,6,7,15)$ | $5 \times A_1, A_2, A_6$ |
| No. 74 | $X_{30} \subset \mathbb{P}(3,4,10,13)$ | $A_3, A_1, A_{12}$ |
| No. 75 | $X_{30} \subset \mathbb{P}(4,5,6,15)$ | $A_3, 2 \times A_1, 2 \times A_4, A_2$ |
| No. 76 | $X_{30} \subset \mathbb{P}(5,6,8,11)$ | $A_1, A_7, A_{10}$ |
| No. 77 | $X_{32} \subset \mathbb{P}(2,5,9,16)$ | $2 \times A_1, A_4, A_8$ |
| No. 78 | $X_{32} \subset \mathbb{P}(4,5,7,16)$ | $2 \times A_3, A_4, A_6$ |
| No. 79 | $X_{33} \subset \mathbb{P}(3,5,11,14)$ | $A_4, A_{13}$ |

| No. 80 | $X_{34} \subset \mathbb{P}(3,4,10,17)$ | $A_2, A_3, A_1, A_9$ |
|---|---|---|
| No. 81 | $X_{34} \subset \mathbb{P}(4,6,7,17)$ | $A_3, 2 \times A_1, A_5, A_6$ |
| No. 82 | $X_{36} \subset \mathbb{P}(1,5,12,18)$ | $A_4, A_5$ |
| No. 83 | $X_{36} \subset \mathbb{P}(3,4,11,18)$ | $2 \times A_2, A_1, A_{10}$ |
| No. 84 | $X_{36} \subset \mathbb{P}(7,8,9,12)$ | $A_6, A_7, A_3, A_2$ |
| No. 85 | $X_{38} \subset \mathbb{P}(3,5,11,19)$ | $A_2, A_4, A_{10}$ |
| No. 86 | $X_{38} \subset \mathbb{P}(5,6,8,19)$ | $A_4, A_5, A_1, A_7$ |
| No. 87 | $X_{40} \subset \mathbb{P}(5,7,8,20)$ | $2 \times A_4, A_6, A_3$ |
| No. 88 | $X_{42} \subset \mathbb{P}(1,6,14,21)$ | $A_1, A_2, A_6$ |
| No. 89 | $X_{42} \subset \mathbb{P}(2,5,14,21)$ | $3 \times A_1, A_4, A_6$ |
| No. 90 | $X_{42} \subset \mathbb{P}(3,4,14,21)$ | $2 \times A_2, A_3, A_1, A_6$ |
| No. 91 | $X_{44} \subset \mathbb{P}(4,5,13,22)$ | $A_1, A_4, A_{12}$ |
| No. 92 | $X_{48} \subset \mathbb{P}(3,5,16,24)$ | $2 \times A_2, A_4, A_7$ |
| No. 93 | $X_{50} \subset \mathbb{P}(7,8,10,25)$ | $A_6, A_7, A_1, A_4$ |
| No. 94 | $X_{54} \subset \mathbb{P}(4,5,18,27)$ | $A_3, A_1, A_4, A_8$ |
| No. 95 | $X_{66} \subset \mathbb{P}(5,6,22,33)$ | $A_4, A_1, A_2, A_{10}$ |

Table 1: Reid's 95 codimension 1 K3 surfaces

However there are not so many dimension 2 weighted hypersurfaces with $\omega_S \simeq \mathcal{O}_S(\pm 1)$:

**13.4 Theorem** *There are exactly three families of dimension 2 weighted hypersurfaces with at worst canonical singularities and $\omega_S \simeq \mathcal{O}_S(1)$, and exactly three families with $\omega_S \simeq \mathcal{O}_S(-1)$,*

| $\alpha = 1$ | $\alpha = -1$ |
|---|---|
| $S_5 \subset \mathbb{P}(1,1,1,1)$ | $S_3 \subset \mathbb{P}(1,1,1,1)$ |
| $S_6 \subset \mathbb{P}(1,1,1,2)$ | $S_4 \subset \mathbb{P}(1,1,1,2)$ |
| $S_8 \subset \mathbb{P}(1,1,1,4)$ | $S_6 \subset \mathbb{P}(1,1,2,3)$ |

**13.5 Note** These families are all nonsingular.

**Proof** Condition (*2ii*) of Theorem 13.1 is very strong when $\alpha = \pm 1$ and forces the $a_i$ to be pairwise coprime. Similarly condition (*1iv*) forces $a_i \mid d$ for each $i$. So $a_0 a_1 a_2 a_3 \mid d$ and $d = a_0 + \cdots + a_3 + \alpha$. Order $a_3 \geq a_2 \geq a_1 \geq a_0 \geq 1$ and let $d = \lambda a_3$. Thus $a_0 a_1 a_2 \mid \lambda$ and $(\lambda - 1)a_3 = a_0 + \cdots + a_2 + \alpha$.

Suppose $\alpha = 1$. Then $2a_3 \leq \lambda a_3 = a_0 + \cdots + a_3 + 1 \leq 5a_3$. So $2 \leq \lambda \leq 5$. Running through the possible values of $\lambda$:

(i) $\lambda = 5$. So $a_0 a_1 a_2 \mid 5$. If $a_2 = 1$ then $a_4 = 1$ (that is, $S_5$ in $\mathbb{P}(1,1,1,1)$). If $a_2 = 5$ then $a_3 = 2$, a contradiction.

(ii) $\lambda = 4$. So $a_0 a_1 a_2 \mid 4$. If $a_2 = 1$ then $a_4 = \frac{4}{3}$, a contradiction. If $a_2 = 2$ then $a_4 = \frac{5}{3}$, a contradiction. If $a_2 = 4$ then $a_4 = \frac{7}{3}$, a contradiction.

(iii) $\lambda = 3$. So $a_0 a_1 a_2 \mid 3$. If $a_2 = 1$ then $a_4 = 2$ (that is, $S_6$ in $\mathbb{P}(1,1,1,2)$). If $a_2 = 3$ then $a_4 = 3$, a contradiction.

(iv) $\lambda = 2$. So $a_0 a_1 a_2 \mid 2$. If $a_2 = 1$ then $a_4 = 4$ (that is, $S_8$ in $\mathbb{P}(1,1,1,4)$). If $a_2 = 2$ then $a_4 = \frac{5}{2}$, a contradiction.

So there are exactly three families.

Suppose that $\alpha = -1$. Then $2a_3 \leq \lambda a_3 = a_0 + \cdots + a_3 - 1 \leq 6a_3$. Thus $2 \leq \lambda \leq 6$. As above this gives rise to the following families: $S_3$ in $\mathbb{P}(1,1,1,1)$ in the case $\lambda = 3$, $S_4$ in $\mathbb{P}(1,1,1,2)$ and $S_6$ in $\mathbb{P}(1,1,2,3)$ in the case $\lambda = 2$. □

Consider the case of codimension 2 complete intersections.

**13.6 Theorem** *Suppose $S = S_{d_1,d_2}$ in $\mathbb{P} = \mathbb{P}(a_0, \ldots, a_4)$ is quasismooth and is not the intersection of a linear cone with another hypersurface. Let $\alpha = \sum d_\lambda - \sum a_i$. $S$ is well formed and has at worst canonical singularities if and only if the following hold:*

(1) *for all $i$, if $a_i \nmid d_1$ and $a_i \nmid d_2$ then $a_i \mid \alpha$.*

(2) *for all distinct $i$ and $j$, with $h = \mathrm{hcf}(a_i, a_j)$, one of the following occurs:*

   (i) *$h \mid d_1$ and $h \mid d_2$,*

   (ii) *$h \mid d_1$, $h \nmid d_2$ and $h \mid \alpha$, or*

   (iii) *$h \nmid d_1$, $h \mid d_2$ and $h \mid \alpha$.*

(3) *for all distinct $i$, $j$ and $k$, with $h = \mathrm{hcf}(a_i, a_j, a_k)$, $h \mid d_1$, $h \mid d_2$ and $h \mid \alpha$.*

(4) *for all distinct $i$, $j$, $k$ and $l$, $h = \mathrm{hcf}(a_i, a_j, a_k, a_l) = 1$*

**13.7 Note** *Since the hypersurface is well formed we have that $\omega_S = \mathcal{O}_S(\alpha)$.*

**Proof** Let $f_1$ and $f_2$ be sufficiently general homogeneous polynomials of degrees $d_1$ and $d_2$ respectively, in the variables $x_0, \ldots, x_4$ with respect to the weights $a_0, \ldots, a_4$. Define $S : (f_1 = 0, f_2 = 0) \subset \mathbb{P}$.

Since $S$ is quasismooth the only singularities are due to the $\mathbb{K}^*$-action and hence are all cyclic quotient singularities occurring on the fundamental simplex $\Delta$.

Assume conditions $(1), \ldots, (4)$ hold. By conditions $(2)$, $(3)$ and $(4)$ $S$ is well formed. By condition $(4)$ only the vertices, edges and faces of $\Delta$ need be considered.

Suppose $P_i \in S$. By renumbering we can assume that $i = 0$. So $a_0 \nmid d_1$ and $a_0 \nmid d_2$. As $S$ is quasismooth (and using $I = \{0\}$ in Theorem 8.7) there exist monomials $x_0^n x_{e_1}$ and $x_0^m x_{e_2}$ of degrees $d_1$ and $d_2$, where $e_1 \neq e_2$. By renumbering we can write $e_1 = 1$ and $e_2 = 2$. So $f_1$ and $f_2$ are of the form:

$$f_1 = x_0^n x_1 + \cdots$$
$$f_2 = x_0^m x_2 + \cdots$$

Thus $\partial f_1 / \partial x_1$ and $\partial f_2 / \partial x_2$ are nonzero at $P_0$. By the Inverse Function Theorem, $x_3$ and $x_4$ are local coordinates around $P_0$. Hence $P_0 \in S$ is of type $\frac{1}{a_0}(a_3, a_4)$. As $d_1 + d_2 = a_0 + \cdots + a_4 + \alpha$ and $a_0 \mid \alpha$ then $a_0 \mid a_3 + a_4$. Let $h = \mathrm{hcf}(a_0, a_3)$. So $h \mid a_4$ and, by condition (3), $h \mid d_1$. Since $\deg(x_0^n x_1) = d_1$, $h \mid a_1$ and so, by condition (4), $h = 1$. Thus $P_0 \in S$ is canonical.

Consider the edge $P_i P_j$. By renumbering we can assume that $i = 0$ and $j = 1$. Let $h = \mathrm{hcf}(a_0, a_1)$. Notice that $P_0 P_1 \subset X_{d_\lambda}$ if and only if $h \nmid d_\lambda$ for $\lambda = 0, 1$. By condition (2), $h \mid d_\lambda$ for some $\lambda$. Without loss of generality assume that $h \mid d_1$. There are 2 cases:

(a) $h \mid d_2$. $P_0 P_1 \cap (f_\lambda = 0)$ is a finite set of points for $\lambda = 0, 1$. Thus $P_0 P_1 \cap S = \emptyset$.

(b) $h \nmid d_2$. In this case no monomial of the form $x_0^n x_1^m$ of degree $d_2$ exists (or else $h \mid d_2$). From Theorem 8.7 (with $I = \{0, 1\}$) there exists a monomial $x_0^n x_1^m x_e$ of degree $d_2$, where $e \neq 0, 1$. By renumbering we can assume that $e = 2$. Thus $f_2$ is of the form:

$$f_2 = x_0^n x_1^m x_2 + \cdots$$

and $\partial f_2 / \partial x_2$ is nonzero on $P_0 P_1 \cap S$. By the Inverse Function Theorem, $x_3$ and $x_4$ are local coordinates around each point of $P_0 P_1 \cap S$ and so each is of type $\frac{1}{h}(a_3, a_4)$. Condition $(2b)$ gives $h \mid \alpha$ and so $h \mid a_3 + a_4$. Let $h' = \mathrm{hcf}(h, a_3)$. So $h' \mid a_4$ and thus by condition (4) $h' = 1$. Thus these points are canonical.

Therefore $S$ has at worst canonical points along $P_0 P_1$.

Consider the face $P_i P_j P_k$. As before assume $i = 0$, $j = 1$ and $k = 2$. By condition (3) $h = \mathrm{hcf}(a_0, a_1, a_2) \mid d_1$ and $h \mid d_2$. So $P_0 P_1 P_2$ intersects $S$ transversally. Each point in the intersection is of type $\frac{1}{h}(a_3, a_4)$. As $h \mid \alpha$, $h \mid a_3 + a_4$. By condition (4) $\mathrm{hcf}(h, a_3) = 1$. Thus these points are canonical.

Therefore conditions $(1), \ldots, (4)$ are sufficient.

Conversely assume that $S$ is well formed and has at worst canonical singularities. Suppose $a_i \nmid d_1$ and $a_i \nmid d_2$. By renumbering assume $i = 0$. Thus $P_0 \in S$. Since $S$ is quasismooth there exist 2 monomials $x_0^n x_{e_1}$ and $x_0^m x_{e_2}$ of degrees $d_1$ and $d_2$, where $e_1 \neq e_2$. Without loss of generality we can assume that $e_1 = 1$ and $e_2 = 2$. As before we find that $P_0 \in S$ is of type $\frac{1}{a_0}(a_3, a_4)$. As this is canonical $a_0 \mid a_3 + a_4$ and so $a_0 \mid \alpha$. This is condition (1).

Suppose $h = \mathrm{hcf}(a_i, a_j)$ for distinct $i$ and $j$. As usual we can renumber such that $i = 0$ and $j = 1$. As $S$ is well formed then $h \mid d_\lambda$ for some $\lambda$. Suppose $h \mid d_1$. If $h \mid d_2$ then this is condition (2a). So assume that $h \nmid d_2$. As above each point of $P_0 P_1 \cap S$ is isolated and of type $\frac{1}{h}(a_3, a_4)$. Thus $h \mid a_3 + a_4$ and so $h \mid \alpha$. This is condition (2b). Likewise for the case when $h \mid d_2$ but $h \nmid d_1$. This gives condition (2c).

Suppose $h = \mathrm{hcf}(a_i, a_j, a_k)$ for distinct $i$, $j$ and $k$. Renumber such that $i = 0$, $j = 1$ and $k = 2$. As $S$ is well formed then $h \mid d_1$ and $h \mid d_2$. Thus $P_0 P_1 P_2 \cap S$ is a finite number of points, all of type $\frac{1}{h}(a_3, a_4)$. As these are canonical $h \mid a_3 + a_4$ and so $h \mid \alpha$. This is condition (3). Also $\mathrm{hcf}(h, a_3) = \mathrm{hcf}(h, a_4) = 1$, which is condition (4).

Therefore these conditions are necessary. $\square$

**13.8 Codimension 2 Weighted K3 Surfaces** There are 84 families of codimension 2 quasismooth, well-formed K3 surfaces with only canonical singularities and $\sum a_i \leq 100$. These were found by means of a computer search program.

| | Weighted K3 | Singularities |
|---|---|---|
| No. 1 | $X_{2,3} \subset \mathbb{P}(1,1,1,1,1)$ | |
| No. 2 | $X_{3,3} \subset \mathbb{P}(1,1,1,1,2)$ | $A_1$ |
| No. 3 | $X_{3,4} \subset \mathbb{P}(1,1,1,2,2)$ | $2 \times A_1$ |
| No. 4 | $X_{4,4} \subset \mathbb{P}(1,1,1,2,3)$ | $A_2$ |
| No. 5 | $X_{4,4} \subset \mathbb{P}(1,1,2,2,2)$ | $4 \times A_1$ |
| No. 6 | $X_{4,5} \subset \mathbb{P}(1,1,2,2,3)$ | $2 \times A_1, A_2$ |
| No. 7 | $X_{4,6} \subset \mathbb{P}(1,1,2,3,3)$ | $2 \times A_2$ |
| No. 8 | $X_{4,6} \subset \mathbb{P}(1,2,2,2,3)$ | $6 \times A_1$ |
| No. 9 | $X_{5,6} \subset \mathbb{P}(1,1,2,3,4)$ | $A_1, A_3$ |
| No. 10 | $X_{5,6} \subset \mathbb{P}(1,2,2,3,3)$ | $3 \times A_1, 2 \times A_2$ |
| No. 11 | $X_{6,6} \subset \mathbb{P}(1,1,2,3,5)$ | $A_4$ |
| No. 12 | $X_{6,6} \subset \mathbb{P}(1,2,2,3,4)$ | $4 \times A_1, A_3$ |
| No. 13 | $X_{6,6} \subset \mathbb{P}(1,2,3,3,3)$ | $4 \times A_2$ |
| No. 14 | $X_{6,6} \subset \mathbb{P}(2,2,2,3,3)$ | $9 \times A_1$ |
| No. 15 | $X_{6,7} \subset \mathbb{P}(1,2,2,3,5)$ | $3 \times A_1, A_4$ |
| No. 16 | $X_{6,7} \subset \mathbb{P}(1,2,3,3,4)$ | $A_1, 2 \times A_2, A_3$ |
| No. 17 | $X_{6,8} \subset \mathbb{P}(1,1,3,4,5)$ | $A_4$ |
| No. 18 | $X_{6,8} \subset \mathbb{P}(1,2,3,3,5)$ | $2 \times A_2, A_4$ |
| No. 19 | $X_{6,8} \subset \mathbb{P}(1,2,3,4,4)$ | $2 \times A_1, 2 \times A_3$ |
| No. 20 | $X_{6,8} \subset \mathbb{P}(2,2,3,3,4)$ | $6 \times A_1, 2 \times A_2$ |
| No. 21 | $X_{6,9} \subset \mathbb{P}(1,2,3,4,5)$ | $A_1, A_3, A_4$ |

| | | |
|---|---|---|
| No. 22 | $X_{7,8} \subset \mathbb{P}(1,2,3,4,5)$ | $2 \times A_1, A_2, A_4$ |
| No. 23 | $X_{6,10} \subset \mathbb{P}(1,2,3,5,5)$ | $2 \times A_4$ |
| No. 24 | $X_{6,10} \subset \mathbb{P}(2,2,3,4,5)$ | $7 \times A_1, A_3$ |
| No. 25 | $X_{8,9} \subset \mathbb{P}(1,2,3,4,7)$ | $2 \times A_1, A_6$ |
| No. 26 | $X_{8,9} \subset \mathbb{P}(1,3,4,4,5)$ | $2 \times A_3, A_4$ |
| No. 27 | $X_{8,9} \subset \mathbb{P}(2,3,3,4,5)$ | $2 \times A_1, 3 \times A_2, A_4$ |
| No. 28 | $X_{8,10} \subset \mathbb{P}(1,2,3,5,7)$ | $A_2, A_6$ |
| No. 29 | $X_{8,10} \subset \mathbb{P}(1,2,4,5,6)$ | $3 \times A_1, A_5$ |
| No. 30 | $X_{8,10} \subset \mathbb{P}(1,3,4,5,5)$ | $A_2, 2 \times A_4$ |
| No. 31 | $X_{8,10} \subset \mathbb{P}(2,3,4,4,5)$ | $4 \times A_1, A_2, 2 \times A_3$ |
| No. 32 | $X_{9,10} \subset \mathbb{P}(1,2,3,5,8)$ | $A_1, A_7$ |
| No. 33 | $X_{9,10} \subset \mathbb{P}(1,3,4,5,6)$ | $A_2, A_3, A_5$ |
| No. 34 | $X_{9,10} \subset \mathbb{P}(2,2,3,5,7)$ | $5 \times A_1, A_6$ |
| No. 35 | $X_{9,10} \subset \mathbb{P}(2,3,4,5,5)$ | $2 \times A_1, A_3, 2 \times A_4$ |
| No. 36 | $X_{8,12} \subset \mathbb{P}(1,3,4,5,7)$ | $A_4, A_6$ |
| No. 37 | $X_{8,12} \subset \mathbb{P}(2,3,4,5,6)$ | $4 \times A_1, 2 \times A_2, A_4$ |
| No. 38 | $X_{9,12} \subset \mathbb{P}(2,3,4,5,7)$ | $3 \times A_1, A_4, A_6$ |
| No. 39 | $X_{10,11} \subset \mathbb{P}(2,3,4,5,7)$ | $2 \times A_1, A_2, A_3, A_6$ |
| No. 40 | $X_{10,12} \subset \mathbb{P}(1,3,4,5,9)$ | $A_2, A_8$ |
| No. 41 | $X_{10,12} \subset \mathbb{P}(1,3,5,6,7)$ | $2 \times A_2, A_6$ |
| No. 42 | $X_{10,12} \subset \mathbb{P}(1,4,5,6,6)$ | $A_1, 2 \times A_5$ |
| No. 43 | $X_{10,12} \subset \mathbb{P}(2,3,4,5,8)$ | $3 \times A_1, A_3, A_7$ |
| No. 44 | $X_{10,12} \subset \mathbb{P}(2,3,5,5,7)$ | $2 \times A_4, A_6$ |
| No. 45 | $X_{10,12} \subset \mathbb{P}(2,4,5,5,6)$ | $5 \times A_1, 2 \times A_4$ |
| No. 46 | $X_{10,12} \subset \mathbb{P}(3,3,4,5,7)$ | $4 \times A_2, A_6$ |
| No. 47 | $X_{10,12} \subset \mathbb{P}(3,4,4,5,6)$ | $2 \times A_2, 3 \times A_3, A_1$ |
| No. 48 | $X_{11,12} \subset \mathbb{P}(1,4,5,6,7)$ | $A_1, A_4, A_6$ |
| No. 49 | $X_{10,14} \subset \mathbb{P}(1,2,5,7,9)$ | $A_8$ |
| No. 50 | $X_{10,14} \subset \mathbb{P}(2,3,5,7,7)$ | $A_2, 2 \times A_6$ |
| No. 51 | $X_{10,14} \subset \mathbb{P}(2,4,5,6,7)$ | $5 \times A_1, A_3, A_5$ |
| No. 52 | $X_{10,15} \subset \mathbb{P}(2,3,5,7,8)$ | $A_1, A_6, A_7$ |
| No. 53 | $X_{12,13} \subset \mathbb{P}(3,4,5,6,7)$ | $2 \times A_2, A_1, A_4, A_6$ |
| No. 54 | $X_{12,14} \subset \mathbb{P}(1,3,4,7,11)$ | $A_{10}$ |
| No. 55 | $X_{12,14} \subset \mathbb{P}(1,4,6,7,8)$ | $A_1, A_3, A_7$ |
| No. 56 | $X_{12,14} \subset \mathbb{P}(2,3,4,7,10)$ | $4 \times A_1, A_9$ |
| No. 57 | $X_{12,14} \subset \mathbb{P}(2,3,5,7,9)$ | $A_2, A_4, A_8$ |
| No. 58 | $X_{12,14} \subset \mathbb{P}(3,4,5,7,7)$ | $A_4, 2 \times A_6$ |
| No. 59 | $X_{12,14} \subset \mathbb{P}(4,4,5,6,7)$ | $3 \times A_3, 2 \times A_1, A_4$ |
| No. 60 | $X_{12,15} \subset \mathbb{P}(1,4,5,6,11)$ | $A_1, A_{10}$ |
| No. 61 | $X_{12,15} \subset \mathbb{P}(3,4,5,6,9)$ | $3 \times A_2, A_1, A_8$ |
| No. 62 | $X_{12,15} \subset \mathbb{P}(3,4,5,7,8)$ | $A_3, A_6, A_7$ |
| No. 63 | $X_{12,16} \subset \mathbb{P}(2,5,6,7,8)$ | $4 \times A_1, A_4, A_6$ |

| No. 64 | $X_{14,15} \subset \mathbb{P}(2,3,5,7,12)$ | $A_1, A_2, A_{11}$ |
|--------|--------------------------------------------|---------------------|
| No. 65 | $X_{14,15} \subset \mathbb{P}(2,5,6,7,9)$ | $2 \times A_1, A_5, A_8$ |
| No. 66 | $X_{14,15} \subset \mathbb{P}(3,4,5,7,10)$ | $A_3, A_4, A_9$ |
| No. 67 | $X_{14,15} \subset \mathbb{P}(3,5,6,7,8)$ | $2 \times A_2, A_5, A_7$ |
| No. 68 | $X_{14,16} \subset \mathbb{P}(1,5,7,8,9)$ | $A_4, A_8$ |
| No. 69 | $X_{14,16} \subset \mathbb{P}(3,4,5,7,11)$ | $A_2, A_4, A_{10}$ |
| No. 70 | $X_{14,16} \subset \mathbb{P}(4,5,6,7,8)$ | $A_1, 2 \times A_3, A_4, A_5$ |
| No. 71 | $X_{15,16} \subset \mathbb{P}(2,3,5,8,13)$ | $2 \times A_1, A_{12}$ |
| No. 72 | $X_{15,16} \subset \mathbb{P}(3,4,5,8,11)$ | $2 \times A_3, A_{10}$ |
| No. 73 | $X_{14,18} \subset \mathbb{P}(2,3,7,9,11)$ | $2 \times A_2, A_{10}$ |
| No. 74 | $X_{14,18} \subset \mathbb{P}(2,6,7,8,9)$ | $5 \times A_1, A_2, A_7$ |
| No. 75 | $X_{12,20} \subset \mathbb{P}(4,5,6,7,10)$ | $2 \times A_1, 2 \times A_4, A_6$ |
| No. 76 | $X_{16,18} \subset \mathbb{P}(1,6,8,9,10)$ | $A_1, A_2, A_9$ |
| No. 77 | $X_{16,18} \subset \mathbb{P}(4,6,7,8,9)$ | $2 \times A_1, 2 \times A_3, A_2, A_6$ |
| No. 78 | $X_{18,20} \subset \mathbb{P}(4,5,6,9,14)$ | $2 \times A_1, A_2, A_{13}$ |
| No. 79 | $X_{18,20} \subset \mathbb{P}(4,5,7,9,13)$ | $A_6, A_{12}$ |
| No. 80 | $X_{18,20} \subset \mathbb{P}(5,6,7,9,11)$ | $A_2, A_6, A_{10}$ |
| No. 81 | $X_{18,22} \subset \mathbb{P}(2,5,9,11,13)$ | $A_4, A_{12}$ |
| No. 82 | $X_{20,21} \subset \mathbb{P}(3,4,7,10,17)$ | $A_1, A_{16}$ |
| No. 83 | $X_{18,30} \subset \mathbb{P}(6,8,9,10,15)$ | $2 \times A_1, 2 \times A_2, A_7, A_4$ |
| No. 84 | $X_{24,30} \subset \mathbb{P}(8,9,10,12,15)$ | $A_1, A_3, A_8, A_2, A_4$ |

Table 2: The codimension 2 weighted K3 surfaces

# 14  Weighted 3-fold complete intersections

This section gives the corresponding conditions and lists for 3-folds.

**14.1 Theorem** *Let $X_d$ be a general hypersurface in $\mathbb{P} = \mathbb{P}(a_0, \ldots, a_4)$ and let $\alpha = d - \sum a_i$. Then $X_d$ is quasismooth with only isolated terminal quotient singularities and is not a linear cone if and only if all the following hold:*

(1) *For all $i$,*

    (i) *$d > a_i$.*

    (ii) *there exists a monomial $x_i^m x_e$ of degree $d$ (that is, there exists $e$ such that $a_i \mid d - a_e$).*

    (iii) *if $a_i \nmid d$, there exists an $m \neq i, e$ such that $a_i \mid a_m + \alpha$.*

(2) *For all distinct $i, j$, with $h = \mathrm{hcf}(a_i, a_j)$, then*

    (i) *$h \mid d$.*

(ii) there exists an $m \neq i, j$ such that $h \mid a_m + \alpha$.

(iii) one of the following holds:

either there exists a monomial $x_i^m x_j^n$ of degree $d$,

or there exist monomials $x_i^{n_1} x_j^{m_1} x_{e_1}$ and $x_i^{n_2} x_j^{m_2} x_{e_2}$ of degree $d$ such that $e_1$ and $e_2$ are distinct.

(iv) there exists a monomial of degree $d$ which does not involve $x_i$ or $x_j$.

(3) For all distinct $i, j, k$, $\mathrm{hcf}(a_i, a_j, a_k) = 1$.

**14.2 Note** Since the hypersurface is quasismooth and of dimension 3 then it is well formed, and so $\omega_X = \mathcal{O}_X(\alpha)$.

**Proof** Let $f$ be a general homogeneous polynomial of degree $d$ in variables $x_0, \ldots, x_3$; define $X_d : (f = 0) \subset \mathbb{P}$.

$X_d$ is quasismooth and not a linear cone (and therefore well-formed) if and only if conditions $(1i)$, $(1ii)$, $(2i)$, $(2iii)$, $(2iv)$ and $(3)$ hold (see Corollary 8.6). By calculating the types of the singularities on $X_d$ we can show that conditions $(1iii)$, $(2i)$, $(2ii)$ and $(3)$ are equivalent to these singularities being terminal; the combinatorial conditions for which are found in Lemma 9.3.

Suppose furthermore that conditions $(1iii)$, $(2i)$, $(2ii)$ and $(3)$ hold. As $X_d$ is quasismooth the only singularities are due to the $\mathbb{K}^*$-action and hence are cyclic quotient singularities on the fundamental simplex $\Delta \subset \mathbb{P}$. By condition $(3)$ only vertices and edges need be checked.

Consider $P_i \in X_d$. By renumbering we can assume that $i = 0$. So $a_0 \nmid d$. By condition $(1ii)$ there exists an $e \neq 0$ such that $a_0 \mid d - a_e$. Without loss of generality we can assume that $e = 1$. So $f$ is of the form $f = x_0^n x_1 + \cdots$. Thus $\partial f / \partial x_1$ is nonzero at $P_0$. By the Inverse Function Theorem $x_2$, $x_3$ and $x_4$ are local coordinates around $P_0$. So $P_0 \in X_d$ is of type $\frac{1}{a_0}(a_2, a_3, a_4)$. However $d = a_0 + \cdots + a_4 + \alpha$ and so $a_0 \mid a_2 + a_4 + \alpha$. By condition $(1iv)$, $a_0 \mid \alpha + a_m$ for some $m = 2, 3, 4$. Without loss of generality assume $m = 2$. By condition $(1iv)$, $a_0 \mid a_3 + a_4$. Let $h = \mathrm{hcf}(a_0, a_3)$. So $h \mid a_3$ and hence, by condition $(3)$, $h = 1$. Therefore $P_0 \in X_d$ is a terminal singularity.

Consider the edge $P_i P_j$. Again by renumbering assume that $i = 0$ and $j = 1$. $f$ restricted to $P_0 P_1$ is:

$$f = \sum x_0^n x_1^m,$$

where the sum is taken over the set $\{(n, m) : na_0 + ma_1 = d\}$. If $a_0 \nmid d$ then $a_0 \mid d - a_e$ for some $e \neq 0$. If $e \neq 1$ then $h = \mathrm{hcf}(a_0, a_1) \mid a_e$ and by condition $(4)$ $h = 1$. Then $P_0 P_1$ is nonsingular. So assume that either $a_0 \mid d$ or $a_0 \mid d - a_1$. Hence $f$ is not identically zero on $P_0 P_1$, and so $X_d \cap P_0 P_1$ is finite. Each point in this intersection is of type $\frac{1}{h}(a_2, a_3, a_4)$. By condition (2$ii$) $h \mid \alpha + a_m$ for some $m = 2, 3, 4$. By renumbering assume $m = 2$. Since $d = a_0 + \cdots + a_4 + \alpha$, then $h \mid a_3 + a_4$. Also $\mathrm{hcf}(h, a_3) = 1$. Thus each point is terminal.

Therefore $X_d$ in $\mathbb{P}$ has at worst terminal singularities.

Conversely assume that $X_d$ is quasismooth, not a linear cone and has at worst only terminal singularities. Suppose $a_i \nmid d$. By renumbering we can assume that $i = 0$. So $P_0 \in X_d$ and $a_0 \mid d - a_e$ for some $e$. Without loss of generality assume that $e = 1$. As above the singularity at $P_0 \in X_d$ is of type $\frac{1}{a_0}(a_2, a_3, a_4)$. Since this is terminal we have, after renumbering, $a_0 \mid a_2 + a_3$ and so $a_0 \mid \alpha + a_m$ for some $m$. This is condition (1$iv$).

Suppose $h = \mathrm{hcf}(a_i, a_j)$. By renumbering assume that $i = 0$ and $j = 1$. As $X_d$ is well formed then $h \mid d$, which is condition (2$i$). So $P_0 P_1 \cap X_d$ is a finite intersection, where each point is of type $\frac{1}{h}(a_2, a_3, a_4)$. This is terminal and so $h \mid \alpha + a_m$ for $m = 2, 3, 4$. This is condition (2$ii$).

Suppose $h = \mathrm{hcf}(a_i, a_j, a_k)$. Without loss of generality assume that $i = 0$, $j = 1$ and $k = 2$. Let $h' = \mathrm{hcf}(a_0, a_1)$. So $h' \mid d$. Hence the line $P_0 P_1$ contains singularities of type $\frac{1}{h'}(a_2, a_3, a_4)$. As these are terminal $h = \mathrm{hcf}(h', a_2) = 1$. This is condition (3). $\square$

**14.3 Theorem** *There are exactly 4 families of quasismooth 3-fold weighted hypersurfaces with only terminal isolated quotient singularities and $\omega_X \simeq \mathcal{O}_X$:*

$$X_5 \subset \mathbb{P}(1, 1, 1, 1, 1)$$
$$X_6 \subset \mathbb{P}(1, 1, 1, 1, 2)$$
$$X_8 \subset \mathbb{P}(1, 1, 1, 1, 4)$$
$$X_{10} \subset \mathbb{P}(1, 1, 1, 2, 5)$$

Notice that the above are all nonsingular.

**Proof** As $K_X \simeq \mathcal{O}_X$ then $\alpha = 0$. Suppose $h = \mathrm{hcf}(a_i, a_j) \neq 1$ for distinct $i$, $j$. By Theorem 14.1 (2$ii$) there exists an $m \neq i, j$ such that $h \mid a_m + \alpha$. However $\alpha = 0$ and so $h \mid a_m$. By (3) $h = 1$, a contradiction. Hence $a_i$ and $a_j$ are coprime for distinct $i$, $j$.

Suppose that $a_i \nmid d$. Then there exists an $m \neq i, e_i$ such that $a_i \mid a_m + \alpha$. Thus $a_i = \mathrm{hcf}(a_i, a_m) = 1$, contradicting $a_i \nmid d$. Thus $a_i \mid d$ for all $i$.

Order the $\{a_i\}$ such that $a_4 \geq \cdots \geq a_0$. So $5a_4 \geq d \geq 2a_4$. Let $d = \lambda a_4$. Thus $\lambda = 2, 3, 4$ or $5$. As the $\{a_i\}$ are pairwise coprime then $a_0 a_1 a_2 a_3 a_4 \mid d$ and so $a_0 a_1 a_2 a_3 \mid \lambda$. Also $a_0 + \cdots + a_3 = (\lambda - 1)a_4$. There are four cases:

(i) $\lambda = 5$. Either $(a_0, a_1, a_2, a_3) = (1, 1, 1, 1)$ giving $a_4 = 1$ (that is, $X_5$ in $\mathbb{P}(1, 1, 1, 1, 1)$) or $(a_0, a_1, a_2, a_3) = (1, 1, 1, 5)$ giving $a_4 = 2 < a_3$.

(ii) $\lambda = 4$. So $\lambda - 1 = 3$ and divides $a_0 + \cdots + a_3$. There are three possibilities:

  (a) $(a_0, a_1, a_2, a_3) = (1, 1, 1, 1)$, giving $3 \mid 4$.

  (b) $(a_0, a_1, a_2, a_3) = (1, 1, 1, 2)$, giving $3 \mid 5$,

  (c) $(a_0, a_1, a_2, a_3) = (1, 1, 1, 4)$, giving $3 \mid 7$.

All of these possibilities give contradictions.

(iii) $\lambda = 3$. Either $(a_0, a_1, a_2, a_3) = (1, 1, 1, 1)$ giving $a_4 = 2$ (that is, $X_6$ in $\mathbb{P}(1, 1, 1, 1, 2)$), or $(a_0, a_1, a_2, a_3) = (1, 1, 1, 3)$ giving $a_4 = 3$, contradicting the coprime condition.

(iv) $\lambda = 2$. Either $(a_0, a_1, a_2, a_3) = (1, 1, 1, 1)$ giving $a_4 = 4$ (that is, $X_8$ in $\mathbb{P}(1, 1, 1, 1, 4)$), or $(a_0, a_1, a_2, a_3) = (1, 1, 1, 2)$ giving $a_4 = 5$ (that is, $X_{10}$ in $\mathbb{P}(1, 1, 1, 2, 5)$).  $\square$

Consider the case of codimension 2 complete intersections.

**14.4 Theorem** *Suppose* $X = X_{d_1, d_2}$ *in* $\mathbb{P} = \mathbb{P}(a_0, \ldots, a_5)$ *is quasismooth and not the intersection of a linear cone with another hypersurface. Let* $\alpha = \sum d_\lambda - \sum a_i$. *X has at worst terminal singularities if and only if the following hold:*

(1) *for all* $i$, *if* $a_i \nmid d_1$ *and* $a_i \nmid d_2$ *then there exists* $e_1$, $e_2$ *and* $m$ *such that* $a_i \mid d_1 - a_{e_1}$, $a_i \mid d_2 - a_{e_2}$ *and* $a_i \mid \alpha + a_m$, *where* $\{i, e_1, e_2, m\}$ *are distinct.*

(2) *for all distinct* $i$ *and* $j$, *with* $h = \mathrm{hcf}(a_i, a_j)$, *at least one of the following occurs:*

  (a) $h \mid d_1$ *and* $h \mid d_2$,

  (b) $h \mid d_1$, $h \nmid d_2$ *and* $h \mid \alpha + a_m$ *for some* $m \neq i, j$, *or*

  (c) $h \nmid d_1$, $h \mid d_2$ *and* $h \mid \alpha + a_m$ *for some* $m \neq i, j$.

(3) *for all distinct* $i$, $j$ *and* $k$, *with* $h = \mathrm{hcf}(a_i, a_j, a_k)$, $h \mid d_1$, $h \mid d_2$ *and* $h \mid \alpha + a_m$ *for some* $m \neq i, j, k$.

(4) *for all distinct* $i$, $j$, $k$ *and* $l$, $h = \mathrm{hcf}(a_i, a_j, a_k, a_l) = 1$.

**14.5 Note** Since $X$ is quasismooth, of dimension 3 and not the intersect of a linear cone with other hypersurfaces then $X$ is well formed. Thus $\omega_X = \mathcal{O}_X(\alpha)$.

**Proof** Let $f_1$ and $f_2$ be sufficiently general homogeneous polynomials of degrees $d_1$ and $d_2$ respectively, in the variables $x_0, \ldots, x_4$ with respect to the weights $a_0, \ldots, a_4$. Define $X : (f_1 = 0, f_2 = 0) \subset \mathbb{P}$.

Since $X$ is quasismooth the only singularities are due to the $\mathbb{K}^*$-action and hence are all cyclic quotient singularities occurring on the fundamental simplex $\Delta$.

Assume conditions $(1), \ldots, (4)$ hold. By condition $(4)$ only the vertices, edges and faces of $\Delta$ need be considered.

Suppose $P_i \in X$. By renumbering we can assume that $i = 0$. So $a_0 \nmid d_1$ and $a_0 \nmid d_2$. By condition $(1)$, there exist monomials $x_0^{n_1} x_{e_1}$ and $x_0^{n_2} x_{e_2}$ of degrees $d_1$ and $d_2$, where $e_1 \neq e_2$. Note that this is really quasismoothness. By renumbering we can write $e_1 = 1$ and $e_2 = 2$. So $f_1$ and $f_2$ are of the form:

$$f_1 = x_0^{n_1} x_1 + \cdots,$$
$$f_2 = x_0^{n_2} x_2 + \cdots.$$

Thus $\partial f_1 / \partial x_1$ and $\partial f_2 / \partial x_2$ are nonzero at $P_0$. By the Inverse Function Theorem, $x_3$, $x_4$ and $x_5$ are local coordinates. Hence $P_0 \in X$ is of type $\frac{1}{a_0}(a_3, a_4, a_5)$. By condition $(1)$ $a_0 \mid \alpha + a_m$ for some $m \neq 0, 1, 2$. Without loss of generality assume $m = 3$. As $d_1 + d_2 = a_0 + \cdots + a_5 + \alpha$ then $a_0 \mid a_4 + a_5$. Let $h = \mathrm{hcf}(a_0, a_4)$. So $h \mid a_5$ and, by condition $(3)$, $h \mid d_1$. Since $\deg x_0^n x_1 = d_1$, $h \mid a_1$ and so, by condition $(4)$, $h = 1$. Thus $P_0 \in X$ is terminal.

Consider the edge $P_i P_j$. By renumbering we can assume that $i = 0$ and $j = 1$. Let $h = \mathrm{hcf}(a_0, a_1)$. Notice that $P_0 P_1 \subset X_{d_\lambda}$ if and only if $h \nmid d_\lambda$ for $\lambda = 0, 1$. By condition $(2)$, $h \mid d_\lambda$ for some $\lambda$. Without loss of generality assume that $h \mid d_1$. There are two cases:

(a) $h \mid d_2$. $P_0 P_1 \cap (f_\lambda = 0)$ is a finite set of points for $\lambda = 0, 1$. Thus $P_0 P_1 \cap X = \emptyset$.

(b) $h \nmid d_2$. In this case no monomial of the form $x_0^{n_0} x_1^{n_1}$ of degree $d_2$ exists (or else $h \mid d_2$). From Theorem 8.7 (with $I = \{0, 1\}$) there exists a monomial $x_0^{n_0} x_1^{n_1} x_e$ of degree $d_2$, where $e \neq 0, 1$. By renumbering we can assume that $e = 2$. Thus $f_2$ is of the form:

$$f_2 = x_0^{n_0} x_1^{n_1} x_2 + \cdots$$

and $\partial f_2 / \partial x_2$ is nonzero on $P_0 P_1 \cap X$. By the Inverse Function Theorem, $x_3$, $x_4$ and $x_5$ are local coordinates and so each point of $P_0 P_1 \cap X$ is of type $\frac{1}{h}(a_3, a_4, a_5)$. Condition $(2b)$ gives $h \mid \alpha + a_m$ for some $m \neq 0, 1, 2$. Assume that $m = 3$, and hence $h \mid a_4 + a_5$. Let $h' = \mathrm{hcf}(h, a_4)$. So $h \mid a_4$ and thus by condition $(4)$ $h = 1$. Thus these points are terminal.

Therefore $X$ has at worst terminal points along $P_0 P_1$.

Consider the face $P_i P_j P_k$. As before assume $i = 0$, $j = 1$ and $k = 2$. By condition (3) $h = \text{hcf}(a_0, a_1, a_2) \mid d_1$ and $h \mid d_2$. So $P_0 P_1 P_2$ intersects $X$ transversally. Each point in the intersection is of type $\frac{1}{h}(a_3, a_4, a_5)$. As $h \mid \alpha + a_m$ for some $m \neq 0, 1, 2$, after renumbering, $h \mid a_3 + a_4$. By condition (4) $\text{hcf}(h, a_3) = 1$. Thus these points are terminal.

Therefore condition $(1), \ldots, (4)$ are sufficient.

Conversely assume that $X$ has at worst terminal singularities. Suppose $a_i \nmid d_1$ and $a_i \nmid d_2$. By renumbering assume $i = 0$. Thus $P_0 \in X$. Since $X$ is quasismooth there exist 2 monomials $x_0^n x_{e_1}$ and $x_0^m x_{e_2}$ of degrees $d_1$ and $d_2$, where $e_1 \neq e_2$. This gives the first part of condition (1). Without loss of generality we can assume that $e_1 = 1$ and $e_2 = 2$. As before we find that $P_0 \in X$ is of type $\frac{1}{a_0}(a_3, a_4, a_5)$. As this is terminal, after renumbering, $a_0 \mid a_3 + a_4$ and so $a_0 \mid \alpha + a_5$. This is condition (1).

Suppose $h = \text{hcf}(a_i, a_j)$ for distinct $i$ and $j$. As usual we can renumber such that $i = 0$ and $j = 1$. As $X$ is well formed then $h \mid d_\lambda$ for some $\lambda$. Suppose $h \mid d_1$. If $h \mid d_2$ then this is condition $(2a)$. So assume that $h \nmid d_2$. As above each point of $P_0 P_1 \cap X$ is isolated and of type $\frac{1}{h}(a_3, a_4, a_5)$. After renumbering, $h \mid a_3 + a_4$ and so $h \mid \alpha + a_5$. This is condition $(2b)$. Likewise for the case when $h \mid d_2$ but $h \nmid d_1$. This gives condition $(2c)$.

Suppose $h = \text{hcf}(a_i, a_j, a_k)$ for distinct $i$, $j$ and $k$. Renumber such that $i = 0$, $j = 1$ and $k = 2$. Since $X$ is well formed $h \mid d_1$ and $h \mid d_2$. $P_0 P_1 P_2 \cap X$ is a finite number of points, all of type $\frac{1}{h}(a_3, a_4, a_5)$. As these are terminal, after renumbering, $h \mid a_3 + a_4$ and so $h \mid \alpha + a_5$. This is condition (3). Condition (4) follows from the fact that $\text{hcf}(h, a_3) = \text{hcf}(h, a_4) = 1$.

Therefore these conditions are necessary.  □

**14.6 Codimension 2 weighted 3-fold complete intersection with trivial canonical bundle**  The four families of 3-fold codimension 2 quasismooth complete intersections with at worst terminal singularities, $\omega_X \simeq \mathcal{O}_X$ and $\sum a_i < 100$ are:

$$X_{2,4} \subset \mathbb{P}(1,1,1,1,1,1) \qquad X_{3,3} \subset \mathbb{P}(1,1,1,1,1,1)$$
$$X_{3,4} \subset \mathbb{P}(1,1,1,1,1,2) \qquad X_{4,4} \subset \mathbb{P}(1,1,1,1,2,2)$$

Again the above are all nonsingular and were found using a computer search based on the conditions of Theorem 14.4.

# 15   Canonically embedded weighted 3-folds

**15.1 Canonically embedded 3-fold weighted hypersurfaces**  There are 23 families of 3-fold quasismooth weighted hypersurfaces with only terminal isolated quotient singularities with $\omega_X \simeq \mathcal{O}_X(1)$ and $\sum a_i \leq 100$.

| | Hypersurface | $K_X^3$ | $p_g$ | Singularities |
|---|---|---|---|---|
| No. 1 | $X_6 \subset \mathbb{P}(1,1,1,1,1)$ | 6 | 5 | |
| No. 2 | $X_7 \subset \mathbb{P}(1,1,1,1,2)$ | 7/2 | 4 | $\frac{1}{2}(1,-1,1)$ |
| No. 3 | $X_8 \subset \mathbb{P}(1,1,1,2,2)$ | 2 | 3 | $4 \times \frac{1}{2}(1,-1,1)$ |
| No. 4 | $X_9 \subset \mathbb{P}(1,1,1,2,3)$ | 3/2 | 3 | $\frac{1}{2}(1,-1,1)$ |
| No. 5 | $X_{10} \subset \mathbb{P}(1,1,1,1,5)$ | 2 | 4 | |
| No. 6 | $X_{10} \subset \mathbb{P}(1,1,2,2,3)$ | 5/6 | 2 | $5 \times \frac{1}{2}(1,-1,1), \frac{1}{3}(1,-1,1)$ |
| No. 7 | $X_{12} \subset \mathbb{P}(1,1,1,2,6)$ | 1 | 3 | $2 \times \frac{1}{2}(1,-1,1)$ |
| No. 8 | $X_{12} \subset \mathbb{P}(1,1,2,3,4)$ | 1/2 | 2 | $3 \times \frac{1}{2}(1,-1,1)$ |
| No. 9 | $X_{12} \subset \mathbb{P}(1,2,2,3,3)$ | 1/3 | 1 | $6 \times \frac{1}{2}(1,-1,1), 4 \times \frac{1}{3}(1,-1,1)$ |
| No. 10 | $X_{14} \subset \mathbb{P}(1,1,2,2,7)$ | 1/2 | 2 | $7 \times \frac{1}{2}(1,-1,1)$ |
| No. 11 | $X_{15} \subset \mathbb{P}(1,2,3,3,5)$ | 1/6 | 1 | $\frac{1}{2}(1,-1,1), 5 \times \frac{1}{3}(1,-1,1)$ |
| No. 12 | $X_{16} \subset \mathbb{P}(1,1,2,3,8)$ | 1/3 | 2 | $2 \times \frac{1}{2}(1,-1,1), \frac{1}{3}(1,-1,1)$ |
| No. 13 | $X_{16} \subset \mathbb{P}(1,2,3,4,5)$ | 2/15 | 1 | $4 \times \frac{1}{2}(1,-1,1), \frac{1}{3}(1,-1,1),$ $\frac{1}{5}(1,-1,2)$ |
| No. 14 | $X_{18} \subset \mathbb{P}(1,2,2,3,9)$ | 1/6 | 1 | $9 \times \frac{1}{2}(1,-1,1), 2 \times \frac{1}{3}(1,-1,1)$ |
| No. 15 | $X_{18} \subset \mathbb{P}(2,3,3,4,5)$ | 1/20 | 0 | $4 \times \frac{1}{2}(1,-1,1), 6 \times \frac{1}{3}(1,-1,1),$ $\frac{1}{4}(1,-1,1), \frac{1}{5}(1,-1,2)$ |
| No. 16 | $X_{20} \subset \mathbb{P}(2,3,4,5,5)$ | 1/30 | 0 | $5 \times \frac{1}{2}(1,-1,1), \frac{1}{3}(1,-1,1),$ $4 \times \frac{1}{5}(1,-1,2)$ |
| No. 17 | $X_{21} \subset \mathbb{P}(1,3,4,5,7)$ | 1/20 | 1 | $\frac{1}{4}(1,-1,1), \frac{1}{5}(1,-1,2)$ |
| No. 18 | $X_{22} \subset \mathbb{P}(1,2,3,4,11)$ | 1/12 | 1 | $5 \times \frac{1}{2}(1,-1,1), \frac{1}{3}(1,-1,1),$ $\frac{1}{4}(1,-1,1)$ |
| No. 19 | $X_{28} \subset \mathbb{P}(1,3,4,5,14)$ | 1/30 | 1 | $\frac{1}{3}(1,-1,1), \frac{1}{2}(1,-1,1),$ $\frac{1}{5}(1,-1,1)$ |
| No. 20 | $X_{28} \subset \mathbb{P}(3,4,5,7,8)$ | 1/120 | 0 | $\frac{1}{3}(1,-1,1), 3 \times \frac{1}{4}(1,-1,1),$ $\frac{1}{5}(1,-1,2), \frac{1}{8}(1,-1,3)$ |
| No. 21 | $X_{30} \subset \mathbb{P}(2,3,4,5,15)$ | 1/60 | 0 | $7 \times \frac{1}{2}(1,-1,1), 2 \times \frac{1}{3}(1,-1,1),$ $\frac{1}{4}(1,-1,1), 2 \times \frac{1}{5}(1,-1,2)$ |
| No. 22 | $X_{40} \subset \mathbb{P}(3,4,5,7,20)$ | 1/210 | 0 | $\frac{1}{3}(1,-1,1), 2 \times \frac{1}{4}(1,-1,1),$ $2 \times \frac{1}{5}(1,-1,2), \frac{1}{7}(1,-1,2)$ |
| No. 23 | $X_{46} \subset \mathbb{P}(4,5,6,7,23)$ | 1/420 | 0 | $\frac{1}{4}(1,-1,1), 3 \times \frac{1}{2}(1,-1,1),$ $\frac{1}{5}(1,-1,2), \frac{1}{6}(1,-1,1),$ $\frac{1}{7}(1,-1,3)$ |

Table 3: 3-fold canonical hypersurfaces

**15.2 Conjecture** *This list was produced using a computer program. In fact the program was run much further but produced no more examples. I* *conjecture that the lists in this section and in Sections 15.4, 16.6, and 16.7 are complete lists, and not limited by $\sum a_i \leq 100$.*

**15.3  Interesting example**  The family $X_{46}$ in $\mathbb{P}(4,5,6,7,23)$ has $p_g$, $P_2$ and $P_3$ all zero. It is interesting to find canonical 3-folds with as many of their first plurigenera equal to zero as possible (see also [F1], Section 4.9). This is the best such weighted complete intersections example found in these lists.

**15.4  Canonically embedded codimension 2 weighted 3-folds**  There are 59 families of 3-fold codimension 2 weighted complete intersections satisfying the conditions of Theorem 14.4 with $\omega_X \simeq \mathcal{O}_X(1)$ and $\sum a_i \leq 100$.

| | Complete Intersection | $K_X^3$ | $p_g$ | Singularities |
|---|---|---|---|---|
| No. 1 | $X_{2,5} \subset \mathbb{P}(1,1,1,1,1,1)$ | 10 | 6 | |
| No. 2 | $X_{3,4} \subset \mathbb{P}(1,1,1,1,1,1)$ | 12 | 6 | |
| No. 3 | $X_{3,5} \subset \mathbb{P}(1,1,1,1,1,2)$ | 15/2 | 5 | $\frac{1}{2}(1,-1,1)$ |
| No. 4 | $X_{4,4} \subset \mathbb{P}(1,1,1,1,1,2)$ | 8 | 5 | |
| No. 5 | $X_{3,6} \subset \mathbb{P}(1,1,1,1,2,2)$ | 9/2 | 4 | $3 \times \frac{1}{2}(1,-1,1)$ |
| No. 6 | $X_{4,5} \subset \mathbb{P}(1,1,1,1,2,2)$ | 5 | 4 | $2 \times \frac{1}{2}(1,-1,1)$ |
| No. 7 | $X_{2,8} \subset \mathbb{P}(1,1,1,1,1,4)$ | 4 | 5 | |
| No. 8 | $X_{4,6} \subset \mathbb{P}(1,1,1,1,2,3)$ | 4 | 4 | |
| No. 9 | $X_{4,6} \subset \mathbb{P}(1,1,1,2,2,2)$ | 3 | 3 | $6 \times \frac{1}{2}(1,-1,1)$ |
| No. 10 | $X_{3,8} \subset \mathbb{P}(1,1,1,1,2,4)$ | 3 | 4 | $2 \times \frac{1}{2}(1,-1,1)$ |
| No. 11 | $X_{4,7} \subset \mathbb{P}(1,1,1,2,2,3)$ | 7/3 | 3 | $\frac{1}{3}(1,-1,1), 2 \times \frac{1}{2}(1,-1,1)$ |
| No. 12 | $X_{5,6} \subset \mathbb{P}(1,1,1,2,2,3)$ | 5/2 | 3 | $3 \times \frac{1}{2}(1,-1,1)$ |
| No. 13 | $X_{6,6} \subset \mathbb{P}(1,1,1,2,3,3)$ | 2 | 3 | |
| No. 14 | $X_{4,8} \subset \mathbb{P}(1,1,2,2,2,3)$ | 4/3 | 2 | $\frac{1}{3}(1,-1,1), 8 \times \frac{1}{2}(1,-1,1)$ |
| No. 15 | $X_{6,6} \subset \mathbb{P}(1,1,2,2,2,3)$ | 3/2 | 2 | $9 \times \frac{1}{2}(1,-1,1)$ |
| No. 16 | $X_{3,10} \subset \mathbb{P}(1,1,1,2,2,5)$ | 3/2 | 3 | $5 \times \frac{1}{2}(1,-1,1)$ |
| No. 17 | $X_{4,9} \subset \mathbb{P}(1,1,2,2,3,3)$ | 1 | 2 | $2 \times \frac{1}{2}(1,-1,1), 3 \times \frac{1}{3}(1,-1,1)$ |
| No. 18 | $X_{6,7} \subset \mathbb{P}(1,1,2,2,3,3)$ | 7/6 | 2 | $3 \times \frac{1}{2}(1,-1,1), 2 \times \frac{1}{3}(1,-1,1)$ |
| No. 19 | $X_{4,10} \subset \mathbb{P}(1,1,1,2,3,5)$ | 4/3 | 3 | $\frac{1}{3}(1,-1,1)$ |
| No. 20 | $X_{4,10} \subset \mathbb{P}(1,1,2,2,2,5)$ | 1 | 2 | $10 \times \frac{1}{2}(1,-1,1)$ |
| No. 21 | $X_{6,8} \subset \mathbb{P}(1,1,2,2,3,4)$ | 1 | 2 | $6 \times \frac{1}{2}(1,-1,1)$ |
| No. 22 | $X_{6,8} \subset \mathbb{P}(1,2,2,2,3,3)$ | 2/3 | 1 | $12 \times \frac{1}{2}(1,-1,1), 2 \times \frac{1}{3}(1,-1,1)$ |
| No. 23 | $X_{6,9} \subset \mathbb{P}(1,1,2,2,3,3,4)$ | 3/4 | 2 | $\frac{1}{4}(1,-1,1), \frac{1}{2}(1,-1,1)$ |
| No. 24 | $X_{6,9} \subset \mathbb{P}(1,2,2,3,3,3)$ | 1/2 | 1 | $3 \times \frac{1}{2}(1,-1,1), 6 \times \frac{1}{3}(1,-1,1)$ |
| No. 25 | $X_{4,12} \subset \mathbb{P}(1,1,2,2,3,6)$ | 2/3 | 2 | $4 \times \frac{1}{2}(1,-1,1), 2 \times \frac{1}{3}(1,-1,1)$ |
| No. 26 | $X_{6,10} \subset \mathbb{P}(1,1,2,3,3,5)$ | 2/3 | 2 | $2 \times \frac{1}{3}(1,-1,1)$ |
| No. 27 | $X_{6,10} \subset \mathbb{P}(1,2,2,2,3,5)$ | 1/2 | 1 | $15 \times \frac{1}{2}(1,-1,1)$ |
| No. 28 | $X_{6,10} \subset \mathbb{P}(1,2,2,3,3,4)$ | 5/12 | 1 | $\frac{1}{4}(1,-1,1), 7 \times \frac{1}{2}(1,-1,1),$ $2 \times \frac{1}{3}(1,-1,1)$ |

No. 29 $X_{4,14} \subset \mathbb{P}(1,2,2,2,3,7)$  1/3  1  $\frac{1}{3}(1,-1,1), 14 \times \frac{1}{2}(1,-1,1)$

No. 30 $X_{6,12} \subset \mathbb{P}(1,2,2,3,4,5)$  3/10  1  $\frac{1}{5}(1,-1,2), 9 \times \frac{1}{2}(1,-1,1)$

No. 31 $X_{8,10} \subset \mathbb{P}(1,2,2,3,4,5)$  1/3  1  $\frac{1}{3}(1,-1,1), 10 \times \frac{1}{2}(1,-1,1)$

No. 32 $X_{6,12} \subset \mathbb{P}(1,2,3,3,4,4)$  1/4  1  $3 \times \frac{1}{2}(1,-1,1), 3 \times \frac{1}{4}(1,-1,1)$

No. 33 $X_{6,12} \subset \mathbb{P}(2,2,3,3,3,4)$  1/6  0  $9 \times \frac{1}{2}(1,-1,1), 8 \times \frac{1}{3}(1,-1,1)$

No. 34 $X_{6,13} \subset \mathbb{P}(1,2,3,3,4,5)$  13/60  1  $\frac{1}{4}(1,-1,1), \frac{1}{5}(1,-1,2),$
$\frac{1}{2}(1,-1,1), 2 \times \frac{1}{3}(1,-1,1)$

No. 35 $X_{9,10} \subset \mathbb{P}(1,2,3,3,4,5)$  1/4  1  $\frac{1}{4}(1,-1,1), 2 \times \frac{1}{2}(1,-1,1),$
$3 \times \frac{1}{3}(1,-1,1)$

No. 36 $X_{6,14} \subset \mathbb{P}(1,2,2,3,4,7)$  1/4  1  $\frac{1}{4}(1,-1,1), 10 \times \frac{1}{2}(1,-1,1)$

No. 37 $X_{8,12} \subset \mathbb{P}(1,2,3,4,4,5)$  1/5  1  $\frac{1}{5}(1,-1,1), 6 \times \frac{1}{2}(1,-1,1)$

No. 38 $X_{6,14} \subset \mathbb{P}(2,2,2,3,3,7)$  1/6  0  $21 \times \frac{1}{2}(1,-1,1), 2 \times \frac{1}{3}(1,-1,1)$

No. 39 $X_{8,12} \subset \mathbb{P}(2,2,3,3,4,5)$  2/15  0  $\frac{1}{5}(1,-1,2), 12 \times \frac{1}{2}(1,-1,1),$
$4 \times \frac{1}{3}(1,-1,1)$

No. 40 $X_{6,15} \subset \mathbb{P}(2,3,3,3,4,5)$  1/12  0  $\frac{1}{4}(1,-1,1), \frac{1}{2}(1,-1,1),$
$10 \times \frac{1}{3}(1,-1,1)$

No. 41 $X_{6,16} \subset \mathbb{P}(1,2,3,3,4,8)$  1/6  1  $2 \times \frac{1}{2}(1,-1,1), 2 \times \frac{1}{3}(1,-1,1),$
$2 \times \frac{1}{4}(1,-1,1)$

No. 42 $X_{10,12} \subset \mathbb{P}(1,2,3,4,5,6)$  1/6  1  $5 \times \frac{1}{2}(1,-1,1), 2 \times \frac{1}{3}(1,-1,1)$

No. 43 $X_{10,12} \subset \mathbb{P}(2,2,3,4,5,5)$  1/10  0  $15 \times \frac{1}{2}(1,-1,1), 2 \times \frac{1}{5}(1,-1,2)$

No. 44 $X_{10,12} \subset \mathbb{P}(2,3,3,4,4,5)$  1/12  0  $6 \times \frac{1}{2}(1,-1,1), 4 \times \frac{1}{3}(1,-1,1),$
$3 \times \frac{1}{4}(1,-1,1)$

No. 45 $X_{8,15} \subset \mathbb{P}(2,3,3,4,5,5)$  1/15  0  $2 \times \frac{1}{2}(1,-1,1), 5 \times \frac{1}{3}(1,-1,1),$
$3 \times \frac{1}{5}(1,-1,2)$

No. 46 $X_{6,18} \subset \mathbb{P}(1,2,3,3,5,9)$  2/15  1  $\frac{1}{5}(1,-1,2), 4 \times \frac{1}{3}(1,-1,1)$

No. 47 $X_{6,18} \subset \mathbb{P}(2,2,3,3,4,9)$  1/12  0  $\frac{1}{4}(1,-1,1), 13 \times \frac{1}{2}(1,-1,1),$
$4 \times \frac{1}{3}(1,-1,1)$

No. 48 $X_{10,14} \subset \mathbb{P}(2,2,3,4,5,7)$  1/12  0  $\frac{1}{3}(1,-1,1), \frac{1}{4}(1,-1,1),$
$17 \times \frac{1}{2}(1,-1,1)$

No. 49 $X_{6,20} \subset \mathbb{P}(1,2,3,4,5,10)$  1/10  1  $3 \times \frac{1}{2}(1,-1,1), 2 \times \frac{1}{5}(1,-1,2)$

No. 50 $X_{12,14} \subset \mathbb{P}(2,3,4,4,5,7)$  1/20  0  $\frac{1}{5}(1,-1,2), 9 \times \frac{1}{2}(1,-1,1),$
$3 \times \frac{1}{4}(1,-1,1)$

No. 51 $X_{12,15} \subset \mathbb{P}(1,3,4,5,6,7)$  1/14  1  $\frac{1}{7}(1,-1,2), \frac{1}{2}(1,-1,1)$

No. 52 $X_{10,18} \subset \mathbb{P}(2,3,4,5,6,7)$  1/28  0  $\frac{1}{4}(1,-1,1), \frac{1}{7}(1,-1,3),$
$7 \times \frac{1}{2}(1,-1,1), 3 \times \frac{1}{3}(1,-1,1)$

No. 53 $X_{12,16} \subset \mathbb{P}(2,3,4,5,6,7)$  4/105  0  $\frac{1}{5}(1,-1,2), \frac{1}{7}(1,-1,2),$
$8 \times \frac{1}{2}(1,-1,1), 2 \times \frac{1}{3}(1,-1,1)$

No. 54 $X_{8,22} \subset \mathbb{P}(2,3,4,4,5,11)$  1/30  0  $\frac{1}{3}(1,-1,1), \frac{1}{5}(1,-1,1),$
$10 \times \frac{1}{2}(1,-1,1), 2 \times \frac{1}{4}(1,-1,1)$

No. 55 $X_{12,18} \subset \mathbb{P}(2,3,4,5,6,9)$  1/30  0  $\frac{1}{5}(1,-1,1), 9 \times \frac{1}{2}(1,-1,1),$
$4 \times \frac{1}{3}(1,-1,1)$

No. 56 $X_{12,18} \subset \mathbb{P}(3,4,4,5,6,7)$    3/140  0  $\frac{1}{5}(1,-1,1), \frac{1}{7}(1,-1,2),$
$3 \times \frac{1}{4}(1,-1,1), 3 \times \frac{1}{2}(1,-1,1)$

No. 57 $X_{10,21} \subset \mathbb{P}(3,4,5,5,6,7)$    1/60  0  $\frac{1}{4}(1,-1,1), \frac{1}{6}(1,-1,1),$
$3 \times \frac{1}{3}(1,-1,1), 2 \times \frac{1}{5}(1,-1,2)$

No. 58 $X_{12,21} \subset \mathbb{P}(3,4,5,6,7,7)$    1/70  0  $\frac{1}{5}(1,-1,2), \frac{1}{2}(1,-1,1),$
$3 \times \frac{1}{7}(1,-1,2)$

No. 59 $X_{12,28} \subset \mathbb{P}(3,4,5,6,7,14)$   1/105  0  $\frac{1}{5}(1,-1,1), 2 \times \frac{1}{3}(1,-1,1),$
$2 \times \frac{1}{2}(1,-1,1), 2 \times \frac{1}{7}(1,-1,2)$

Table 4: Canonical codimension 2 complete intersections

# 16   Q-Fano 3-folds

In [R4], Section 4.3 Reid conjectures that if $X$ is a $\mathbb{Q}$-Fano 3-fold then $\mathcal{O}_X(-K_X)$ has a global section. This is false as shown by the following example:

**16.1 Example** The family $X_{12,14}$ in $\mathbb{P}(2,3,4,5,6,7)$ is an anticanonically embedded Fano 3-fold with only the following isolated terminal singularities: 1 of type $\frac{1}{5}(4,1,2)$, 2 of type $\frac{1}{3}(2,1,1)$ and 7 of type $\frac{1}{2}(1,1,1)$. These singularities were determined in 10.3 above.

Since it is quasismooth and of dimension 3, $\omega_X \simeq \mathcal{O}_X(-1)$ and $K_X^3 = -\frac{1}{30}$. By an unpublished result of Barlow (see [R4], Corollary 10.3) we have

$$K_X \cdot c_2(X) = \sum_{\text{singularities } Q} \frac{r_Q^2 - 1}{r_Q} - 24\chi(\mathcal{O}_X)$$

where $r_Q$ is the index of the singularity $Q$ of type $\frac{1}{r_Q}(1,-1,b_Q)$. So $K_X \cdot c_2 = -\frac{101}{30} < 0$. However $\mathcal{O}_X(-K_X) \simeq \mathcal{O}_X(1)$ has no global sections.

Experimentation leads to the following:

**16.2 Conjecture** *Every weighted hypersurface $\mathbb{Q}$-Fano 3-fold $X$, with canonical singularities, has a global section of $\omega_X^{-1}$.*

This is clear in one particular case.

**16.3 Lemma** *Consider $X_d$ in $\mathbb{P}(a_0,\ldots,a_4)$ be a family of $\mathbb{Q}$-Fano 3-folds with only isolated terminal singularities. Suppose also that $a_0 \leq \cdots \leq a_4$ and $a_4 \nmid d$. Then $\omega_X^{-1}$ has a global section.*

**Proof** As $a_4 \nmid d$, the vertex $P_4$ is contained in $X$. The condition for a terminal singularity at $P_i$ gives that there exists an $a_m$ such that $a_4 \mid a_m + \alpha$. So $a_m = \mu a_4 + (-\alpha)$ for some integer $\mu$. Since $\alpha < 0$ and $a_4 \geq a_m$, then $\mu \leq 0$. Thus $\deg(x_4^{(-\mu)} x_m) = -\alpha$ and so $\dim \mathrm{H}^0(\mathcal{O}_X(-\alpha)) \geq 1$. But $\mathrm{H}^0(\omega_X^{-1}) \simeq \mathrm{H}^0(\mathcal{O}_X(-\alpha))$, and so $\omega_X^{-1}$ has a global section. $\square$

Notice that when $\alpha = -1$, there exists a generator $x_i$ with $\deg(x_i) = 1$, that is, $a_0 = 1$.

**16.4 Lemma** *There is a bijection between the following:*

(i) *the set of families of quasismooth, well-formed weighted surface hypersurfaces $S_d$ in $\mathbb{P}(a_1, \ldots, a_4)$ with only canonical singularities and trivial canonical class.*

(ii) *the set of families of quasismooth weighted 3-folds hypersurfaces $X_d$ in $\mathbb{P}(1, a_1, \ldots, a_4)$ with only terminal singularities and $\omega_X \simeq \mathcal{O}_X(-1)$.*

**Proof** Suppose that $S_d$ in $\mathbb{P} = \mathbb{P}(a_1, \ldots, a_4)$ is a K3 surface, with at worst canonical singularities. By comparing the conditions in Theorems 13.1 and 14.1 it is clear that the conditions of the latter are satisfied for $X = X_d$ in $\mathbb{P}(1, a_1, \ldots, a_n)$. Thus $X$ is quasismooth with at worst terminal singularities.

Conversely suppose $X_d$ in $\mathbb{P}(1, a_1, \ldots, a_n)$ is quasismooth and has at worst terminal singularities. It can be seen from Theorems 13.1 and 14.1 that only condition (1ii) of Theorem 13.1 needs proof (the others being either trivially satisfied or equivalent in both the surface and the 3-fold case).

Set $a_0 = 1$ and consider $a_i$ for $i \neq 0$. Suppose that condition (1ii) does not hold. So $a_i \nmid d - a_e$ for all $e = 1, \ldots, 4$. In particular $a_i \nmid d$. Thus $a_i \mid d - a_0$, that is, $a_i \mid d - 1$. Since $a_i \nmid d$ then Theorem 14.1 (1iv) gives that there exists an $m \neq 0, i$ such that $a_i \mid a_m - 1$. Hence $a_i \mid (d - 1) - (a_m - 1)$, that is, $a_i \mid d - a_m$, a contradiction. So $a_i \mid d - a_e$ for some $e \neq 0, i$, which is condition (1ii) of Theorem 13.1. $\square$

**16.5 Note** Each singularity on the K3 surface is of type $\frac{1}{r}(a, -a)$ for some $r$ and $a$, with respect to some pair of the coordinates $x_1, \ldots, x_4$. Forming the corresponding $\mathbb{Q}$-Fano 3-fold results in an extra local coordinate $x_0$ at each singularity, which is thus of type $\frac{1}{r}(a, -a, 1)$. A similar result holds for higher codimensions.

**16.6 List of anticanonically embedded ($\mathbb{Q}$-Fano) weighted 3-folds**

The previous lemma gives a bijection between Reid's list of 95 families of weighted K3 surfaces (see Section 13.3 or [R4], Section 4.5) and the 95 families of quasismooth weighted hypersurface $\mathbb{Q}$-Fano 3-folds, with $\alpha = -1$ and $\sum a_i \leq 100$. These were found by a computer search and are listed below.

| | Hypersurface | $-K_X^3$ | Singularities |
|---|---|---|---|
| No. 1 | $X_4 \subset \mathbb{P}(1,1,1,1,1)$ | 4 | |
| No. 2 | $X_5 \subset \mathbb{P}(1,1,1,1,2)$ | 5/2 | $\frac{1}{2}(1,-1,1)$ |
| No. 3 | $X_6 \subset \mathbb{P}(1,1,1,1,3)$ | 2 | |
| No. 4 | $X_6 \subset \mathbb{P}(1,1,1,2,2)$ | 3/2 | $3 \times \frac{1}{2}(1,-1,1)$ |
| No. 5 | $X_7 \subset \mathbb{P}(1,1,1,2,3)$ | 7/6 | $\frac{1}{2}(1,-1,1), \frac{1}{3}(1,-1,1)$ |
| No. 6 | $X_8 \subset \mathbb{P}(1,1,1,2,4)$ | 1 | $2 \times \frac{1}{2}(1,-1,1)$ |
| No. 7 | $X_8 \subset \mathbb{P}(1,1,2,2,3)$ | 2/3 | $4 \times \frac{1}{2}(1,-1,1), \frac{1}{3}(1,-1,1)$ |
| No. 8 | $X_9 \subset \mathbb{P}(1,1,1,3,4)$ | 3/4 | $\frac{1}{4}(1,-1,1)$ |
| No. 9 | $X_9 \subset \mathbb{P}(1,1,2,3,3)$ | 1/2 | $\frac{1}{2}(1,-1,1), 3 \times \frac{1}{3}(1,-1,1)$ |
| No. 10 | $X_{10} \subset \mathbb{P}(1,1,1,3,5)$ | 2/3 | $\frac{1}{3}(1,-1,1)$ |
| No. 11 | $X_{10} \subset \mathbb{P}(1,1,2,2,5)$ | 1/2 | $5 \times \frac{1}{2}(1,-1,1)$ |
| No. 12 | $X_{10} \subset \mathbb{P}(1,1,2,3,4)$ | 5/12 | $2 \times \frac{1}{2}(1,-1,1), \frac{1}{3}(1,-1,1), \frac{1}{4}(1,-1,1)$ |
| No. 13 | $X_{11} \subset \mathbb{P}(1,1,2,3,5)$ | 11/30 | $\frac{1}{2}(1,-1,1), \frac{1}{3}(1,-1,1), \frac{1}{5}(1,-1,2)$ |
| No. 14 | $X_{12} \subset \mathbb{P}(1,1,1,4,6)$ | 1/2 | $\frac{1}{2}(1,-1,1)$ |
| No. 15 | $X_{12} \subset \mathbb{P}(1,1,2,3,6)$ | 1/3 | $2 \times \frac{1}{2}(1,-1,1), 2 \times \frac{1}{3}(1,-1,1)$ |
| No. 16 | $X_{12} \subset \mathbb{P}(1,1,2,4,5)$ | 3/10 | $3 \times \frac{1}{2}(1,-1,1), \frac{1}{5}(1,-1,1)$ |
| No. 17 | $X_{12} \subset \mathbb{P}(1,1,3,4,4)$ | 1/4 | $3 \times \frac{1}{4}(1,-1,1)$ |
| No. 18 | $X_{12} \subset \mathbb{P}(1,2,2,3,5)$ | 1/5 | $6 \times \frac{1}{2}(1,-1,1), \frac{1}{5}(1,-1,2)$ |
| No. 19 | $X_{12} \subset \mathbb{P}(1,2,3,3,4)$ | 1/6 | $3 \times \frac{1}{2}(1,-1,1), 4 \times \frac{1}{3}(1,-1,1)$ |
| No. 20 | $X_{13} \subset \mathbb{P}(1,1,3,4,5)$ | 13/60 | $\frac{1}{3}(1,-1,1), \frac{1}{4}(1,-1,1), \frac{1}{5}(1,-1,1)$ |
| No. 21 | $X_{14} \subset \mathbb{P}(1,1,2,4,7)$ | 1/4 | $3 \times \frac{1}{2}(1,-1,1), \frac{1}{4}(1,-1,1)$ |
| No. 22 | $X_{14} \subset \mathbb{P}(1,2,2,3,7)$ | 1/6 | $7 \times \frac{1}{2}(1,-1,1), \frac{1}{3}(1,-1,1)$ |
| No. 23 | $X_{14} \subset \mathbb{P}(1,2,3,4,5)$ | 7/60 | $3 \times \frac{1}{2}(1,-1,1), \frac{1}{3}(1,-1,1),$ $\frac{1}{4}(1,-1,1), \frac{1}{5}(1,-1,2)$ |
| No. 24 | $X_{15} \subset \mathbb{P}(1,1,2,5,7)$ | 3/14 | $\frac{1}{2}(1,-1,1), \frac{1}{7}(1,-1,3)$ |
| No. 25 | $X_{15} \subset \mathbb{P}(1,1,3,4,7)$ | 5/28 | $\frac{1}{4}(1,-1,1), \frac{1}{7}(1,-1,2)$ |
| No. 26 | $X_{15} \subset \mathbb{P}(1,1,3,5,6)$ | 1/6 | $2 \times \frac{1}{3}(1,-1,1), \frac{1}{6}(1,-1,1)$ |
| No. 27 | $X_{15} \subset \mathbb{P}(1,2,3,5,5)$ | 1/10 | $\frac{1}{2}(1,-1,1), 3 \times \frac{1}{5}(1,-1,2)$ |
| No. 28 | $X_{15} \subset \mathbb{P}(1,3,3,4,5)$ | 1/12 | $5 \times \frac{1}{3}(1,-1,1), \frac{1}{4}(1,-1,1)$ |
| No. 29 | $X_{16} \subset \mathbb{P}(1,1,2,5,8)$ | 1/5 | $2 \times \frac{1}{2}(1,-1,1), \frac{1}{5}(1,-1,2)$ |
| No. 30 | $X_{16} \subset \mathbb{P}(1,1,3,4,8)$ | 1/6 | $\frac{1}{3}(1,-1,1), 2 \times \frac{1}{4}(1,-1,1)$ |
| No. 31 | $X_{16} \subset \mathbb{P}(1,1,4,5,6)$ | 2/15 | $\frac{1}{2}(1,-1,1), \frac{1}{5}(1,-1,1), \frac{1}{6}(1,-1,1)$ |
| No. 32 | $X_{16} \subset \mathbb{P}(1,2,3,4,7)$ | 2/21 | $4 \times \frac{1}{2}(1,-1,1), \frac{1}{3}(1,-1,1), \frac{1}{7}(1,-1,2)$ |
| No. 33 | $X_{17} \subset \mathbb{P}(1,2,3,5,7)$ | $\frac{17}{210}$ | $\frac{1}{2}(1,-1,1), \frac{1}{3}(1,-1,1), \frac{1}{5}(1,-1,2),$ $\frac{1}{7}(1,-1,3)$ |
| No. 34 | $X_{18} \subset \mathbb{P}(1,1,2,6,9)$ | 1/6 | $3 \times \frac{1}{2}(1,-1,1), \frac{1}{3}(1,-1,1)$ |
| No. 35 | $X_{18} \subset \mathbb{P}(1,1,3,5,9)$ | 2/15 | $2 \times \frac{1}{3}(1,-1,1), \frac{1}{5}(1,-1,1)$ |
| No. 36 | $X_{18} \subset \mathbb{P}(1,1,4,6,7)$ | 3/28 | $\frac{1}{4}(1,-1,1), \frac{1}{2}(1,-1,1), \frac{1}{7}(1,-1,1)$ |
| No. 37 | $X_{18} \subset \mathbb{P}(1,2,3,4,9)$ | 1/12 | $4 \times \frac{1}{2}(1,-1,1), 2 \times \frac{1}{3}(1,-1,1),$ $\frac{1}{4}(1,-1,1)$ |

No. 38  $X_{18} \subset \mathbb{P}(1,2,3,5,8)$  3/40  $2 \times \frac{1}{2}(1,-1,1), \frac{1}{5}(1,-1,2), \frac{1}{8}(1,-1,3)$

No. 39  $X_{18} \subset \mathbb{P}(1,3,4,5,6)$  1/20  $3 \times \frac{1}{3}(1,-1,1), \frac{1}{4}(1,-1,1),$
$\frac{1}{2}(1,-1,1), \frac{1}{5}(1,-1,1)$

No. 40  $X_{19} \subset \mathbb{P}(1,3,4,5,7)$  $\frac{19}{420}$  $\frac{1}{3}(1,-1,1), \frac{1}{4}(1,-1,1), \frac{1}{5}(1,-1,2),$
$\frac{1}{7}(1,-1,2)$

No. 41  $X_{20} \subset \mathbb{P}(1,1,4,5,10)$  1/10  $\frac{1}{2}(1,-1,1), 2 \times \frac{1}{5}(1,-1,1)$

No. 42  $X_{20} \subset \mathbb{P}(1,2,3,5,10)$  1/15  $2 \times \frac{1}{2}(1,-1,1), \frac{1}{3}(1,-1,1),$
$2 \times \frac{1}{5}(1,-1,2)$

No. 43  $X_{20} \subset \mathbb{P}(1,2,4,5,9)$  1/18  $5 \times \frac{1}{2}(1,-1,1), \frac{1}{9}(1,-1,2)$

No. 44  $X_{20} \subset \mathbb{P}(1,2,5,6,7)$  1/21  $3 \times \frac{1}{2}(1,-1,1), \frac{1}{6}(1,-1,1), \frac{1}{7}(1,-1,3)$

No. 45  $X_{20} \subset \mathbb{P}(1,3,4,5,8)$  1/24  $\frac{1}{3}(1,-1,1), 2 \times \frac{1}{4}(1,-1,1), \frac{1}{8}(1,-1,3)$

No. 46  $X_{21} \subset \mathbb{P}(1,1,3,7,10)$  1/10  $\frac{1}{10}(1,-1,3)$

No. 47  $X_{21} \subset \mathbb{P}(1,1,5,7,8)$  3/40  $\frac{1}{5}(1,-1,2), \frac{1}{8}(1,-1,1)$

No. 48  $X_{21} \subset \mathbb{P}(1,2,3,7,9)$  1/18  $\frac{1}{2}(1,-1,1), 2 \times \frac{1}{3}(1,-1,1), \frac{1}{9}(1,-1,4)$

No. 49  $X_{21} \subset \mathbb{P}(1,3,5,6,7)$  1/30  $3 \times \frac{1}{3}(1,-1,1), \frac{1}{5}(1,-1,2), \frac{1}{6}(1,-1,1)$

No. 50  $X_{22} \subset \mathbb{P}(1,1,3,7,11)$  2/21  $\frac{1}{3}(1,-1,1), \frac{1}{7}(1,-1,2)$

No. 51  $X_{22} \subset \mathbb{P}(1,1,4,6,11)$  1/12  $\frac{1}{4}(1,-1,1), \frac{1}{2}(1,-1,1), \frac{1}{6}(1,-1,1)$

No. 52  $X_{22} \subset \mathbb{P}(1,2,4,5,11)$  1/20  $5 \times \frac{1}{2}(1,-1,1), \frac{1}{4}(1,-1,1), \frac{1}{5}(1,-1,1)$

No. 53  $X_{24} \subset \mathbb{P}(1,1,3,8,12)$  1/12  $2 \times \frac{1}{3}(1,-1,1), \frac{1}{4}(1,-1,1)$

No. 54  $X_{24} \subset \mathbb{P}(1,1,6,8,9)$  1/18  $\frac{1}{2}(1,-1,1), \frac{1}{3}(1,-1,1), \frac{1}{9}(1,-1,1)$

No. 55  $X_{24} \subset \mathbb{P}(1,2,3,7,12)$  1/21  $2 \times \frac{1}{2}(1,-1,1), 2 \times \frac{1}{3}(1,-1,1),$
$\frac{1}{7}(1,-1,3)$

No. 56  $X_{24} \subset \mathbb{P}(1,2,3,8,11)$  1/22  $3 \times \frac{1}{2}(1,-1,1), \frac{1}{11}(1,-1,4)$

No. 57  $X_{24} \subset \mathbb{P}(1,3,4,5,12)$  1/30  $2 \times \frac{1}{3}(1,-1,1), 2 \times \frac{1}{4}(1,-1,1),$
$\frac{1}{5}(1,-1,2)$

No. 58  $X_{24} \subset \mathbb{P}(1,3,4,7,10)$  1/35  $\frac{1}{2}(1,-1,1), \frac{1}{7}(1,-1,2), \frac{1}{10}(1,-1,3)$

No. 59  $X_{24} \subset \mathbb{P}(1,3,6,7,8)$  1/42  $4 \times \frac{1}{3}(1,-1,1), \frac{1}{2}(1,-1,1), \frac{1}{7}(1,-1,1)$

No. 60  $X_{24} \subset \mathbb{P}(1,4,5,6,9)$  1/45  $2 \times \frac{1}{2}(1,-1,1), \frac{1}{5}(1,-1,1),$
$\frac{1}{3}(1,-1,1), \frac{1}{9}(1,-1,2)$

No. 61  $X_{25} \subset \mathbb{P}(1,4,5,7,9)$  $\frac{5}{252}$  $\frac{1}{4}(1,-1,1), \frac{1}{7}(1,-1,3), \frac{1}{9}(1,-1,2)$

No. 62  $X_{26} \subset \mathbb{P}(1,1,5,7,13)$  2/35  $\frac{1}{5}(1,-1,2), \frac{1}{7}(1,-1,1)$

No. 63  $X_{26} \subset \mathbb{P}(1,2,3,8,13)$  1/24  $3 \times \frac{1}{2}(1,-1,1), \frac{1}{3}(1,-1,1), \frac{1}{8}(1,-1,3)$

No. 64  $X_{26} \subset \mathbb{P}(1,2,5,6,13)$  1/30  $4 \times \frac{1}{2}(1,-1,1), \frac{1}{5}(1,-1,2), \frac{1}{6}(1,-1,1)$

No. 65  $X_{27} \subset \mathbb{P}(1,2,5,9,11)$  $\frac{3}{110}$  $\frac{1}{2}(1,-1,1), \frac{1}{5}(1,-1,1), \frac{1}{11}(1,-1,5)$

No. 66  $X_{27} \subset \mathbb{P}(1,5,6,7,9)$  1/70  $\frac{1}{5}(1,-1,1), \frac{1}{6}(1,-1,1), \frac{1}{3}(1,-1,1),$
$\frac{1}{7}(1,-1,3)$

No. 67  $X_{28} \subset \mathbb{P}(1,1,4,9,14)$  1/18  $\frac{1}{2}(1,-1,1), \frac{1}{9}(1,-1,2)$

No. 68  $X_{28} \subset \mathbb{P}(1,3,4,7,14)$  1/42  $\frac{1}{3}(1,-1,1), \frac{1}{2}(1,-1,1), 2 \times \frac{1}{7}(1,-1,2)$

No. 69  $X_{28} \subset \mathbb{P}(1,4,6,7,11)$  1/66  $2 \times \frac{1}{2}(1,-1,1), \frac{1}{6}(1,-1,1),$
$\frac{1}{11}(1,-1,3)$

No. 70  $X_{30} \subset \mathbb{P}(1,1,4,10,15)$  1/20  $\frac{1}{4}(1,-1,1), \frac{1}{2}(1,-1,1), \frac{1}{5}(1,-1,1)$

No. 71 $X_{30} \subset \mathbb{P}(1,1,6,8,15)$    1/24    $\frac{1}{2}(1,-1,1), \frac{1}{3}(1,-1,1), \frac{1}{8}(1,-1,1)$

No. 72 $X_{30} \subset \mathbb{P}(1,2,3,10,15)$    1/30    $3 \times \frac{1}{2}(1,-1,1), 2 \times \frac{1}{3}(1,-1,1),$
$\frac{1}{5}(1,-1,2)$

No. 73 $X_{30} \subset \mathbb{P}(1,2,6,7,15)$    1/42    $5 \times \frac{1}{2}(1,-1,1), \frac{1}{3}(1,-1,1), \frac{1}{7}(1,-1,1)$

No. 74 $X_{30} \subset \mathbb{P}(1,3,4,10,13)$    1/52    $\frac{1}{4}(1,-1,1), \frac{1}{2}(1,-1,1), \frac{1}{13}(1,-1,4)$

No. 75 $X_{30} \subset \mathbb{P}(1,4,5,6,15)$    1/60    $\frac{1}{4}(1,-1,1), 2 \times \frac{1}{2}(1,-1,1),$
$2 \times \frac{1}{5}(1,-1,1), \frac{1}{3}(1,-1,1)$

No. 76 $X_{30} \subset \mathbb{P}(1,5,6,8,11)$    1/88    $\frac{1}{2}(1,-1,1), \frac{1}{8}(1,-1,3), \frac{1}{11}(1,-1,2)$

No. 77 $X_{32} \subset \mathbb{P}(1,2,5,9,16)$    1/45    $2 \times \frac{1}{2}(1,-1,1), \frac{1}{5}(1,-1,1), \frac{1}{9}(1,-1,4)$

No. 78 $X_{32} \subset \mathbb{P}(1,4,5,7,16)$    1/70    $2 \times \frac{1}{4}(1,-1,1), \frac{1}{5}(1,-1,1), \frac{1}{7}(1,-1,3)$

No. 79 $X_{33} \subset \mathbb{P}(1,3,5,11,14)$    1/70    $\frac{1}{5}(1,-1,1), \frac{1}{14}(1,-1,5)$

No. 80 $X_{34} \subset \mathbb{P}(1,3,4,10,17)$    1/60    $\frac{1}{3}(1,-1,1), \frac{1}{4}(1,-1,1), \frac{1}{2}(1,-1,1),$
$\frac{1}{10}(1,-1,3)$

No. 81 $X_{34} \subset \mathbb{P}(1,4,6,7,17)$    1/84    $\frac{1}{4}(1,-1,1), 2 \times \frac{1}{2}(1,-1,1),$
$\frac{1}{6}(1,-1,1), \frac{1}{7}(1,-1,2)$

No. 82 $X_{36} \subset \mathbb{P}(1,1,5,12,18)$    1/30    $\frac{1}{5}(1,-1,2), \frac{1}{6}(1,-1,1)$

No. 83 $X_{36} \subset \mathbb{P}(1,3,4,11,18)$    1/66    $2 \times \frac{1}{3}(1,-1,1), \frac{1}{2}(1,-1,1),$
$\frac{1}{11}(1,-1,3)$

No. 84 $X_{36} \subset \mathbb{P}(1,7,8,9,12)$    $\frac{1}{168}$    $\frac{1}{7}(1,-1,3), \frac{1}{8}(1,-1,1), \frac{1}{4}(1,-1,1),$
$\frac{1}{3}(1,-1,1)$

No. 85 $X_{38} \subset \mathbb{P}(1,3,5,11,19)$    $\frac{2}{165}$    $\frac{1}{3}(1,-1,1), \frac{1}{5}(1,-1,1), \frac{1}{11}(1,-1,4)$

No. 86 $X_{38} \subset \mathbb{P}(1,5,6,8,19)$    $\frac{1}{120}$    $\frac{1}{5}(1,-1,1), \frac{1}{6}(1,-1,1), \frac{1}{2}(1,-1,1),$
$\frac{1}{8}(1,-1,3)$

No. 87 $X_{40} \subset \mathbb{P}(1,5,7,8,20)$    $\frac{1}{140}$    $2 \times \frac{1}{5}(1,-1,2), \frac{1}{7}(1,-1,1), \frac{1}{4}(1,-1,1)$

No. 88 $X_{42} \subset \mathbb{P}(1,1,6,14,21)$    1/42    $\frac{1}{2}(1,-1,1), \frac{1}{3}(1,-1,1), \frac{1}{7}(1,-1,1)$

No. 89 $X_{42} \subset \mathbb{P}(1,2,5,14,21)$    1/70    $3 \times \frac{1}{2}(1,-1,1), \frac{1}{5}(1,-1,1), \frac{1}{7}(1,-1,3)$

No. 90 $X_{42} \subset \mathbb{P}(1,3,4,14,21)$    1/84    $2 \times \frac{1}{3}(1,-1,1), \frac{1}{4}(1,-1,1),$
$\frac{1}{2}(1,-1,1), \frac{1}{2}(1,-1,2)$

No. 91 $X_{44} \subset \mathbb{P}(1,4,5,13,22)$    $\frac{1}{130}$    $\frac{1}{2}(1,-1,1), \frac{1}{5}(1,-1,2), \frac{1}{13}(1,-1,3)$

No. 92 $X_{48} \subset \mathbb{P}(1,3,5,16,24)$    $\frac{1}{120}$    $2 \times \frac{1}{3}(1,-1,1), \frac{1}{5}(1,-1,1), \frac{1}{8}(1,-1,3)$

No. 93 $X_{50} \subset \mathbb{P}(1,7,8,10,25)$    $\frac{1}{280}$    $\frac{1}{7}(1,-1,2), \frac{1}{8}(1,-1,1), \frac{1}{2}(1,-1,1),$
$\frac{1}{5}(1,-1,2)$

No. 94 $X_{54} \subset \mathbb{P}(1,4,5,18,27)$    $\frac{1}{180}$    $\frac{1}{4}(1,-1,1), \frac{1}{2}(1,-1,1), \frac{1}{5}(1,-1,2),$
$\frac{1}{9}(1,-1,2)$

No. 95 $X_{66} \subset \mathbb{P}(1,5,6,22,33)$    $\frac{1}{330}$    $\frac{1}{5}(1,-1,2), \frac{1}{2}(1,-1,1), \frac{1}{3}(1,-1,1),$
$\frac{1}{11}(1,-1,2)$

Table 5: The 95 anticanonical 3-folds hypersurfaces

**16.7  Codimension 2 Q-Fano weighted complete intersections**  There are 85 codimension 2 quasismooth Q-Fano weighted complete intersections which satisfy the conditions of Theorem 14.4, $\alpha = -1$ and $\sum a_i \leq 100$.

| | Complete intersection | $-K_X^3$ | Singularities |
|---|---|---|---|
| No. 1 | $X_{2,3} \subset \mathbb{P}(1,1,1,1,1,1)$ | 6 | |
| No. 2 | $X_{3,3} \subset \mathbb{P}(1,1,1,1,1,2)$ | 9/2 | $\frac{1}{2}(1,-1,1)$ |
| No. 3 | $X_{3,4} \subset \mathbb{P}(1,1,1,1,2,2)$ | 3 | $2 \times \frac{1}{2}(1,-1,1)$ |
| No. 4 | $X_{4,4} \subset \mathbb{P}(1,1,1,1,2,3)$ | 8/3 | $\frac{1}{3}(1,-1,1)$ |
| No. 5 | $X_{4,4} \subset \mathbb{P}(1,1,1,2,2,2)$ | 2 | $4 \times \frac{1}{2}(1,-1,1)$ |
| No. 6 | $X_{4,5} \subset \mathbb{P}(1,1,1,2,2,3)$ | 5/3 | $\frac{1}{3}(1,-1,1), 2 \times \frac{1}{2}(1,-1,1)$ |
| No. 7 | $X_{4,6} \subset \mathbb{P}(1,1,1,2,3,3)$ | 4/3 | $2 \times \frac{1}{3}(1,-1,1)$ |
| No. 8 | $X_{4,6} \subset \mathbb{P}(1,1,2,2,2,3)$ | 1 | $6 \times \frac{1}{2}(1,-1,1)$ |
| No. 9 | $X_{5,6} \subset \mathbb{P}(1,1,1,2,3,4)$ | 5/4 | $\frac{1}{4}(1,-1,1), \frac{1}{2}(1,-1,1)$ |
| No. 10 | $X_{5,6} \subset \mathbb{P}(1,1,2,2,3,3)$ | 5/6 | $3 \times \frac{1}{2}(1,-1,1), 2 \times \frac{1}{3}(1,-1,1)$ |
| No. 11 | $X_{6,6} \subset \mathbb{P}(1,1,1,2,3,5)$ | 6/5 | $\frac{1}{5}(1,-1,2)$ |
| No. 12 | $X_{6,6} \subset \mathbb{P}(1,1,2,2,3,4)$ | 3/4 | $\frac{1}{4}(1,-1,1), 4 \times \frac{1}{2}(1,-1,1)$ |
| No. 13 | $X_{6,6} \subset \mathbb{P}(1,1,2,3,3,3)$ | 2/3 | $4 \times \frac{1}{3}(1,-1,1)$ |
| No. 14 | $X_{6,6} \subset \mathbb{P}(1,2,2,2,3,3)$ | 1/2 | $9 \times \frac{1}{2}(1,-1,1)$ |
| No. 15 | $X_{6,7} \subset \mathbb{P}(1,1,2,2,3,5)$ | 7/10 | $\frac{1}{5}(1,-1,2), 3 \times \frac{1}{2}(1,-1,1)$ |
| No. 16 | $X_{6,7} \subset \mathbb{P}(1,1,2,3,3,4)$ | 7/12 | $\frac{1}{4}(1,-1,1), \frac{1}{2}(1,-1,1),$ $2 \times \frac{1}{3}(1,-1,1)$ |
| No. 17 | $X_{6,8} \subset \mathbb{P}(1,1,1,3,4,5)$ | 4/5 | $\frac{1}{5}(1,-1,1)$ |
| No. 18 | $X_{6,8} \subset \mathbb{P}(1,1,2,3,3,5)$ | 8/15 | $\frac{1}{5}(1,-1,2), 2 \times \frac{1}{3}(1,-1,1)$ |
| No. 19 | $X_{6,8} \subset \mathbb{P}(1,1,2,3,4,4)$ | 1/2 | $2 \times \frac{1}{2}(1,-1,1), 2 \times \frac{1}{4}(1,-1,1)$ |
| No. 20 | $X_{6,8} \subset \mathbb{P}(1,2,2,3,3,4)$ | 1/3 | $6 \times \frac{1}{2}(1,-1,1), 2 \times \frac{1}{3}(1,-1,1)$ |
| No. 21 | $X_{6,9} \subset \mathbb{P}(1,1,2,3,4,5)$ | 9/20 | $\frac{1}{4}(1,-1,1), \frac{1}{5}(1,-1,2),$ $\frac{1}{2}(1,-1,1)$ |
| No. 22 | $X_{7,8} \subset \mathbb{P}(1,1,2,3,4,5)$ | 7/15 | $\frac{1}{3}(1,-1,1), \frac{1}{5}(1,-1,1),$ $2 \times \frac{1}{2}(1,-1,1)$ |
| No. 23 | $X_{6,10} \subset \mathbb{P}(1,1,2,3,5,5)$ | 2/5 | $2 \times \frac{1}{5}(1,-1,2)$ |
| No. 24 | $X_{6,10} \subset \mathbb{P}(1,2,2,3,4,5)$ | 1/4 | $\frac{1}{4}(1,-1,1), 7 \times \frac{1}{2}(1,-1,1)$ |
| No. 25 | $X_{8,9} \subset \mathbb{P}(1,1,2,3,4,7)$ | 3/7 | $\frac{1}{7}(1,-1,2), 2 \times \frac{1}{2}(1,-1,1)$ |
| No. 26 | $X_{8,9} \subset \mathbb{P}(1,1,3,4,4,5)$ | 3/10 | $\frac{1}{5}(1,-1,1), 2 \times \frac{1}{4}(1,-1,1)$ |
| No. 27 | $X_{8,9} \subset \mathbb{P}(1,2,3,3,4,5)$ | 1/5 | $\frac{1}{5}(1,-1,2), 2 \times \frac{1}{2}(1,-1,1),$ $3 \times \frac{1}{3}(1,-1,1)$ |
| No. 28 | $X_{8,10} \subset \mathbb{P}(1,1,2,3,5,7)$ | 8/21 | $\frac{1}{3}(1,-1,1), \frac{1}{7}(1,-1,3)$ |
| No. 29 | $X_{8,10} \subset \mathbb{P}(1,1,2,4,5,6)$ | 1/3 | $\frac{1}{6}(1,-1,1), 3 \times \frac{1}{2}(1,-1,1)$ |
| No. 30 | $X_{8,10} \subset \mathbb{P}(1,1,3,4,5,5)$ | 4/15 | $\frac{1}{3}(1,-1,1), 2 \times \frac{1}{5}(1,-1,1)$ |
| No. 31 | $X_{8,10} \subset \mathbb{P}(1,2,3,4,4,5)$ | 1/6 | $\frac{1}{3}(1,-1,1), 4 \times \frac{1}{2}(1,-1,1),$ $2 \times \frac{1}{4}(1,-1,1)$ |
| No. 32 | $X_{9,10} \subset \mathbb{P}(1,1,2,3,5,8)$ | 3/8 | $\frac{1}{8}(1,-1,3), \frac{1}{2}(1,-1,1)$ |
| No. 33 | $X_{9,10} \subset \mathbb{P}(1,1,3,4,5,6)$ | 1/4 | $\frac{1}{4}(1,-1,1), \frac{1}{6}(1,-1,1),$ $\frac{1}{3}(1,-1,1)$ |

No. 34  $X_{9,10} \subset \mathbb{P}(1,2,2,3,5,7)$    $3/14$   $\frac{1}{7}(1,-1,3), 5 \times \frac{1}{2}(1,-1,1)$

No. 35  $X_{9,10} \subset \mathbb{P}(1,2,3,4,5,5)$    $3/20$   $\frac{1}{4}(1,-1,1), 2 \times \frac{1}{2}(1,-1,1),$
$2 \times \frac{1}{5}(1,-1,2)$

No. 36  $X_{8,12} \subset \mathbb{P}(1,1,3,4,5,7)$    $8/35$   $\frac{1}{5}(1,-1,1), \frac{1}{7}(1,-1,2)$

No. 37  $X_{8,12} \subset \mathbb{P}(1,2,3,4,5,6)$    $2/15$   $\frac{1}{5}(1,-1,1), 4 \times \frac{1}{2}(1,-1,1),$
$2 \times \frac{1}{3}(1,-1,1)$

No. 38  $X_{9,12} \subset \mathbb{P}(1,2,3,4,5,7)$    $9/70$   $\frac{1}{5}(1,-1,2), \frac{1}{7}(1,-1,2),$
$3 \times \frac{1}{2}(1,-1,1)$

No. 39  $X_{10,11} \subset \mathbb{P}(1,2,3,4,5,7)$    $\frac{11}{84}$   $\frac{1}{3}(1,-1,1), \frac{1}{4}(1,-1,1),$
$\frac{1}{7}(1,-1,3), 2 \times \frac{1}{2}(1,-1,1)$

No. 40  $X_{10,12} \subset \mathbb{P}(1,1,3,4,5,9)$    $2/9$   $\frac{1}{9}(1,-1,2), \frac{1}{3}(1,-1,1)$

No. 41  $X_{10,12} \subset \mathbb{P}(1,1,3,5,6,7)$    $4/21$   $\frac{1}{7}(1,-1,1), 2 \times \frac{1}{3}(1,-1,1)$

No. 42  $X_{10,12} \subset \mathbb{P}(1,1,4,5,6,6)$    $1/6$   $\frac{1}{5}(1,-1,1), 2 \times \frac{1}{6}(1,-1,1)$

No. 43  $X_{10,12} \subset \mathbb{P}(1,2,3,4,5,8)$    $1/8$   $\frac{1}{8}(1,-1,3), 3 \times \frac{1}{2}(1,-1,1),$
$\frac{1}{4}(1,-1,1)$

No. 44  $X_{10,12} \subset \mathbb{P}(1,2,3,5,5,7)$    $4/35$   $\frac{1}{7}(1,-1,3), 2 \times \frac{1}{5}(1,-1,2)$

No. 45  $X_{10,12} \subset \mathbb{P}(1,2,4,5,5,6)$    $1/10$   $5 \times \frac{1}{2}(1,-1,1), 2 \times \frac{1}{5}(1,-1,1)$

No. 46  $X_{10,12} \subset \mathbb{P}(1,3,3,4,5,7)$    $2/21$   $\frac{1}{7}(1,-1,2), 4 \times \frac{1}{3}(1,-1,1)$

No. 47  $X_{10,12} \subset \mathbb{P}(1,3,4,4,5,6)$    $1/12$   $2 \times \frac{1}{3}(1,-1,1), 3 \times \frac{1}{4}(1,-1,1),$
$\frac{1}{2}(1,-1,1)$

No. 48  $X_{11,12} \subset \mathbb{P}(1,1,4,5,6,7)$    $\frac{11}{70}$   $\frac{1}{5}(1,-1,1), \frac{1}{7}(1,-1,1),$
$\frac{1}{2}(1,-1,1)$

No. 49  $X_{10,14} \subset \mathbb{P}(1,1,2,5,7,9)$    $2/9$   $\frac{1}{9}(1,-1,4)$

No. 50  $X_{10,14} \subset \mathbb{P}(1,2,3,5,7,7)$    $2/21$   $\frac{1}{3}(1,-1,1), 2 \times \frac{1}{7}(1,-1,3)$

No. 51  $X_{10,14} \subset \mathbb{P}(1,2,4,5,6,7)$    $1/12$   $\frac{1}{4}(1,-1,1), \frac{1}{6}(1,-1,1),$
$5 \times \frac{1}{2}(1,-1,1)$

No. 52  $X_{10,15} \subset \mathbb{P}(1,2,3,5,7,8)$    $5/56$   $\frac{1}{7}(1,-1,3), \frac{1}{8}(1,-1,3),$
$\frac{1}{2}(1,-1,1)$

No. 53  $X_{12,13} \subset \mathbb{P}(1,3,4,5,6,7)$    $\frac{13}{210}$   $\frac{1}{5}(1,-1,1), \frac{1}{7}(1,-1,2),$
$2 \times \frac{1}{3}(1,-1,1), \frac{1}{2}(1,-1,1)$

No. 54  $X_{12,14} \subset \mathbb{P}(1,1,3,4,7,11)$    $2/11$   $\frac{1}{11}(1,-1,3)$

No. 55  $X_{12,14} \subset \mathbb{P}(1,1,4,6,7,8)$    $1/8$   $\frac{1}{8}(1,-1,1), \frac{1}{2}(1,-1,1),$
$\frac{1}{4}(1,-1,1)$

No. 56  $X_{12,14} \subset \mathbb{P}(1,2,3,4,7,10)$    $1/10$   $\frac{1}{10}(1,-1,3), 4 \times \frac{1}{2}(1,-1,1)$

No. 57  $X_{12,14} \subset \mathbb{P}(1,2,3,5,7,9)$    $4/45$   $\frac{1}{5}(1,-1,2), \frac{1}{9}(1,-1,4),$
$\frac{1}{3}(1,-1,1)$

No. 58  $X_{12,14} \subset \mathbb{P}(1,3,4,5,7,7)$    $2/35$   $\frac{1}{5}(1,-1,2), 2 \times \frac{1}{7}(1,-1,2)$

No. 59  $X_{12,14} \subset \mathbb{P}(1,4,4,5,6,7)$    $1/20$   $\frac{1}{5}(1,-1,1), 3 \times \frac{1}{4}(1,-1,1),$
$2 \times \frac{1}{2}(1,-1,1)$

No. 60  $X_{12,14} \subset \mathbb{P}(2,3,4,5,6,7)$    $1/30$   $\frac{1}{5}(1,-1,2), 7 \times \frac{1}{2}(1,-1,1),$
$2 \times \frac{1}{3}(1,-1,1)$

No. 61  $X_{12,15} \subset \mathbb{P}(1,1,4,5,6,11)$   $3/22$   $\frac{1}{11}(1,-1,2), \frac{1}{2}(1,-1,1)$

No. 62  $X_{12,15} \subset \mathbb{P}(1,3,4,5,6,9)$   $1/18$   $\frac{1}{9}(1,-1,2), 3 \times \frac{1}{3}(1,-1,1),$
$\frac{1}{2}(1,-1,1)$

No. 63  $X_{12,15} \subset \mathbb{P}(1,3,4,5,7,8)$   $3/56$   $\frac{1}{7}(1,-1,2), \frac{1}{8}(1,-1,3),$
$\frac{1}{4}(1,-1,1)$

No. 64  $X_{12,16} \subset \mathbb{P}(1,2,5,6,7,8)$   $2/35$   $\frac{1}{5}(1,-1,2), \frac{1}{7}(1,-1,1),$
$4 \times \frac{1}{2}(1,-1,1)$

No. 65  $X_{14,15} \subset \mathbb{P}(1,2,3,5,7,12)$   $1/12$   $\frac{1}{12}(1,-1,5), \frac{1}{2}(1,-1,1),$
$\frac{1}{3}(1,-1,1)$

No. 66  $X_{14,15} \subset \mathbb{P}(1,2,5,6,7,9)$   $1/18$   $\frac{1}{6}(1,-1,1), \frac{1}{9}(1,-1,4),$
$2 \times \frac{1}{2}(1,-1,1)$

No. 67  $X_{14,15} \subset \mathbb{P}(1,3,4,5,7,10)$   $1/20$   $\frac{1}{4}(1,-1,1), \frac{1}{10}(1,-1,3),$
$\frac{1}{5}(1,-1,2)$

No. 68  $X_{14,15} \subset \mathbb{P}(1,3,5,6,7,8)$   $1/24$   $\frac{1}{6}(1,-1,1), \frac{1}{8}(1,-1,3),$
$2 \times \frac{1}{3}(1,-1,1)$

No. 69  $X_{14,16} \subset \mathbb{P}(1,1,5,7,8,9)$   $4/45$   $\frac{1}{5}(1,-1,2), \frac{1}{9}(1,-1,1)$

No. 70  $X_{14,16} \subset \mathbb{P}(1,3,4,5,7,11)$   $\frac{8}{165}$   $\frac{1}{3}(1,-1,1), \frac{1}{5}(1,-1,2),$
$\frac{1}{11}(1,-1,3)$

No. 71  $X_{14,16} \subset \mathbb{P}(1,4,5,6,7,8)$   $1/30$   $\frac{1}{5}(1,-1,2), \frac{1}{6}(1,-1,1),$
$\frac{1}{2}(1,-1,1), 2 \times \frac{1}{4}(1,-1,1)$

No. 72  $X_{15,16} \subset \mathbb{P}(1,2,3,5,8,13)$   $1/13$   $\frac{1}{13}(1,-1,5), 2 \times \frac{1}{2}(1,-1,1)$

No. 73  $X_{15,16} \subset \mathbb{P}(1,3,4,5,8,11)$   $1/22$   $\frac{1}{11}(1,-1,4), 2 \times \frac{1}{4}(1,-1,1)$

No. 74  $X_{14,18} \subset \mathbb{P}(1,2,3,7,9,11)$   $2/33$   $\frac{1}{11}(1,-1,5), 2 \times \frac{1}{3}(1,-1,1)$

No. 75  $X_{14,18} \subset \mathbb{P}(1,2,6,7,8,9)$   $1/24$   $\frac{1}{8}(1,-1,1), 5 \times \frac{1}{2}(1,-1,1),$
$\frac{1}{3}(1,-1,1)$

No. 76  $X_{12,20} \subset \mathbb{P}(1,4,5,6,7,10)$   $1/35$   $\frac{1}{7}(1,-1,2), 2 \times \frac{1}{2}(1,-1,1),$
$2 \times \frac{1}{5}(1,-1,1)$

No. 77  $X_{16,18} \subset \mathbb{P}(1,1,6,8,9,10)$   $1/15$   $\frac{1}{10}(1,-1,1), \frac{1}{2}(1,-1,1),$
$\frac{1}{3}(1,-1,1)$

No. 78  $X_{16,18} \subset \mathbb{P}(1,4,6,7,8,9)$   $1/42$   $\frac{1}{7}(1,-1,1), 2 \times \frac{1}{2}(1,-1,1),$
$2 \times \frac{1}{4}(1,-1,1), \frac{1}{3}(1,-1,1)$

No. 79  $X_{18,20} \subset \mathbb{P}(1,4,5,6,9,14)$   $1/42$   $\frac{1}{14}(1,-1,3), 2 \times \frac{1}{2}(1,-1,1),$
$\frac{1}{3}(1,-1,1)$

No. 80  $X_{18,20} \subset \mathbb{P}(1,4,5,7,9,13)$   $2/91$   $\frac{1}{7}(1,-1,3), \frac{1}{13}(1,-1,3)$

No. 81  $X_{18,20} \subset \mathbb{P}(1,5,6,7,9,11)$   $\frac{4}{231}$   $\frac{1}{7}(1,-1,3), \frac{1}{11}(1,-1,2),$
$\frac{1}{3}(1,-1,1)$

No. 82  $X_{18,22} \subset \mathbb{P}(1,2,5,9,11,13)$   $2/65$   $\frac{1}{5}(1,-1,1), \frac{1}{13}(1,-1,6)$

No. 83  $X_{20,21} \subset \mathbb{P}(1,3,4,7,10,17)$   $1/34$   $\frac{1}{17}(1,-1,5), \frac{1}{2}(1,-1,1)$

No. 84  $X_{18,30} \subset \mathbb{P}(1,6,8,9,10,15)$   $\frac{1}{120}$   $\frac{1}{8}(1,-1,1), 2 \times \frac{1}{2}(1,-1,1),$
$2 \times \frac{1}{3}(1,-1,1), \frac{1}{5}(1,-1,1)$

No. 85  $X_{24,30} \subset \mathbb{P}(1,8,9,10,12,15)$  $\frac{1}{180}$   $\frac{1}{9}(1,-1,1), \frac{1}{2}(1,1,1),$
$\frac{1}{4}(1,-1,1), \frac{1}{3}(1,-1,1),$
$\frac{1}{5}(1,-1,2)$

Table 6: Anticanonical 3-fold codimension 2 complete intersections

**16.8 Note** $X_{12,14}$ in $\mathbb{P}(2,3,4,5,6,7)$ is the only element in the above list with $a_i \geq 2$ for all $i$ (see Example 16.1).

# 17   The plurigenus formula

Before we describe Reid's table method for producing examples of weighted complete intersection we must state the plurigenus formula for canonical and $\mathbb{Q}$-Fano 3-folds.

**17.1 Definition** For a singularity $Q$ of type $\frac{1}{r}(1,-1,b)$ define:

$$l(Q,n) = \begin{cases} 0 & \text{if } n = 0, 1, \\ \sum\limits_{k=1}^{n-1} \frac{\overline{bk}(r-\overline{bk})}{2r} & \text{if } n \geq 2, \end{cases}$$

where $\overline{x}$ denotes the smallest nonnegative residue of $x$ modulo $r$. This is extended to negative integers via:

$$l(-n) = -l(n+1)$$

for all $n \geq 0$. This is for consistency with Serre duality. For a collection (or *basket*) $\mathcal{B}$ of singularities define:

$$l(n) = \sum_{Q \in \mathcal{B}} l(q,n)$$

for all $n \in \mathbb{Z}$.

From [F1], Theorem 2.5, equation (4) (see also [R4], Chapter III) we have the following:

**17.2 Theorem** *For any projective 3-fold $X$, with at worst canonical singularities, there exists a basket $\mathcal{B}$ of singularities such that*

$$\chi(\mathcal{O}_X(nK_X)) = \frac{(2n-1)n(n-1)}{12r}K_X^3 - (2n-1)\chi(\mathcal{O}_X) + l(n)$$

*for all $n \in \mathbb{Z}$.*

**17.3 Canonical 3-folds** Let $X$ be a canonical 3-fold. Then $K_X$ is ample and we have:

$$P_n = \chi(\mathcal{O}_X(nK_X)) = \frac{(2n-1)n(n-1)}{12r}K_X^3 - (2n-1)\chi(\mathcal{O}_X) + l(n)$$

for all $n \geq 2$. This formula is Reid's exact plurigenus formula.

**17.4 Q-Fano 3-fold complete intersections** If $X$ is a Q-Fano 3-fold then $-K_X$ is ample. Moreover if $X$ is also a complete intersection then $\chi(\mathcal{O}_X) = 1$. So:

$$P_{-n} = \chi(\mathcal{O}_X(-nK_X)) = \frac{(2n+1)n(n+1)}{12r}(-K_X)^3 + (2n+1) - l(n+1)$$

for all $n \geq 1$.

# 18    The Reid table method

Consider a complete intersection $X_{d_1,\dots,d_c}$ in $\mathbb{P}(a_0,\dots,a_n)$. The Poincaré series (see [WPS], Section 3.4 and compare [AM], 11.1) corresponding to the coordinate ring $R$ of $X$ is:

$$\mathcal{P}(t) = \sum_{n=0}^{\infty} h^0(X, \mathcal{O}_X(n))t^n$$

$$= \frac{\prod_{i=1}^{i=c}(1-t^{d_i})}{\prod_{i=0}^{i=n}(1-t^{a_i})}.$$

Moreover, if $\omega_X \simeq \mathcal{O}_X(1)$ then $\mathcal{P}(t) = \sum_{n=0}^{\infty} P_n(X)t^n$, where $P_n(X)$ are the plurigenera of $X$. In the case of a Q-Fano 3-fold with $\omega_X \simeq \mathcal{O}_X(-1)$ then $\mathcal{P}(t) = \sum_{n=0}^{\infty} P_{-n}(X)t^n$, where $P_{-n}(X)$ are the anti-plurigenera of $X$.

**18.1 Example** $X_6$ in $\mathbb{P}^4$ has Poincaré series

$$\mathcal{P}(t) = \frac{(1-t^6)}{(1-t)^5} = 1 + t + 5t^2 + 15t^3 + \cdots$$

So $p_g = 1$, $P_2 = 5$, $P_3 = 15$, etc.

**18.2 Question** Given a list of plurigenera (which could arise from a record of pluridata) does there exist a complete intersection with $\omega_X \simeq \mathcal{O}_X(\pm 1)$?

The following lemma due to Reid helps answer the above.

**18.3 Lemma** *Given a sequence $p_0 = 1, p_1, p_2, \ldots$ such that*

$$\sum_{i=0}^{\infty} p_i t^i = \frac{\prod_{i=1}^{i=c} (1 - t^{d_i})}{\prod_{i=0}^{i=n} (1 - t^{a_i})}$$

*for some $\{d_i, a_i\}$. Then these $\{d_i, a_i\}$ are unique up to $a_i \neq d_j$ and are determinable.*

**Proof**   The following is a constructive proof. Let $q_i^0 = p_i$. So

$$\sum_{i=0}^{\infty} q_i^0 t^i = \frac{\prod (1 - t^{d_i})}{\prod (1 - t^{a_i})}.$$

Without loss of generality assume that $d_c \geq \cdots \geq d_1$ and $a_n \geq \cdots \geq a_0$. Clearly we may assume $a_0 \neq d_1$ or else these two terms would cancel. There are two cases:

(i) $a_0 < d_1$. Let $a_0$ occur with multiplicity $\mu$. Then $\mathcal{P}(t) = 1 + \mu t^{a_0} +$higher order terms. So the first nonzero $q_i^0$ is $q_{a_0}^0 = \mu < 0$. Define $q_i^1 = q_i^0 - q_{i-a_0}^0$, where $q_i^0 = 0$ if $i < 0$. Then $q_{a_0}^1 = q_{a_0}^0 - 1$. Thus

$$\sum_{i=0}^{\infty} q_i^1 t^i = \sum_{i=0}^{\infty} (q_i^0 - q_{i-a_0}^0) t^i$$

$$= (1 - t^{a_0}) \sum_{i=0}^{\infty} q_i^0 t^i$$

$$= \frac{\prod_{i=1}^{c} (1 - t^{d_i})}{\prod_{i=1}^{n} (1 - t^{a_i})}.$$

This involves one less $a_i$.

(ii) $d_1 < a_0$. Let $d_1$ occur with multiplicity $\mu$. Then $\mathcal{P}(t) = 1 - \mu t^{d_1} +$higher order terms. So the first nonzero $q_i^0$ is $q_{d_1}^0 = -\mu < 0$. Define $q_i^1 = q_i^0 + q_{i-d_1}^1$, for $i = 1, 2, \ldots$ where $q_i^1 = 0$ if $i ¡ 0$. This corresponds to:

$$\sum_{i=0}^{\infty} q_i^1 t^i = \sum_{i=0}^{\infty} (q_i^0 + q_{i-d_1}^1) t^i$$

$$= \sum_{i=0}^{\infty} (q_i^0 + q_{i-d_1}^0 + q_{i-2d_1}^0 + \cdots) t^i$$

$$= \frac{\prod_{i=2}^{c} (1 - t^{d_i})}{\prod_{i=0}^{n} (1 - t^{a_i})}.$$

This involves one less $d_i$.

Repetition of the above steps clearly terminates when

$$\sum_{i=0}^{\infty} q_i^b t^i = 1$$

By induction on the number of $a_i$ and $d_j$ it is clear that the process uniquely determines the $a_i$ and $d_j$. $\square$

**18.4  The table method** The proof of the above lemma allows us to construct a weighted complete intersection from a list of 'plurigenera'. This construction is easily set out in the form of a table. In the first column write down the integers $\{0, 1, 2, \dots\}$ and in the second the list $\{1, P_1, P_2, \dots\}$. Let the $n$th column be denoted by $q_i^n$ for $i = 0, 1, \dots$. Each successive column is obtained as follows. Look down the list $\{q_i^n\}$ of the $n$th column to find the position of the first nonzero entry (disregard the initial 1 at the top of the column). Suppose this is in row $r$. There are 2 cases:

(i) this entry is positive. First enter $(r)$ at the head of this column. This will keep a record of the degrees of the generators. The $(n+1)$th column is obtained by the rule:

$$q_i^{n+1} = q_i^n - q_{i-r}^n,$$

assuming that $q_i^n = 0$ for all $i < 0$.

(ii) this entry is negative. First enter $(-r)$ at the head of this column. This will keep a record of the degrees of the relations. The $(n+1)$st column is obtained by the rule:

$$q_i^{n+1} = q_i^n - q_{i-r}^{n+1},$$

assuming that $q_i^{n+1} = 0$ and for all $i < 0$.

The process is clearly defined and the integers at the head of each column keep track of the $a_i$ and $-d_i$.

**18.5  Example** Consider the record of pluridata $K^3 = \frac{1}{6}$, $\chi = 1$, $p_g = 0$, 9 singularities of type $\frac{1}{2}(1, 1, 1)$ and 8 singularities of type $\frac{1}{3}(2, 1, 1)$. Using Reid's plurigenus formula (see Section 17) the plurigenera $P_n$ corresponding to this record was calculated and is given below. The table obtained is the following:

| $n$ | $P_n$ | (2) | (2) | (3) | (3) | (3) | (4) | (−6) | (−12) |
|---|---|---|---|---|---|---|---|---|---|
| 0 | 1 | 1 | 1 | 1 | 1 | 1 | 1 | 1 | 1 |
| 1 | 0 | 0 | 0 | 0 | 0 | 0 | 0 | 0 | 0 |
| 2 | 2 | 1 | 0 | 0 | 0 | 0 | 0 | 0 | 0 |
| 3 | 3 | 3 | 3 | 2 | 1 | 0 | 0 | 0 | 0 |
| 4 | 4 | 2 | 1 | 1 | 1 | 1 | 0 | 0 | 0 |
| 5 | 6 | 3 | 0 | 0 | 0 | 0 | 0 | 0 | 0 |
| 6 | 11 | 7 | 5 | 2 | 0 | −1 | −1 | 0 | 0 |
| 7 | 12 | 6 | 3 | 2 | 1 | 0 | 0 | 0 | 0 |
| 8 | 19 | 8 | 1 | 1 | 1 | 1 | 0 | 0 | 0 |
| 9 | 25 | 13 | 7 | 2 | 0 | 0 | 0 | 0 | 0 |
| 10 | 32 | 13 | 5 | 2 | 0 | −1 | 0 | 0 | 0 |
| 11 | 41 | 16 | 3 | 2 | 1 | 0 | 0 | 0 | 0 |
| 12 | 54 | 22 | 9 | 2 | 0 | 0 | −1 | −1 | 0 |
| 13 | 64 | 23 | 7 | 2 | 0 | 0 | 0 | 0 | 0 |
| 14 | 81 | 27 | 5 | 2 | 0 | −1 | 0 | 0 | 0 |
| 15 | 98 | 34 | 11 | 2 | 0 | 0 | 0 | 0 | 0 |
| 16 | 117 | 36 | 9 | 2 | 0 | 0 | 0 | 0 | 0 |
| 17 | 139 | 41 | 7 | 2 | 0 | 0 | 0 | 0 | 0 |
| 18 | 166 | 49 | 13 | 2 | 0 | 0 | 1 | 0 | 0 |
| 19 | 191 | 52 | 11 | 2 | 0 | 0 | 0 | 0 | 0 |
| 20 | 224 | 58 | 9 | 2 | 0 | 0 | 0 | 0 | 0 |

This gives $X_{6,12}$ in $\mathbb{P}(2,2,3,3,3,4)$, which has the above record.

**18.6 Note** Of course this method cannot tell the difference between $X_6$ in $\mathbb{P}(1,1,1,2)$ and the example of V. Iliev $X_{3,6}$ in $\mathbb{P}(1,1,1,2,3)$, in which the cubic relation does not involve the degree 3 generator.

However in this section we are only interested in the general member of a family of weighted complete intersections and so Iliev's example does not occur.

**18.7 Warning** Although in general it is clear when this process stops, it is not clear when it is worth continuing with a particular list of integers.

**18.8 The analysis** This process is basically the same as that in Section 12.6 on the coordinate ring

$$R = \bigoplus_{m \geq 0} R_m.$$

Starting from the dimensions of each $R_m$ the degrees of the generators and relations can be found. At each stage it is assumed that the monomials are linearly independent unless

either there already exist some relations of a lower degree,

or a relation is forced by the dimension not being large enough.

For the above example we have the following analysis:

| deg | dim | monomials |
|-----|-----|-----------|
| 0 | 1 | 1 |
| 1 | 0 | $\emptyset$ |
| 2 | 2 | $x_0, x_1$ |
| 3 | 3 | $y_0, y_1, y_2$ |
| 4 | 4 | $x_0^2, x_0 x_1, x_1^2, z$ |
| 5 | 6 | $x_0 y_0, x_0 y_1, x_0 y_2, x_1 y_0, x_1 y_1, x_1 y_2$ |
| 6 | 11 | $x_0^3, x_0^2 x_1, x_0 x_1^2, x_1^3, y_0^2, y_0 y_1, y_0 y_2, y_1^2, y_1 y_2, y_2^2, x_0 z, x_1 z$ |
|   |   | 1 relation. |

If this calculation is continued only one more relation is found, of degree 12

**18.9  Canonical 3-fold complete intersections**  The formula:

$$P_2 = \frac{1}{2} K_X^3 - 3(1 - p_g) + l(2)$$

limits the value of $p_g$ (since $K_X^3 > 0$) and defines $K_X^3$ in terms of a particular basket of singularities and $P_2$.

**18.10  Q-Fano complete intersections**  The formula:

$$P_{-1} = -\frac{1}{2} K_X^3 + 3 - l(2)$$

defines $K_X^3$ in terms of a particular basket of singularities and $P_{-1}$.

**18.11  The search**  The search through all combinations of $P \geq 0$ ($P_2 = P$ for canonical 3-folds and $P_{-1} = P$ for the Fano case) and baskets will give every possible list of plurigenera (respectively anti-plurigenera). Hence a list of canonically (respectively anticanonically) embedded complete intersections can be found. Of course this is not a finite search, and requires a computer to make any reasonable progress.

The order of the search was as follows. Let $Q_i$ for $i = 0, 1, \ldots$ be a list of the types of 3-fold cyclic quotient singularity $\frac{1}{r}(1, -1, a)$ in order of increasing index $r$ and increasing $a$ within each index. So $Q_0 = \frac{1}{2}(1, 1, 1)$, $Q_1 = \frac{1}{3}(1, -1, 1)$, etc. The program took 2 integer arguments $l$ and $u$, and searched through all baskets $\{n_i \times Q_i\}$ such that $l \leq \sum_{i=0}^{\infty} n_i(i + 2) < u$.

**18.12   The raw list**   Here is the first part of the list produced by the search program (with arguments $l = 0$, $u = 8$).

| | |
|---|---|
| $X_6 \subset \mathbb{P}(1,1,1,1,3)$ | $X_{12} \subset \mathbb{P}(1,1,1,4,6)$ |
| $X_4 \subset \mathbb{P}(1,1,1,1,1)$ | $X_5 \subset \mathbb{P}(1,1,1,1,2)$ |
| $X_8 \subset \mathbb{P}(1,1,1,2,4)$ | $X_{10} \subset \mathbb{P}(1,1,1,3,5)$ |
| $X_{2,3} \subset \mathbb{P}(1,1,1,1,1,1)$ | $X_{3,3} \subset \mathbb{P}(1,1,1,1,1,2)$ |
| $X_{3,4} \subset \mathbb{P}(1,1,1,1,2,2)$ | $X_6 \subset \mathbb{P}(1,1,1,2,2)$ |
| $X_{4,4} \subset \mathbb{P}(1,1,1,1,2,3)$ | $X_7 \subset \mathbb{P}(1,1,1,2,3)$ |
| $X_9 \subset \mathbb{P}(1,1,1,3,4)$ | $X_{2,2,2} \subset \mathbb{P}(1,1,1,1,1,1,1)$ |
| $X_{6,6} \subset \mathbb{P}(1,1,1,2,3,3)$ | $X_{12} \subset \mathbb{P}(1,1,2,3,4)$ |
| $X_{4,4} \subset \mathbb{P}(1,1,1,2,2,2)$ | $X_{10} \subset \mathbb{P}(1,1,2,2,5)$ |
| $X_{4,5} \subset \mathbb{P}(1,1,1,2,2,3)$ | $X_{18} \subset \mathbb{P}(1,1,2,6,9)$ |
| $X_{4,6} \subset \mathbb{P}(1,1,1,2,3,3)$ | $X_{5,6} \subset \mathbb{P}(1,1,1,2,3,4)$ |
| $X_{6,8} \subset \mathbb{P}(1,1,1,3,4,5)$ | |

**18.13   Refinement**   Of course this list contains complete intersections already obtained in other ways (see Sections 15 and 16) and some intersections which do not meet the requirements; that is,

(1) dimension 3,

(2) quasismooth but not the intersection of a linear cone with other hypersurfaces,

(3) canonically or anticanonically embedded,

(4) and have at worst terminal singularities.

The example $X_{6,22}$ in $\mathbb{P}(2,2,3,4,5,11)$ from the raw list is not quasismooth, since the polynomial of degree 6 does not involve the generator of weight 5.

We use the following lemma to cut out a large number of elements from the raw list produced by the search program.

**18.14   Lemma**   *Let $X_{d_1,\dots,d_c}$ in $\mathbb{P}(a_0,\dots,a_n)$ be quasismooth but not an intersection of a linear cone with other hypersurfaces. Suppose also that $d_1,\dots,d_c$ and $a_0,\dots,a_n$ are in increasing order. Then:*

*(i) $d_c > a_n$, $d_{c-1} > a_{n-1}$, $\dots$, and $d_1 > a_{n-c+1}$.*

*(ii) if $d_{c-1} < a_n$ then $a_n \mid d_c$.*

**Proof** (*i*). Suppose $d_c > a_n, \ldots, d_{c-k+1} > a_{n-k+1}$ and $d_{c-k} < a_{n-k}$ for some $k = 0, \ldots, c - 1$. So $d_i < a_{n-k}$ for all $i \leq c - k$. Therefore the polynomials $f_1, \ldots, f_{n-k}$ do not involve the variables $x_{n-k}, \ldots, x_n$.

Let $\Pi$ be the coordinate $(k + 1)$-plane in $\mathbb{A}^n + 1$ given by $x_0 = \cdots = x_{n-k-1} = 0$. So $f_1, \ldots, f_{n-k}$ are identically zero on $\Pi$. Define $Z = (f_{c-k+1} = \cdots = f_c = 0) \cap \Pi$. Thus $\dim Z \geq 1$ and so $Z - \underline{0}$ is nonempty. Let $Q \in Z - \underline{0}$. Then $\partial f_i / \partial x_j$ are zero at $Q$ for all $i \leq c - k$ and for all $j$. Therefore

$$
\text{rank} \begin{pmatrix} \partial f_1 / \partial x_0(Q) & \cdots & \partial f_1 / \partial x_n(Q) \\ \vdots & & \vdots \\ \partial f_c / \partial x_0(Q) & \cdots & \partial f_c / \partial x_n(Q) \end{pmatrix} \leq k - c.
$$

Thus $Q \in C_X^*$ is singular and so $X$ is not quasismooth.

(*ii*) is treated likewise. $\square$

**18.15 Examples** Therefore a codimension 2 complete intersection $X_{d_1,d_2}$ in $\mathbb{P}(a_1, \ldots, a_n)$, which is quasismooth and not the intersection of a linear cone with another hypersurface, satisfies:

(i) $d_2 > a_n$ and $d_1 > a_{n-1}$.

(ii) if $d_1 < a_n$ then $a_n \mid d_2$.

So this lemma gives extra combinatoric conditions to help remove *nasty* complete intersections.

**18.16 The final list** The program was run between the limits 0 and 32 and gave the following list (after cutting out repetitions and nasty complete intersections):

| | Complete intersection | $K_X^3$ | $p_g$ | Singularities |
|---|---|---|---|---|
| 1. | $X_{2,2,2} \subset \mathbb{P}(1,1,1,1,1,1,1)$ | $-8$ | 0 | |
| 2. | $X_{2,2,4} \subset \mathbb{P}(1,1,1,1,1,1,1)$ | 16 | 7 | |
| 3. | $X_{2,2,6} \subset \mathbb{P}(1,1,1,1,1,1,3)$ | 8 | 6 | |
| 4. | $X_{2,3,3} \subset \mathbb{P}(1,1,1,1,1,1,1)$ | 18 | 7 | |
| 5. | $X_{3,3,3} \subset \mathbb{P}(1,1,1,1,1,1,2)$ | 27/2 | 6 | $\frac{1}{2}(1,-1,1)$ |
| 6. | $X_{3,3,4} \subset \mathbb{P}(1,1,1,1,1,2,2)$ | 9 | 5 | $2 \times \frac{1}{2}(1,-1,1)$ |
| 7. | $X_{3,4,4} \subset \mathbb{P}(1,1,1,1,2,2,2)$ | 6 | 4 | $4 \times \frac{1}{2}(1,1,1)$ |
| 8. | $X_{4,4,4} \subset \mathbb{P}(1,1,1,1,2,2,3)$ | 16/3 | 4 | $\frac{1}{3}(1,-1,1)$ |
| 9. | $X_{4,4,4} \subset \mathbb{P}(1,1,1,2,2,2,2)$ | 4 | 3 | $8 \times \frac{1}{2}(1,-1,1)$ |
| 10. | $X_{4,4,5} \subset \mathbb{P}(1,1,1,2,2,2,3)$ | 10/3 | 3 | $\frac{1}{3}(1,-1,1), 4 \times \frac{1}{2}(1,1,1)$ |
| 11. | $X_{4,4,6} \subset \mathbb{P}(1,1,1,2,2,3,3)$ | 8/3 | 3 | $2 \times \frac{1}{3}(1,1-,1)$ |
| 12. | $X_{4,4,6} \subset \mathbb{P}(1,1,2,2,2,2,3)$ | 2 | 2 | $12 \times \frac{1}{2}(1,1,1)$ |

| | | | |
|---|---|---|---|
| 13. | $X_{4,5,6} \subset \mathbb{P}(1,1,2,2,2,3,3)$ | $5/3$ | $2$ | $2 \times \frac{1}{3}(1,-1,1), 6 \times \frac{1}{2}(1,1,1)$ |
| 14. | $X_{4,6,6} \subset \mathbb{P}(1,1,2,2,3,3,3)$ | $4/3$ | $2$ | $4 \times \frac{1}{3}(1,-1,1)$ |
| 15. | $X_{4,6,6} \subset \mathbb{P}(1,2,2,2,2,3,3)$ | $1$ | $1$ | $18 \times \frac{1}{2}(1,1,1)$ |
| 16. | $X_{5,6,6} \subset \mathbb{P}(1,1,2,2,3,3,4)$ | $5/4$ | $2$ | $\frac{1}{4}(1,-1,1), 4 \times \frac{1}{2}(1,1,1)$ |
| 17. | $X_{5,6,6} \subset \mathbb{P}(1,2,2,2,3,3,3)$ | $5/6$ | $1$ | $4 \times \frac{1}{3}(1,-1,1), 9 \times \frac{1}{2}(1,1,1)$ |
| 18. | $X_{6,6,10} \subset \mathbb{P}(2,2,2,3,3,4,5)$ | $1/4$ | $0$ | $\frac{1}{4}(1,-1,1), 22 \times \frac{1}{2}(1,1,1)$ |
| 19. | $X_{6,6,6} \subset \mathbb{P}(1,2,2,2,3,3,4)$ | $3/4$ | $1$ | $\frac{1}{4}(1,-1,1), 13 \times \frac{1}{2}(1,1,1)$ |
| 20. | $X_{6,6,6} \subset \mathbb{P}(1,2,2,3,3,3,3)$ | $2/3$ | $1$ | $8 \times \frac{1}{3}(1,-1,1)$ |
| 21. | $X_{6,6,6} \subset \mathbb{P}(2,2,2,2,3,3,3)$ | $1/2$ | $0$ | $27 \times \frac{1}{2}(1,1,1)$ |
| 22. | $X_{6,6,7} \subset \mathbb{P}(1,2,2,3,3,3,4)$ | $7/12$ | $1$ | $\frac{1}{4}(1,-1,1), 4 \times \frac{1}{3}(1,-1,1),$<br>$4 \times \frac{1}{2}(1,1,1)$ |
| 23. | $X_{6,6,8} \subset \mathbb{P}(1,1,2,3,3,4,5)$ | $4/5$ | $2$ | $\frac{1}{5}(1,-1,2)$ |
| 24. | $X_{6,6,8} \subset \mathbb{P}(1,2,2,3,3,4,4)$ | $1/2$ | $1$ | $\frac{1}{4}(1,-1,1), 8 \times \frac{1}{2}(1,1,1)$ |
| 25. | $X_{6,6,8} \subset \mathbb{P}(2,2,2,3,3,3,4)$ | $1/3$ | $0$ | $18 \times \frac{1}{2}(1,1,1), 4 \times \frac{1}{3}(1,-1,1)$ |
| 26. | $X_{6,7,8} \subset \mathbb{P}(1,2,2,3,3,4,5)$ | $7/15$ | $1$ | $\frac{1}{5}(1,-1,2), 2 \times \frac{1}{3}(1,-1,1),$<br>$6 \times \frac{1}{2}(1,1,1)$ |
| 27. | $X_{6,8,10} \subset \mathbb{P}(1,2,3,3,4,5,5)$ | $4/15$ | $1$ | $2 \times \frac{1}{5}(1,-1,2), 2 \times \frac{1}{3}(1,-1,1)$ |
| 28. | $X_{6,8,10} \subset \mathbb{P}(2,2,3,3,4,4,5)$ | $1/6$ | $0$ | $2 \times \frac{1}{4}(1,-1,1), 2 \times \frac{1}{3}(1,-1,1)$<br>$14 \times \frac{1}{2}(1,1,1)$ |
| 29. | $X_{6,8,9} \subset \mathbb{P}(1,2,3,3,4,4,5)$ | $3/10$ | $1$ | $\frac{1}{5}(1,-1,2), 2 \times \frac{1}{4}(1,-1,1),$<br>$2 \times \frac{1}{2}(1,1,1)$ |
| 30. | $X_{8,10,12} \subset \mathbb{P}(2,3,4,4,5,5,6)$ | $1/15$ | $0$ | $2 \times \frac{1}{5}(1,-1,1), 2 \times \frac{1}{3}(1,-1,1)$<br>$10 \times \frac{1}{2}(1,1,1)$ |
| 31. | $X_{8,9,10} \subset \mathbb{P}(2,3,3,4,4,5,5)$ | $1/10$ | $0$ | $2 \times \frac{1}{5}(1,-1,2), 2 \times \frac{1}{4}(1,-1,1)$<br>$3 \times \frac{1}{3}(1,-1,1), 4 \times \frac{1}{2}(1,1,1)$ |
| 32. | $X_{9,10,12} \subset \mathbb{P}(2,3,3,4,5,6,7)$ | $1/14$ | $0$ | $\frac{1}{7}(1,-1,2), 6 \times \frac{1}{3}(1,-1,1),$<br>$5 \times \frac{1}{2}(1,1,1)$ |
| 33. | $X_{10,11,12} \subset \mathbb{P}(2,3,4,5,5,6,7)$ | $\frac{11}{210}$ | $0$ | $5 \times \frac{1}{2}(1,1,1), 2 \times \frac{1}{3}(1,-1,1),$<br>$2 \times \frac{1}{5}(1,-1,2), \frac{1}{7}(1,-1,3)$ |
| 34. | $X_{10,12,14} \subset \mathbb{P}(2,3,4,5,6,7,8)$ | $1/24$ | $0$ | $\frac{1}{8}(1,-1,3), \frac{1}{4}(1,-1,1),$<br>$2 \times \frac{1}{3}(1,-1,1), 8 \times \frac{1}{2}(1,1,1)$ |
| 35. | $X_{10,12,18} \subset \mathbb{P}(3,4,5,5,6,7,9)$ | $\frac{2}{105}$ | $0$ | $\frac{1}{7}(1,-1,1), 2 \times \frac{1}{5}(1,-1,1),$<br>$4 \times \frac{1}{3}(1,-1,1)$ |
| 36. | $X_{12,14,15} \subset \mathbb{P}(3,4,5,6,7,7,8)$ | $1/56$ | $0$ | $\frac{1}{2}(1,1,1), \frac{1}{4}(1,-1,1),$<br>$2 \times \frac{1}{7}(1,-1,2), \frac{1}{8}(1,-1,3)$ |
| 37. | $X_{12,15,16} \subset \mathbb{P}(3,4,5,6,7,8,9)$ | $1/63$ | $0$ | $2 \times \frac{1}{2}(1,1,1), 3 \times \frac{1}{3}(1,-1,1),$<br>$\frac{1}{7}(1,-1,2), \frac{1}{9}(1,-1,2)$ |
| 38. | $X_{12,16,18} \subset \mathbb{P}(4,5,6,6,7,8,9)$ | $\frac{1}{105}$ | $0$ | $\frac{1}{7}(1,-1,1), \frac{1}{5}(1,-1,1),$<br>$2 \times \frac{1}{3}(1,-1,1), 6 \times \frac{1}{2}(1,1,1)$ |

Table 7: Some codimension 3 complete intersections

**18.17 Note** After refinement there are no codimension 2 or 1 complete intersections left in the list.

**18.18 Extra example** The family of intersections $X_{2,2,2,2,2}$ in $\mathbb{P}^8$ is smooth, $K_X^3 = 16$, $p_g = 9$ and $\chi(\mathcal{O}_X) = -8$. If the above search were continued this family would eventually appear; however my implementation becomes painfully slow.

**18.19 Conjecture**   *(1) There are no canonical complete intersections with codimension greater than 5.*

   *(2) There are no $\mathbb{Q}$-Fano complete intersections with codimension greater than 3.*

**18.20 K3 surfaces** Reid has done a similar search to produce lists of K3 surface weighted complete intersections; using Riemann–Roch for $\mathcal{O}_S(1)$ (see [R4], Theorem 9.1).

   This time the search is finite due to the following theorem pointed out by Reid:

**18.21 Theorem** *Let $S$ be a K3 surface with canonical (Du Val) singularities of types $A_{n_i}$, $D_{n_i}$ or $E_{n_i}$ for $i = 1,\ldots,n$. So $\sum n_i \leq 19$. This limits the singularities present on the K3 surface to a finite list.*

**Proof** Let $f: T \to S$ be a minimal resolution. $T$ is still a K3 surface. By [BPV], Proposition VIII.3.3, $h^{1,1} = h^1(\Omega_T^1) = 20$. By the Signature Theorem [BPV], Theorem IV.2.13, the cup product restricted to $H^2(T,\mathbb{R})$ is nondegenerate of type $(1, h^{1,1} - 1) = (1, 19)$. Via the Néron–Severi group, the exceptional $-2$-curves of the resolution $f$ are linearly independent in $H^{1,1}$, each with negative self-intersection.
   It is well known that a Du Val singularity of type $A_n$, $D_n$ or $E_n$ contributes exactly $n$ $-2$-curves to $T$. Thus $\sum n_i \leq 19$. □

# References

[ACGH] E. Arbarello, M. Cornalba, P. A. Griffiths, J. Harris: *Geometry of algebraic curves,* Vol. 1, Comprehensive studies in Math, **267**, (1985) Springer Verlag.

[AM] M. F. Atiyah, I. G. MacDonald: *Introduction to commutative algebra,* Addison Wesley Publishing Co. 1969.

172    *Working with weighted complete intersections*

[BPV]  W. Barth, C. Peters, A. Van de Ven: *Compact complex surfaces*, (1984) Springer Verlag.

[Be]  D. N. Bernshtein: *The number of roots of a system of equations*, Funk. Anal **9** no. 3 (1975) pp. 1–4.

[Da]  V. I. Danilov: *The geometry of toric varieties*, Uspekhi Mat. Nauk. **33** no. 2 (1978) pp. 85–134, English transl. Russian Math Survey **33** no. 2 (1978) pp. 97–154.

[De]  C. Delorme: *Espaces projectifs anisotropes*, Bull. Soc. math. France, **103** (1975) pp. 203–223.

[Di]  A. Dimca: *Singularities and coverings of weighted complete intersections*, Journal reine u. angew. Math. **366** (1986) pp. 184–193.

[Du]  A. Durfee: *Fifteen characterizations of rational double points and simple critical points*, Ens. math **25** no. 2 (1979) pp. 131–163.

[EGA]  A. Grothendieck: *Elements de géométrie algébrique* Chap. II. Publ. Math. de l'IHES **8** (1961).

[F1]  A. R. Fletcher: *Contributions to Riemann–Roch on projective 3-folds with only canonical singularities and applications*, Proc. Symposia in Pure Math **46** (1987) Vol 1, pp. 221–231.

[F2]  A. R. Fletcher: *Plurigenera of 3-folds and weighted hypersurfaces*, Thesis submitted to the University of Warwick for the degree of Ph.D. (1988).

[Hart]  R. Hartshorne: *Algebraic geometry*, Grad. Texts in Mathematics **52**, Springer Verlag 1977.

[Ku]  A. G. Kushnirenko: *Newton polytopes and the Bezout's theorem*, Funk. Anal. **10** no. 3 (1976) pp. 82–83

[M]  D. Mumford: *Curves and their Jacobians*, University of Michigan Press, 1975.

[OW]  P. Orlick, P. Wagreich: *Equivariant resolution of singularities with $\mathbb{C}^*$-actions*, Proc. Second Conference on Compact Transformation Groups, Part I, LNM **298** (1971), Springer Verlag.

[R1]  M. Reid: *Canonical 3-folds*, Proc. Alg. Geom. Anger 1979. Sijthoff and Nordhoff, pp. 273–310.

[R2]   M. Reid: *Minimal models of canonical 3-folds*, Algebraic varieties and analytic varieties, Adv. Stud. Pure Math., North Holland, Amsterdam and Kinokuniya Book Co., Tokyo, Math vol. 1, 1983 pp. 131–180.

[R3]   M. Reid: *Tendencious survey of 3-folds*, Proc. Symposia in Pure Math **46** (1987) Vol 1, pp. 333–344.

[R4]   M. Reid: *Young person's guide to canonical singularities*, Proc. Symposia in Pure Math **46** (1987) Vol 1, pp. 345–414.

[S]    J. H. M. Steenbrink: *Mixed Hodge structures and singularities: a survey*, Géométrie algébrique et applications, III (La Rábida, 1984), Travaux en Cours, **24**, Hermann, Paris, 1987, pp. 99–123

[WPS] I. Dolgachev: *Weighted projective spaces*, Group actions and vector fields, Proc. Vancouver 1981, LNM **956**, pp. 34–71 Springer Verlag.

Anthony Iano-Fletcher,
Computational Bioscience and Engineering Laboratory,
Centre for Information Technology, Building 12A, Room 2033,
National Institutes of Health, 9000 Rockville Pike,
Bethesda, MD 20892-5624, USA.
e-mail: Anthony.Iano-Fletcher@cbel.cit.nih.gov
web: cbel.cit.nih.gov/~arif/cv.html

# Fano 3-fold hypersurfaces

Alessio Corti       Aleksandr Pukhlikov       Miles Reid

### Abstract

We study the birational geometry of the 95 families of Fano 3-fold weighted hypersurfaces $X = X_d \subset \mathbb{P}(1, a_1, a_2, a_3, a_4)$, corresponding to the famous 95 families of K3 surfaces $X_d \subset \mathbb{P}(a_1, a_2, a_3, a_4)$ of Reid and Fletcher ([C3-f, §4] and [Fl, 13.3]). Our main aim is to prove a rigidity theorem for the general $X_d$ in each family, by analogy with the famous theorem of Iskovskikh and Manin on the rigidity of the quartic 3-fold; on the way, we derive as much instruction and amusement as possible on topics in biregular and birational geometry of Fano 3-folds and Mori fibre spaces. While this paper uncovers an amazing wealth of new phenomena and methods of calculation, many of the basic questions remain open, and we spell some of these out.

## Contents

1 **Simple-minded introduction**                                   177

2 **Cubic surface**                                                 179
   2.2  Set-up . . . . . . . . . . . . . . . . . . . . . . . . . . . 180
   2.4  Exclusion  . . . . . . . . . . . . . . . . . . . . . . . . . 180
   2.6  Construction of Geiser and Bertini involutions . . . . . . . 181
        2.6.1  "Italian" . . . . . . . . . . . . . . . . . . . . . . 181
        2.6.2  "Russian"  . . . . . . . . . . . . . . . . . . . . . . 181
        2.6.3  Our approach: calculating the anticanonical ring of $Y$ . 182
   2.9  Geiser and Bertini involutions as Sarkisov links, untwisting . . 183
   2.10 Conclusion of the proof of Theorem 2.1 . . . . . . . . . . . 185
   2.11 Discussion: special position and bad links . . . . . . . . . 186

3 **Plan of proof of Theorem 1.3**                                  187
   3.1  Basic notions  . . . . . . . . . . . . . . . . . . . . . . . 187
   3.3  Theorem 3.2 implies Theorem 1.3 . . . . . . . . . . . . . . . 189
   3.4  Divisorial contractions  . . . . . . . . . . . . . . . . . . 190

**4  Untwisting**                                                                 **193**

   4.1   Introduction to involutions . . . . . . . . . . . . . . . . . . . . 193

   4.4   Quadratic involutions . . . . . . . . . . . . . . . . . . . . . . . 194

   4.10  Elliptic involutions  . . . . . . . . . . . . . . . . . . . . . . . . 198

**5  Excluding**                                                                  **202**

   5.2   The test class method, and curves . . . . . . . . . . . . . . . . 204

   5.3   Two variations on the test class method, and nonsingular points207

   5.4   Determination of $\overline{NE}\,Y$, and singular points with $B^3 \le 0$ . . . 212

      5.4.1   Singular points with $B^3 \le 0$ . . . . . . . . . . . . . . 216

   5.5   Bad links, and singular points with $B^3 > 0$ . . . . . . . . . . 217

   5.6   Proof of Theorem 5.3.1: construction of the test class for non-
      singular points  . . . . . . . . . . . . . . . . . . . . . . . . . . 219

   5.7   Singular points: construction of $T$ . . . . . . . . . . . . . . . . 222

      5.7.1   Hyperplane sections and easy cases . . . . . . . . . . . 222

      5.7.2   Harder cases . . . . . . . . . . . . . . . . . . . . . . . . 225

      5.7.3   The amazing example  . . . . . . . . . . . . . . . . . . . 229

**6  The Big Table**                                                              **232**

**7  Final remarks**                                                             **243**

   7.1   Iskovskikh–Manin by our methods . . . . . . . . . . . . . . . . 244

   7.2   Numerics  . . . . . . . . . . . . . . . . . . . . . . . . . . . . . 244

      7.2.1   Predicting the degrees  . . . . . . . . . . . . . . . . . . 244

      7.2.2   Hilbert function . . . . . . . . . . . . . . . . . . . . . . 245

      7.2.3   Restriction from Fano to K3 . . . . . . . . . . . . . . . 245

   7.3   Involutions and projections . . . . . . . . . . . . . . . . . . . . 246

   7.4   Comments on the results . . . . . . . . . . . . . . . . . . . . . 248

      7.4.1   Mini-definition of rigidity . . . . . . . . . . . . . . . . . 248

      7.4.2   The key monomials . . . . . . . . . . . . . . . . . . . . 248

      7.4.3   Comments on Bir $X$  . . . . . . . . . . . . . . . . . . . 249

      7.4.4   Inverse flips . . . . . . . . . . . . . . . . . . . . . . . . 249

   7.5   The Sarkisov category and the rigid boundary . . . . . . . . . 250

   7.6   The spectrum: all the arguments of the rainbow . . . . . . . . 251

   7.7   Remarks relating to the proof of exclusion . . . . . . . . . . . 253

   7.8   Generality . . . . . . . . . . . . . . . . . . . . . . . . . . . . . 254

   7.9   Eccentricity  . . . . . . . . . . . . . . . . . . . . . . . . . . . . 255

**References**                                                                    **257**

# 1 Simple-minded introduction

**Definition 1.1** In this paper a *Fano variety* (formerly ℚ-*Fano variety*) is a projective variety $X$ satisfying

(a) $X$ has at worst ℚ-factorial terminal singularities;

(b) $-K_X$ is ample;

(c) Pic $X$ has rank 1.

The "Italian" and "Russian" traditions usually study Fano 3-folds with (a) replaced by nonsingularity (or assumptions in projective geometry that imply Gorenstein). The class of varieties just defined arises in Mori theory as one of the possible end products of classification: it is the particular case $S =$ pt. of Mori fibre space $X \to S$; the other cases (for 3-folds) are fibrations of del Pezzo surfaces over a curve and conic bundles over a surface (see for example [CKM], Chapter 5, [Ko] or [R]).

Our study does not aspire to this generality; we usually work under the following concrete assumptions, in addition to (a–c):

(d) The plurianticanonical ring of $X$ is a hypersurface:

$$R(X, -K_X) = \bigoplus_{n \geq 0} H^0(X, -nK_X) = k[x_0, \ldots, x_4]/(F).$$

In other words, $X$ is a member of one of the famous 95 families of weighted projective hypersurfaces $X_d \subset \mathbb{P}(1, a_1, a_2, a_3, a_4)$ of Fletcher and Reid (see [Fl], II.6.6).

(e) $X$ has at worst the terminal quotient singularities $\frac{1}{r}(1, a, r-a)$ of [YPG], 5.2, where $r \geq 2$ and $a$ is coprime to $r$; that is, $X$ is quasismooth (see [Fl], I.3.3 and I.5).

**Definition 1.2** A Fano variety $X$ is *rigid* (or *birationally rigid*) if for every Mori fibre space $Y \to S$, the existence of a birational equivalence $\varphi \colon X \dashrightarrow Y$ implies that $Y \cong X$.

Although we do not include it as part of the definition, the only way of proving such a statement that we can conceive of is to show that any $\varphi \colon X \dashrightarrow Y$ factors as a chain of suitable selfmaps $X \dashrightarrow X$ followed by an isomorphism $X \cong Y$ (compare 7.4.1 below).

When it holds, rigidity provides a uniqueness statement in Mori theory: any minimal model program run on any variety birational to $X$ terminates with $X$ itself (up to isomorphism); in other words, the model $X \to$ pt. of $X$

as a Mori fibre space is unique up to isomorphism. In birational geometry, the striking thing about rigidity is that it replaces the *category* of birational maps from $X$ to something else by the *group* of birational selfmaps. Note that Iskovskikh and Manin [IM] prove that a nonsingular quartic hypersurface $X_4 \subset \mathbb{P}^4$ is rigid (compare 7.1 below), and Iskovskikh and his school have extended this to many other classes of nonsingular Fano varieties.

The definition generalises to the strict Mori fibre spaces (with some pain, see [Co2], Definition 1.3), and the results of Sarkisov on conic bundles (see [Sa1]–[Sa2]) and Pukhlikov on del Pezzo fibre spaces (see [Pu2]) are rigidity statements. Our main aim is the following.

**Main Theorem–Conjecture 1.3** *Let $X$ be a Fano 3-fold satisfying (a–e) above. Then $X$ is rigid.*

We prove this for the *general* member of each of the famous 95 families, and conjecture it for *all quasismooth* $X$ (and prove some cases of it). It is definitely false if we allow other terminal singularities, even quite mild ones (see Example 7.5.1 below). We believe that our methods should also prove the remaining conjectural part, but the volume of calculations required for this seems disproportionate to the interest of the result; see 7.4.2 for an example of interesting nongeneral behaviour, and 7.8 for a discussion of what generality means in the theorem.

**Remark 1.4** Our proof includes, and in fact depends on, a description of birational involutions $i \colon X \dashrightarrow X$ that generate the birational group Bir $X$. A remarkable and quite beautiful fact is that these divide into just two types: Type I or *quadratic involutions* and Type II or *elliptic involutions*; see 2.6, Figure 1 for the Geiser and Bertini involutions of a cubic surface, 4.4 and 4.10 for the involutions of 3-folds, and the final discussion in 7.3. This dichotomy seems to extend also to other classes of rigid Fano 3-folds not treated here, and we speculate that it may be a completely general phenomenon of birational geometry.

The proof of the theorem involves hundreds of calculations, including many interesting special cases. A lot of this detail is summarised in the Big Table of Chapter 6. Rather than the statement of the theorem, which takes up just a single line, it is this wealth of fascinating information that we consider to be the essential content of the paper.

Our introductory discussion continues at a technically more sophisticated level in Chapter 3. However, we want to set the scene first with the case of a "minimal" cubic surface, which illuminates almost all the main ideas and methods of this paper in an elementary setting.

# 2    Cubic surface

A minimal nonsingular cubic surface $X = X_3 \subset \mathbb{P}^3$ is an example of a Fano variety as defined in Chapter 1: the conditions (a–c) specialise to the following:

(a)  $X$ is nonsingular;

(b)  $-K_X = \mathcal{O}_X(1)$;

(c)  $\operatorname{Pic} X = \mathbb{Z}$.

Here the final condition (c) means that $X$ is defined over an algebraically nonclosed field $K$, and every curve $C$ on $X$ defined over $K$ is the intersection of $X$ with a hypersurface, $C = F \cap X$. We write $\overline{K}$ for the algebraic closure of $K$.

**Theorem 2.1** *A minimal nonsingular cubic surface $X$ is rigid. More precisely, given*

- *any nonsingular surface $Y$ defined over $K$ which is rational over $\overline{K}$ and minimal over $K$;*

- *any birational map $\varphi \colon X \dashrightarrow Y$ defined over $K$.*

*Then $\varphi$ is a composite of a chain of Geiser and Bertini involutions (defined in 2.6 below) followed by an isomorphism $X \cong Y$.*

The result is essentially due to Castelnuovo around 1900, and has been reworked several times since then, most systematically by Iskovskikh and his school. We give a detailed treatment here, in order to illustrate our methods and ideas, and at the same time to compare and contrast different approaches to birational geometry, from classical projective geometry through to Mori theory and the Sarkisov program. For dramatic tension, we mention here that our argument has a little sting in its tail (see 2.11). The structure of the proof (and that of Main Theorem 1.3 for 3-folds in Chapters 3–5) consists of 4 parts.

(1)  Set-up and the Noether–Fano–Iskovskikh inequalities (Theorem 2.3);

(2)  Exclusion (see 2.4);

(3)  Untwisting: construction of Geiser and Bertini involutions (see 2.6);

(4)  Conclusion (see 2.10): untwisting decreases degree (Lemma 2.9.3) so that the theorem follows by induction.

Of these, the two outer steps apply to very general situations in birational geometry, while the inner two depend on specific properties of $X_3$; these two steps certainly do not work to give the same conclusion, say, in the case of a del Pezzo surface $X_4$ of degree 4, and indeed, $X_4$ is usually not rigid.

## 2.2   Set-up

A birational map $\varphi\colon X \dashrightarrow Y \subset \mathbb{P}^N$ is given by a mobile linear system $\mathcal{H}$ (that is, a linear system without fixed part), defined over $K$. Write $A = -K_X = \mathcal{O}_X(1)$ for the positive generator of $\operatorname{Pic} X$ (see assumption (c)), to avoid confusion about signs. Suppose $\mathcal{H} \subset |nA|$. If $n = 1$ then $\varphi$ is linear, and hence must be an isomorphism; this is the basis of an induction.

**Theorem 2.3 (Noether–Fano–Iskovskikh inequalities)** *If $\varphi$ is not an isomorphism then $K_X + \frac{1}{n}\mathcal{H}$ is not canonical.*

This result is completely general (see [Co1], Theorem 4.2 and compare [IM], §2): if a mobile linear system $\mathcal{H}$ on a Fano variety $X$ defines a birational map $\varphi\colon X \dashrightarrow Y$ to another Mori fibre space, not an isomorphism, then $K_X + \lambda\mathcal{H}$ has a noncanonical singularity, where $\lambda \in \mathbb{Q}$ is defined by $K_X + \lambda\mathcal{H} = 0$: roughly, *not canonical* says that the base locus of $\mathcal{H}$ has some points with big multiplicity compared to the class of $\mathcal{H}$ (see Definition 3.1.1 below for the correct statement). In the case of surfaces, it takes on the very concrete form:

$$K_X + \frac{1}{n}\mathcal{H} \text{ is not canonical} \iff \begin{cases} \mathcal{H} \text{ has a base point of} \\ \text{multiplicity } m > n. \end{cases}$$

## 2.4   Exclusion

The aim is to impose restrictions on the maximal singularities of $\mathcal{H}$. Since $\mathcal{H}$ is defined over $K$, its base locus is also defined over $K$. Now, whereas a point over $\overline{K}$ is simply a geometric point, a point over $K$ has a degree, which is the number of points into which it splits up over $\overline{K}$; for example, a point $P$ with $\deg P = 2$ is a pair $P = P_1 + P_2$ of points with coefficients in $\overline{K}$, but conjugate over $K$.

**Lemma 2.5** *A base point of $\mathcal{H}$ of multiplicity $m > n$ has degree 1 or 2.*

**Proof**   Since $A^2 = \deg X = 3$ and $\mathcal{H} \subset |nA|$, we have $\mathcal{H}^2 = 3n^2$. On the other hand, since $\mathcal{H}$ is a mobile linear system, its selfintersection is at least its local selfintersection at the base points, so that $3n^2 = \mathcal{H}^2 \geq \deg P \cdot m^2 > \deg P \cdot n^2$. Therefore $\deg P < 3$. Q.E.D.

## 2.6   Construction of Geiser and Bertini involutions

The aim is to find some birational selfmap $X \dashrightarrow X$ to blow up any point $P$ with $\deg P = 1$ or $2$. We want to describe these constructions in three different languages.

### 2.6.1   "Italian"

We only need a birational map; the construction is geometric (see Figure 1). First, linear projection from a point $P$ of degree 1 expresses $X$ as a rational double cover of $\mathbb{P}^2$. The birational involution exchanging the sheets in this double cover is called the *Geiser involution* $i_P$.

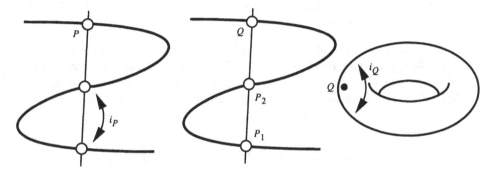

Figure 1: Geiser and Bertini involutions

Next, if $P$ is a point of degree 2 in $X$ then there is a unique line $L$ of $\mathbb{P}^3$ through $P$, meeting $X$ at a third point $Q$. Thus the linear system of plane sections of $X$ through $P$ is a pencil of elliptic curves, each with the marked point $Q$. The generic fibre $F$ of the pencil has a group law with $Q$ as the zero, which defines a reflection $x \mapsto -x$ of $F$, and hence of $X$. This involution is called the *Bertini involution* $j_P$.

(The terms Geiser and Bertini involution are traditionally used to describe birational involutions of $\mathbb{P}^2$ corresponding to the system of cubics through 7 or 8 marked points. For cubic surfaces, the Geiser involution is an obvious projective construction, but the Bertini involution is considerably more subtle. Geiser was a German working around 1860, whereas Bertini was an Italian who continued writing into the 1920s.)

### 2.6.2   "Russian"

We want to add to the birational map information telling us which linear system defines it, and its effect on the curves and points of $X$. The Geiser involution $i_P$ is defined by the linear system $\mathcal{H} = |2A - 3P| = |\mathcal{I}_P^3 \mathcal{O}_X(2)|$.

Let $Y \to X$ be the blowup of $P$, and $E$ the exceptional $-1$-curve. Then $i_P$ defines a biregular automorphism of $Y$, interchanging $E$ and the birational transform of the curve $C = T_P X \cap X$, the unique curve in $|A - 2P|$. Thus the effect of $i_P$ on $\operatorname{Pic} Y$ is given by

$$\begin{array}{l} A \mapsto 2A - 3E, \\ E \mapsto A - 2E, \end{array} \quad \text{that is, by the matrix} \quad \begin{pmatrix} 2 & -3 \\ 1 & -2 \end{pmatrix}.$$

**Exercise 1** Check the following:

$$(2A - 3E)^2 = 3, \quad (A - 2E)^2 = -1, \quad (2A - 3E)(A - 2E) = 0,$$

a curve in $|2A - 3E|$ is elliptic, a curve in $A - 2E$ is rational, and finally, the matrix $\begin{pmatrix} 2 & -3 \\ 1 & -2 \end{pmatrix}$ has square $= 1$.

The Bertini involution $j_P$ is just a bit more tricky. It is given by $\mathcal{H} = |5A - 6E|$, and takes $P$ into $C = |4A - 5E|$, which is a curve over $K$. Over $\overline{K}$, it splits as $C = C_1 + C_2$, where $C_1, C_2$ are rational curves respectively in $|2A - 3P_1 - 2P_2|$ and $|2A - 2P_1 - 3P_2|$, conjugate over $K$, and meeting only at $P = P_1 + P_2$. The corresponding action on $\operatorname{Pic} Y$ is given by

$$\begin{array}{l} A \mapsto 5A - 6E, \\ E \mapsto 4A - 5E, \end{array} \quad \text{that is, by the matrix} \quad \begin{pmatrix} 5 & -6 \\ 4 & -5 \end{pmatrix}.$$

**Exercise 2** Make up your own exercise by analogy with the above.

### 2.6.3 Our approach: calculating the anticanonical ring of $Y$

To express our constructions in their natural generality, in this section 2.6.3 only, we suppose that $X$ is a (not necessarily minimal) cubic surface, and $P \in X$ a point of degree 1 or 2, or $P = P_1 + P_2$ a pair of points; if $\deg P = 2$, we assume that $P$ is not contained in a line of $X$. Write $\sigma \colon Y \to X$ for the blowup of $P$. Consider the *anticanonical ring* (or plurianticanonical ring) of $Y$, defined by

$$R(Y, -K_Y) = \bigoplus_{n \geq 0} H^0(Y, -nK_Y).$$

**Lemma 2.7** *The anticanonical linear system* $|A| = |\mathcal{O}_X(1)|$ *contains an irreducible curve $E$ passing through $P$ and nonsingular there. Moreover:*

*If $\deg P = 1$ then $R(Y, -K_Y) = k[x_1, x_2, x_3, y]/F_4$, where $\deg x_i = 1$ and $\deg y = 2$; that is, $\operatorname{Proj} R(Y, -K_Y)$ is a weighted hypersurface $Z_4 \subset \mathbb{P}(1, 1, 1, 2)$.*

*If* $\deg P = 2$ *then* $R(Y, -K_Y) = k[x_1, x_2, y, z]/F_6$, *where* $\deg x_i = 1$, $\deg y = 2$ *and* $\deg z = 3$; *that is,* $\operatorname{Proj} R(Y, -K_Y)$ *is a weighted hypersurface* $Z_6 \subset \mathbb{P}(1,1,2,3)$.

*Thus in either case, a multiple* $|nA|$ *of the anticanonical system defines a birational map* $f\colon Y \dashrightarrow Z$ *(in fact, a morphism) to a weighted hypersurface with a naturally occurring biregular automorphism* $i_Z$. *We deduce a birational involution* $i_X$ *of* $X$, *the conjugate of* $i_Z$ *by* $X \dashrightarrow Y \dashrightarrow Z$; *this is the Geiser or Bertini involution of* $X$ *corresponding to* $P$. *How to use this involution* $i_X$ *to untwist any birational map* $\varphi\colon X \dashrightarrow Y$ *is explained in 2.10 below.*

**Proof**  The linear system $|A - P| = |\mathcal{I}_P \mathcal{O}_X(1)|$ of hyperplane sections of $X$ through $P$ is mobile, so by Bertini's theorem its general element is irreducible. For any point $P'$ of $X(\overline{K})$ defined over $\overline{K}$, there is exactly one element of $|A|$ singular at $P'$, namely the tangent hyperplane section $T_{P',X} \cap X$. Since $|A-P|$ has dimension $\geq 1$, its general element $E$ is not the tangent hyperplane section for any $P' \in P$, and so $E$ is nonsingular at $P$.

The divisor $A_{|E}$ is a divisor of degree $3 - \deg P = 2$ or $1$ on an elliptic curve, so we know how to calculate the graded ring

$$R(E, A_{|E}).$$

Also $R(Y, A) \twoheadrightarrow R(E, A_{|E})$ is surjective with kernel the principal ideal generated by the equation $x_1$ of $E$ in $Y$. This completes the proof.  Q.E.D.

**Remark 2.8**  In complete generality, we don't know too much about an anticanonical ring – it is not birationally invariant, and there is no particular reason for it to be finitely generated. However, in this paper, the only method we use to get a birational involution is to calculate the anticanonical ring of a suitable blowup $Y \to X$, and to observe that, as here, it naturally involves a quadratic extension of rings. We obtain birational maps $X \leftarrow Y \to Z$, where $Z$ is a Fano hypersurface having a biregular involution. Compare Chapter 4 for many examples.

## 2.9  Geiser and Bertini involutions as Sarkisov links, untwisting

Let $X$ be a minimal cubic surface and $P$ an irreducible point of degree 1 or 2. If $\deg P = 2$ then the minimality assumption (c) implies that the line of $\mathbb{P}^3$ through $P$ is not contained in $X$, so that the construction of the Geiser or Bertini involution in 2.6 applies. Write $Y \to X$ for the blowup of $P$. We now describe the involutions in terms of Mori theory on $Y$, and explain in what sense they *untwists* $P$ as a maximal centre of a linear system $\mathcal{H}$.

**Lemma 2.9.1** $-K_Y$ *is ample.*

It follows that $Y = Z$ (in the notation of Lemma 2.7), and has a biregular involution $i_Z$.

**Proof**  First, $-K_Y^2 = 2$ or $1$. Also $K_Y = -A + E$ and $\mathcal{H}_1 = nA - mE$ gives

$$-K_Y = A - E = \frac{m-n}{m}A + \frac{1}{m}\mathcal{H}_1.$$

Now $A$ and $\mathcal{H}_1$ are both nef, $A\Gamma = 0$ if and only if $\Gamma = E$ is the exceptional curve of $Y \to X$, and $\mathcal{H}_1 E > 0$.   Q.E.D.

**Lemma 2.9.2** $i_Z(E)$ *is an extremal rational curve distinct from $E$, so that the Mori cone $\overline{\mathrm{NE}}\,Y$ is spanned by two extremal rays as in Figure 2.*

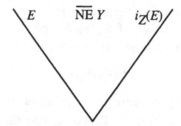

Figure 2: The Mori cone $\overline{\mathrm{NE}}\,Y$

**Proof**  Because $i_Z$ is a biregular involution, and $E$ is an extremal rational curve, so is $i_Z(E)$. The divisor $E + i_Z(E)$ is a pullback from an ample divisor of weighted projective space, so has positive selfintersection. Thus $i_Z(E) = E$ is impossible.   Q.E.D.

**Lemma 2.9.3** *Let $\mathcal{H} \subset |nA|$ be a mobile linear system on $X$ having $P$ as a base point of multiplicity $m > n$, and write $\mathcal{H}' = (i_P)_*\mathcal{H}$ for its birational transform under the involution $i_P$. Then $\mathcal{H}' \subset |n'A|$ with $n' < n$, or in other words, $\deg \mathcal{H}' < \deg \mathcal{H}$.*

**Proof**  Write $\mathcal{H}_1$ for the mobile linear system on $Y$ obtained as the birational transform of $\mathcal{H}$. Then $K_X + \frac{1}{n}\mathcal{H} = 0$ on $X$, but

$$K_Y + \frac{1}{n}\mathcal{H}_1 = -\frac{m-n}{n}E < 0$$

is not quasieffective. In other words, the quasieffective threshold for $\mathcal{H}_1$ is certainly $> \frac{1}{n}$:

$$\lambda_0 := \inf\{\lambda \in \mathbb{R} \mid K_Y + \lambda\mathcal{H}_1 \text{ is quasieffective}\} > \frac{1}{n}. \tag{1}$$

We may view our involution $X \dashrightarrow X$ as the end product of a MMP on $Y$ directed by $\mathcal{H}_1$. For this, start from $K_Y + \lambda\mathcal{H}_1$ with $\lambda \gg 0$, and decrease $\lambda$ until we hit a ray $Q$ of $\overline{NE}\,Y$ with $(K_Y + \lambda\mathcal{H}_1)Q = 0$; necessarily $Q \neq R$, since

$$(K_Y + \frac{1}{m}\mathcal{H}_1)R = 0 \quad \text{and} \quad \lambda > \frac{1}{n} > \frac{1}{m}.$$

In the present case, this ray is the exceptional curve $i_Z(E)$, and we contract it to get back to the surface $X$. This contraction does not change the quasieffective threshold $\lambda_0$, so that $K_X + \lambda_0\mathcal{H}' = 0$, or $\mathcal{H}' \subset |n'A|$ with $n' = 1/\lambda_0 < n$. Q.E.D.

**Remark 2.9.4** Of course, the lemma also follows from the Russian matrix of 2.6.2: it is an easy exercise to see that the degree $n'$ of $\mathcal{H}'$ and its multiplicity $m'$ at the contracted point $P$ are

$$\begin{cases} n' = 2n - m \text{ and } m' = 3n - 2m & \text{for the Geiser involution,} \\ n' = 5n - 4m \text{ and } m' = 6n - 5m & \text{for the Bertini involution;} \end{cases}$$

here $m' = \mathcal{H}_1 \cdot i_Z(E)$, where $i_Z(E) = A - 2E$ (respectively $4A - 5E$) is the exceptional curve determining the nef threshold $\lambda = 1/m'$ in the above argument. Note that $m > n > n' > m'$, so that after the involution, $\mathcal{H}'$ is canonical at $P$: the involution untwists $\mathcal{H}$ both by lowering its degree, and by diluting its noncanonical singularity.

We want to take this opportunity to stress that this traditional and oft-repeated numerical argument is completely unnecessary. While coarser and less precise, the purely adjunction theoretic argument of Lemma 2.9.3 uses essentially no specific geometric properties of $X$, is more general, and extends essentially without change to Sarkisov links in higher dimensions. Compare 4.2 for the higher dimensional version and 7.6 for more discussion.

## 2.10 Conclusion of the proof of Theorem 2.1

Under the assumptions of Theorem 2.1, a birational map $\varphi\colon X \dashrightarrow Y$ is given by a linear system $\mathcal{H} \subset |nA|$. By the NFI inequalities (Theorem 2.3), if $\varphi$ is not an isomorphism then $\mathcal{H}$ has multiplicity $m > n$ at a base point $P$ of degree 1 or 2. Write $i_P$ for the Geiser or Bertini involution. The composite $\varphi \circ i_P\colon X \dashrightarrow Y$ is defined by the linear system $(i_P)_*\mathcal{H}$, which by Lemma 2.9.3 has degree $n' < n$. By induction on $n$, we can assume that $\varphi \circ i_P$ is a composite of Geiser or Bertini involutions, hence so is $\varphi$. Q.E.D.

## 2.11    Discussion: special position and bad links

The Geiser or Bertini involutions in 2.6 are defined as rational maps even for $P$ in *special position*, for example, if $P$ is a geometric point lying on a line of $X_3$; compare, for example, [Isk]. Special position usually contradicts minimality (say, by singling out one or more of the lines of $X_3$), but not always: it can happen, for example, that $X$ is minimal over $K$, but has an Eckardt point $P$ defined over $K$, with 3 conjugate lines out of it. At first sight, it might appear that our proof goes wrong in such cases; at least, it certainly does not work to give the same conclusion: indeed, $-K_Y$ is not ample, the anticanonical model $Y \to Z$ contracts an extremal ray spanned by $-2$-curves, and the two rays of $\overline{NE}\,Y$ are not at all symmetric. Nevertheless, $i_Z$ is defined, and hence it induces a birational selfmap of $Y$.

The solution to our puzzle is striking: for a minimal cubic surface $X$, a Geiser or Bertini involution centred at a point $P$ in special position is *biregular*. The point is that the composite rational map

$$i_Y \colon Y \to Z \to Z \leftarrow Y$$

is clearly biregular outside the $-2$-curves of $Y$, and it is also biregular at the $-2$-curves by the usual argument from absolute minimal models of surfaces. The involution $i_Y$ acts by the identity on $\operatorname{Pic} Y$ and on $\overline{NE}\,Y$.

The fact that the Geiser involution centred at an Eckardt point is biregular is easily verified by a direct calculation in $\mathbb{P}^3$, and was probably known classically; the earliest reference we know is B. Segre [S], pp. 145 ffoll. (Section XIV, Autoprojective nonsingular cubic surfaces). It was rediscovered by Iskovskikh in the 1960s [Isk]. However, we first became aware of it in the context of the current argument via Mori theory.

**Remark 2.12** In this case, our proof implies that such a point $P$ is not a maximal singular centre of any linear system $\mathcal{H}$: if it were, by the Sarkisov program, we could start untwisting it by taking the blowup $Y \to X$ and running a MMP on $Y$; however, the shape of $\overline{NE}$ means that this can only lead back to $X$ via the identity or the Geiser or Bertini involution $i_P$, which is biregular, so doesn't untwist anything. This is a contradiction.

The argument here is an "exclusion" argument quite different from that of 2.4. Recall that in 2.4, if $P$ is a maximal singularity, we calculated intersection numbers to prove that $\deg P \le 2$. In contrast, the argument just given is not at all numerical; it proceeds by attempting to construct a Sarkisov link, then arriving at a contradiction. The only possible ray of $\overline{NE}\,Y$ to contract is a $-2$-curve, which is not a Mori extremal ray; contracting it would take us outside the Mori category of varieties with terminal singularities. We will see this principle in action in much less trivial cases in the "bad link" section 5.5 below. Compare the discussion in 7.6

# 3   Plan of proof of Theorem 1.3

We now give a concise statement that implies our main result Theorem 1.3. We first recall some definitions from the Sarkisov program [Co1]–[Co2] so far only hinted at in Chapters 1–2.

## 3.1   Basic notions

**Definition 3.1.1 (canonical threshold)** $X$ is a variety, $\mathcal{H}$ a mobile linear system, and $\tilde{X} \to X$ a resolution with exceptional divisors $E_i$. As usual, we write

$$K_{\tilde{X}} = K_X + \sum a_i E_i,$$
$$\tilde{\mathcal{H}} = \mathcal{H} - \sum m_i E_i,$$

to define the *discrepancies* $a_i = a_{E_i}(K_X)$ of the exceptional divisors $E_i$ and their *multiplicities* $m_i$ in the base locus of $\mathcal{H}$. For $\lambda \in \mathbb{Q}$, we say that $K_X + \lambda \mathcal{H}$ is *canonical* if all $\lambda m_i \le a_i$, so that $K_{\tilde{X}} + \lambda \tilde{\mathcal{H}} - (K_X + \lambda \mathcal{H})$ is effective ($\ge 0$). Then we define the *canonical threshold* to be

$$c = c(X, \mathcal{H}) = \max\{\lambda \mid K + \lambda \mathcal{H} \text{ is canonical}\}$$
$$= \min_{E_i}\{a_i/m_i\}.$$

This is well defined, independently of the resolution $\tilde{X}$. In all the cases we're interested in, $K_X + \frac{1}{n}\mathcal{H} = 0$ and $c < \frac{1}{n}$.

**Definition 3.1.2 (maximal singularity)** Now suppose that $K_X + \frac{1}{n}\mathcal{H} = 0$ and $K_X + \frac{1}{n}\mathcal{H}$ is not canonical, so that $c < \frac{1}{n}$. We make the following definitions:

(1) a *weak maximal singularity* of $\mathcal{H}$ is a valuation $v_E$ of $k(X)$ for which $m_E(\mathcal{H}) \ge n a_E(K_X)$;

(2) a *strong maximal singularity* is an extremal extraction $Y \to X$ in the Mori category (see Definition 3.4.1) having exceptional divisor $E$ with $c = a_E(K_X)/m_E(\mathcal{H})$.

In either case, the image of $E$ in $X$, or the centre $C(X, v_E)$ of the valuation $v_E$, is called the *centre* of the maximal singularity $E$.

**Remark 3.1.3**   (a) Weak maximal singularity is the traditional definition used by the Moscow school, and is birational in spirit, whereas strong maximal singularity is a biregular construction. It is a seed that always grows into a Sarkisov link (by applying the MMP as in [Co1], Section 5).

The logic of Definition 3.1.2, and the part it plays in the paper is as follows:

(b) If a birational map $\varphi\colon X \dashrightarrow Y$ between Mori fibre spaces is not an isomorphism, the NFI inequalities give the existence of a *weak* maximal singularity.

(c) If a weak maximal singularity exists, the argument of [Co1], Proposition 2.10 (based on running a MMP on a resolution $\tilde{X}$) constructs a *strong* maximal singularity.

(d) The definition of strong maximal singularity contains precise biregular information, in particular on the Mori cone $\overline{NE}\,Y$ of the extraction $Y \to X$, that can be used as the basis of biregular methods of argument.

(e) To summarise, we distinguish three points of view: a *centre* is a point or curve $\Gamma \subset X$; a *weak singularity* is a valuation of $k(X)$; a *strong singularity* is an extremal extraction $(E \subset Y) \to (\Gamma \subset X)$.

**Definition 3.1.4 (Sarkisov link)** A *Sarkisov link of Type II* between two Fano 3-folds $X$ and $X'$ (Definition 1.1) is a birational map $f\colon X \dashrightarrow X'$ that factors as

$$
\begin{array}{ccc}
X_1 & \dashrightarrow & X_1' \\
\downarrow & & \downarrow \\
X & \overset{f}{\dashrightarrow} & X'
\end{array}
$$

where

(a) $X_1 \to X$ and $X_1' \to X'$ are extremal divisorial contractions in the Mori category, and

(b) $X_1 \dashrightarrow X_1'$ is a composite of inverse flips, flops and flips (in that order), and in particular, is an isomorphism in codimension 1.

**Definition 3.1.5 (degree of $\varphi$)** Suppose that $X$ is a Fano 3-fold with the property that $A = -K_X$ generates the Weil divisor class group: $\mathrm{WCl}\,X = \mathbb{Z} \cdot A$ (this holds in our case by Lemma 3.5). Let $\varphi\colon X \dashrightarrow Y$ be a birational map to a given Mori fibre space $Y \to S$, and fix a very ample linear system $\mathcal{H}_Y$ on $Y$; write $\mathcal{H} = \mathcal{H}_X$ for the birational transform $\varphi_*^{-1}(\mathcal{H}_Y)$.

The *degree* of $\varphi$, relative to the given $Y$ and $\mathcal{H}_Y$, is the natural number $n = \deg \varphi$ defined by $\mathcal{H} = nA$, or equivalently $K_X + \frac{1}{n}\mathcal{H} = 0$.

**Definition 3.1.6 (untwisting)** Let $\varphi\colon X \dashrightarrow Y$ be a birational map as above, and $f\colon X \dashrightarrow X'$ a Sarkisov link of Type II. We say that $f$ *untwists* $\varphi$ if $\varphi' = \varphi \circ f^{-1}\colon X' \to Y$ has degree smaller than $\varphi$.

**Remark 3.1.7** The Sarkisov program factors an arbitrary birational map between Mori fibres spaces as a chain of more general types of links, using a more complicated inductive framework. See [Co1], Definition 3.4 for the general definition of a *Sarkisov link* $f\colon X \dashrightarrow X'$, and [Co1], Definition 5.1 for the *Sarkisov degree* of a birational map $\varphi\colon X \dashrightarrow Y$ between Mori fibre spaces. The above Definitions 3.1.4–3.1.6 are special cases that are sufficient for our purposes in this paper. We can get away with this because we start from a Fano 3-fold $X$, and blessed fortune has decreed that we only perform untwistings returning always to $X$ (our aim is to prove rigidity), so that we never really venture beyond the first link of a Sarkisov chain.

The following result is our concise statement:

**Theorem 3.2** $X$ *is a* general *element of one of the famous 95 families of Fano hypersurfaces ("general" is discussed in 7.8). For every point $P$ or curve $C$ of $X$, we can do one of the following:*

*(I)* Either: *construct a birational involution $X \dashrightarrow X$ that is a Sarkisov link of Type II, untwisting $P$ (or $C$) (see Definitions 3.1.4–3.1.6) as a strong maximal centre of any linear system on $X$.*

*(II)* Or: *exclude the possibility that any linear system $\mathcal{H}$ on $X$ can have a strong maximal singularity $E$ centred at $P$ or $C$.*

*All curves and nonsingular points of $X$ are excluded. The Big Table of Chapter 6 documents the fate of the singular points (which by assumption are terminal quotient singularities $\frac{1}{r}(1, a, r - a)$): those marked Q.I. or E.I. are untwisted, and the remainder are excluded. For more details, see the introduction to Chapter 6.*

This theorem summarises the conclusions of a whole series of calculations carried out for Part I in Chapter 4 and for Part II in Chapter 5. Its proof ends on page 232. Each of these chapters has an introduction expanding on a number of points concealed by the concise nature of our definitions and statement.

## 3.3 Theorem 3.2 implies Theorem 1.3

This is standard, and is the same as the proof in Chapter 2. If $X$ is Fano and $Y \to S$ a Mori fibre space, a birational map $\varphi\colon X \dashrightarrow Y$ is defined by a mobile linear system $\mathcal{H}$. By the NFI inequalities ([Co1], Theorem 4.2

or [Pu1], Theorem 2.11), if $\varphi$ is not an isomorphism then $\mathcal{H}$ has a maximal centre $P$ or $C$ (see Definition 3.1.2), hence a strong maximal centre by [Co1], Proposition 2.10. This centre can't be in Part II of Theorem 3.2, so by Part I there is a birational involution $i\colon X \dashrightarrow X$ that is a Sarkisov link, and untwists the maximal centre $P$ or $C$, so that $\varphi \circ i$ has smaller degree. Thus after a number of steps, we arrive at $X \cong Y$.   Q.E.D.

## 3.4   Divisorial contractions

We start with a short preliminary discussion of 3-fold divisorial contractions, fixing the terminology, and stating a theorem of Kawamata used repeatedly in Chapters 4–5. A maximal singularity (Definition 3.1.2) has centre $\Gamma \subset X$ that is one of

    (1) a curve;   (2) a nonsingular point;   (3) a singular point;

our treatment and results are quite different in the three cases.

**Definition 3.4.1**  First, let $P \in X$ be the germ of a 3-fold terminal singularity. A *divisorial contraction* is a proper birational morphism $f\colon Y \to X$ such that

  (1)  $Y$ has terminal singularities,

  (2)  the exceptional set of $f$ is an irreducible divisor $E \subset Y$,

  (3)  $-K_Y$ is relatively ample for $f$.

An *extremal* divisorial contraction $f\colon Y \to X$ is an extremal divisorial contraction in the Mori category. In other words, $Y$ has $\mathbb{Q}$-factorial terminal singularities, $f$ is the contraction of an extremal ray $R$ of $\overline{NE}\,Y$ satisfying $K_Y \cdot R < 0$, and the exceptional set $\operatorname{Exc} f = E \subset Y$ is a divisor in $Y$. Its image $\Gamma = f(E)$ is a closed point or a curve of $X$, and we usually write $f\colon (E \subset Y) \to (\Gamma \subset X)$. Here $X$ is not necessarily a germ, but $Y \to X$ is a divisorial contraction in the above sense above the germ around any point $P \in \Gamma$. Viewed from $X$, we also say that $f$ is an *extremal extraction*, or that it *extracts* the valuation $v = v_E$ of $k(X)$ from its *centre* $\Gamma = C(X, v_E) \subset X$.

The classification of 3-fold divisorial contractions to a point is a beautiful open problem, the answer to which is only known in a few special cases, of which the following is important to us. The result is simple to state and prove, and very useful.

**Theorem–Definition 3.4.2 (Kawamata [Ka])** *Let*

$$P \in X \cong \frac{1}{r}(1, a, r - a) \quad \text{(with } r \geq 2 \text{ and } a \text{ coprime to } r)$$

*be the germ of a 3-fold terminal quotient singularity, and*

$$f \colon (E \subset Y) \to (\Gamma \subset X)$$

*a divisorial contraction such that $P \in \Gamma$. Then $\Gamma = P$ and $f$ is the weighted blowup with weights $(1, a, r - a)$. (Note that we don't need to assume that the contraction $f$ is extremal here.)* $\square$

We refer to it as the *extremal divisorial contraction*, the *extremal blowup* or the *Kawamata blowup* of $P$.

**Corollary 3.4.3** *Suppose that $X$ is a 3-fold with only terminal quotient singularities. If a curve $\Gamma \subset X$ is the centre of a divisorial extraction $f \colon (E \subset Y) \to (\Gamma \subset X)$ then $\Gamma \subset \operatorname{NonSing} X$ (and $f$ is the blowup of $\mathcal{I}_\Gamma$ over the generic point of $\Gamma$).*

For if $\Gamma$ passed through a terminal quotient point $P$, Theorem 3.4.2 would imply that $\Gamma = P$, a contradiction. Q.E.D.

**Remark 3.4.4 (curve as centre)** The trinity of viewpoints discussed in Remark 3.1.3, (e) all boil down to more-or-less the same thing for a curve centre. Let $C \subset X$ be a curve in a 3-fold. The valuation of $k(X)$ above $C$ that could be a weak maximal singularity is obviously unique. There is *at most* one extraction $Y \to X$ which is the blowup above the generic point of $C$. In our case, when $X$ has only terminal quotient singularities, this extraction only exists in the Mori category if $C \subset \operatorname{NonSing} X$ and $C$ has restricted singularities, for example, planar singularities.

We use the following notation and basic results many times in what follows, sometimes without explicit warning.

**Notation 3.4.5** For a divisorial contraction $f \colon (E \subset Y) \to (\Gamma \subset X)$, we adopt the shorthand notation $A = -K_X$, $B = -K_Y$. As before, this avoids confusion about signs and unnecessary subscripts. When $f$ is the Kawamata blowup of a terminal quotient singularity $P \in X \cong \frac{1}{r}(1, a, r - a)$, that is, the $(1, a, r - a)$-blowup, the adjunction formula for $f$ is

$$B = A - \frac{1}{r}E.$$

In this context, we sometimes use $B$ without referring to $f$ explicitly.

**Proposition 3.4.6**   *(1) Let $(E \subset Y) \to (P \in X)$ be the Kawamata blowup of a 3-fold terminal quotient singularity $P \in X \cong \frac{1}{r}(1, a, r - a)$. Then $K_Y = K_X + \frac{1}{r}E$ (or $B = A - \frac{1}{r}E$), and*

$$\frac{1}{r^2}E^3 = \frac{1}{a(r - a)} .$$

*(2) Write $\mathbb{C}^3$ for the local index 1 cover of $P \in X$ with $\xi, \eta, \zeta$ eigen-coordinates for the $\mathbb{Z}/r$ action. Then $\xi, \eta, \zeta$ can be viewed as local analytic sections of $\mathcal{O}_X(1), \mathcal{O}_X(a), \mathcal{O}_X(r - a)$ respectively, and lifted to $Y$, they vanish on $E$ with multiplicities exactly $\frac{1}{r}, \frac{a}{r}, \frac{r-a}{r}$, so that*

$$\xi \in \mathcal{O}_Y(A - \frac{1}{r}E) = \mathcal{O}_Y(B), \quad \eta \in \mathcal{O}_Y(aA - \frac{a}{r}E) = \mathcal{O}_Y(aB),$$

$$\zeta \in \mathcal{O}_Y\left((r - a)A - \frac{r - a}{r}E\right) = \mathcal{O}_Y((r - a)B)$$

*in a neighbourhood of $E \subset Y$.*

**Proof**   More-or-less standard exercises in toric blowups. For (1), it is easy to see that $E \cong \mathbb{P}(1, a, r - a)$ and $\mathcal{O}_E(-E) \cong \mathcal{O}_{\mathbb{P}(1,a,r-a)}(-r)$. To make sense of (2), one can for example take pullback of the invariant functions $\xi^r, \eta^r, \zeta^r \in \mathcal{O}_X$.   $\square$

**Lemma 3.5**   *Let $X_d \subset \mathbb{P}(a_0, a_1, \ldots, a_n)$ be a quasismooth weighted hypersurface of dimension $\geq 3$ (that is, $n \geq 4$). Then every Weil divisor class on $X_d$ is linearly equivalent to $\mathcal{O}_X(i)$ for some $i \in \mathbb{Z}$.*

**Proof**   Write $C_X \subset \mathbb{C}^5$ for the affine cone over $X$, that is, the hypersurface singularity given by the same equation as $X_d$. Then by definition of quasismooth, $0 \in C_X$ is an isolated singularity, and is factorial by a famous theorem of Grothendieck[1] (see Call and Lyubeznik [CL] or [SGA2], Exp. XI, Théorème 3.13). Thus every Weil divisor on $X$ is the $\mathbb{C}^*$ quotient of a $\mathbb{C}^*$-invariant principal divisor on $C_X$.   Q.E.D.

---

[1]In topology, this result is completely elementary: by Lefschetz theory, the link of an isolated hypersurface singularity $0 \in V \subset \mathbb{C}^n$ with $n \geq 5$ is 2-connected. Grothendieck's theorem and its reworking in [CL] concerns arbitrary local rings; the case of graded rings that we need is easier.

# 4   Untwisting

## 4.1   Introduction to involutions

Part I of Theorem 3.2 involves two types of untwisting: 54 quadratic involutions (see 4.4–4.9) and 8 elliptic involutions (see 4.10–4.13). In either case we construct a birational involution $i_P \colon X \dashrightarrow X$ that factors as follows:

Here

(1) $f \colon Y \to X$ is the Kawamata blowup of a terminal quotient singularity $P \in X$ of type $\frac{1}{r}(1, a, r - a)$;

(2) $-K_Y$ is nef, and the anticanonical model of $Y$ defines a morphism $g \colon Y \to Z$ contracting finitely many flopping curves of $Y$ to points of $Z$;

(3) as an anticanonical model, $Z$ is a 3-fold with terminal singularities and ample $-K_Z$, but be warned: it is *not in the Mori category*, since it has local class group $\mathbb{Z}$ at each flopping point (thus $X \dashrightarrow Z$ is *not* a morphism of the Sarkisov category);

(4) $Z$ has a biregular involution $i_Z$ that acts by $-1$ on the local class group at every flopping singularity of $Z$;

(5) the birational involution $i_Y \colon Y \dashrightarrow Y$ is the standard flop of the curves contracted by $Y \to Z$;

(6) finally,

$$i_P = f \circ g^{-1} \circ i_Z \circ g \circ f^{-1}.$$

The key point of the proof is thus the calculation of the anticanonical ring of the Kawamata blowup $Y \to X$. Here $R(Y, B) \subset R(X, A)$ is the subring whose graded elements are sections $H^0(\mathcal{O}_X(dA))$ that vanish $d/r$ times along $E$ (in the sense of Proposition 3.4.6), so that they remain plurianticanonical on $Y$. (Recall that $A = -K_X$ and $B = -K_Y = A - \frac{1}{r}E$.) Note the close parallel with the treatment of Geiser and Bertini involutions in 2.6.3, and compare 7.3 for passing to the subring $R(Y, B)$ as a "projection".

**Lemma 4.2** *A Type II link $X \dashrightarrow X$ untwists $P$ as a strong maximal centre of any linear system $\mathcal{H}$.*

**Proof**  This is essentially the same as the proof of Lemma 2.9.3. Suppose that $P$ is a maximal centre of a linear system $\mathcal{H} \subset |nA|$. Then the Kawamata blowup $Y \to X$ is a strong maximal singularity of $|\mathcal{H}|$, because it is the only divisorial extraction centred at $P$ (see Theorem 3.4.2). Hence

$$K_Y + \frac{1}{n}\mathcal{H}_Y = -\frac{m - na}{n}E < 0$$

The rest is as for Lemma 2.9.3.    Q.E.D.

**Remark 4.3 (Russian matrix)**  Since $i_Z$ takes $B = A - \frac{1}{r}E$ to itself, it is not hard to see that $i_Z$ acts on divisors of $Y$ by $A \mapsto sA - \frac{s+1}{r}E$ and $E \mapsto r(s-1)A - sE$ for some $s$, that is, by the matrix

$$\begin{pmatrix} s & -(s+1)/r \\ r(s-1) & -s \end{pmatrix}.$$

Since we know the image of $E$, it is an easy exercise to calculate $s$ (but we have never felt any inclination to do it). Compare 2.6.2 and Remark 2.9.4.

## 4.4    Quadratic involutions

The untwisting column of the Big Table contains 54 quadratic involutions (indicated by Q.I.). When properly understood, they all follow the same simple pattern: in suitable coordinates, the defining equation can be written

$$F = \eta\xi^2 + a\xi + b, \tag{2}$$

where $\xi = x_i$, $\eta = x_j$ for some $i \neq j$, and $\xi$ does not appear in $a, b$. Here the centre we want to untwist is $P_i = (0, \ldots, 1, 0, \ldots)$. As a birational involution, $i_{P_i}$ just interchanges the two roots of (2). We work this out in biregular terms in several examples and in general in Theorem 4.9. We use slightly different manoeuvres to achieve the form (2):

(i)  In most cases, no extra assumption is needed (beyond quasismooth), and the choice of coordinates is obvious.

(ii)  In 9 of our cases, the presence of the monomial $x_j x_i^2$ is an extra "generality" assumption on $X_d$, indicated by the asterix $*x_j x_i^2$ in the Big Table (see Example 4.6); as we explain in 7.4.2, when it does not appear, the corresponding point $P_i$ is excluded as a maximal centre.

(iii)  In a further 10 cases, the choice of coordinates is an obvious little trick (see Example 4.8).

The conclusion of the construction (and the information documented in the Big Table of Chapter 6) is stated as Theorem 4.9. *Let us play:*

**Example 4.5** No. 2, $X_5 \subset \mathbb{P}(1,1,1,1,2)$, centre $P = P_4 = \frac{1}{2}(1,1,1)$

In this case the preliminary massaging of the equation $F_5$ is very familiar. Write $x_0, \ldots, x_3, y$ for homogeneous coordinates, of which only $y$ is nonzero at $P_4$. Suppose that the equation of $X_5$ is $F_5$; then $y$ only appears quadratically in $F_5$, accompanied by a nonzero linear form, which we can take to be $x_0$. Thus

$$F_5 = x_0 y^2 + a_3 y + b_5 = 0, \tag{3}$$

where $a_3, b_5 \in k[x_0, x_1, x_2, x_3]$ are homogeneous of the indicated degrees. The birational involution $i_P$ just exchanges the roots of $F_5$.

Now to see this as the biregular procedure of 4.1. To specify the midpoint $Z$, we eliminate $y$, replacing it with the new variable $w = x_0 y$, which is integral over $k[x_0, \ldots, x_3]$: indeed, multiplying equation (3) by $x_0$ gives

$$g_6 = w^2 + a_3 w + b_5 x_0 = 0.$$

This is the equation of a Fano hypersurface $Z_6 \subset \mathbb{P}(1,1,1,1,3)$.

The transformation $X_5 \dashrightarrow Z_6$ is obtained by performing the Kawamata blowup $Y \to X$ of $P_4$, followed by a flopping contraction morphism $Y \to Z_6$. Indeed, $P_4$ is a singularity of type $\frac{1}{2}(1,1,1)$ with local coordinate $x_1, x_2, x_3$, and locally around $P_4$, the Kawamata blowup $Y \to X$ is the graph of the rational map to $\mathbb{P}^2$ given by $x_1, x_2, x_3$. The divisor $B = A - \frac{1}{2}E$ is eventually free and defines the morphism $Y \to Z_6$; in fact, in the present case, $|B| = |x_0, x_1, x_2, x_3|$ is itself a free linear system.

What does $Y \to Z_6$ contract? It is clear from what we have just said that the exceptional locus $E$ of the blowup $Y \to X$ is mapped biregularly; thus any contracted curves are present in the original $X$ given by (3), and they are the lines $L_i$ through $P$ given by $x_0 = a_3 = b_5 = 0$. We now prove that this is a finite set of lines (15 for generic $X$): write $\bar{a}_3, \bar{b}_5 \in k[x_1, x_2, x_3]$ for reduction modulo $x_0$; if these two have a common factor $c$ (say) then the original hypersurface $X_5$ contains the surface $x_0 = c = 0$, contradicting quasismooth (because $F$ can be rewritten $\alpha x_0 + \beta c$, which is singular at $\alpha = \beta = x_0 = c = 0$).

**Example 4.6** No. 5, $X_7 \subset \mathbb{P}(1,1,1,2,3)$, centre $P = P_3 = \frac{1}{2}(1,1,1)$

Write $x_0, x_1, x_2, y, z$ for homogeneous coordinates; only $y$ is nonzero at $P$. In this case, $P$ is untwisted by a quadratic involution $i_P$ if and only if the monomial $zy^2$ appears in $F$.

The point is the following: quasismoothness at $P$ involves a monomial $\eta y^m$ with $\eta = x_i$ or $z$. Because $\mathrm{wt}\, y = 2$, a priori, $F_7$ is a cubic in $y$, say

$x_0 y^3 + \gamma z y^2 + \cdots$. However, if $\gamma \neq 0$, a change of coordinates $z \mapsto z' = \gamma z + x_0 y + \cdots$ kills any other monomials divisible by $y^2$ in $F$. After this, we arrive at the quadratic equation

$$F_7 = zy^2 + a_5 y + b_7, \qquad (4)$$

in $y$, where $a_5, b_7 \in k[x_0, x_1, x_2, z]$. The remainder of the construction is word-for-word as in Example 4.5: the involution $i_P$ interchanges the two roots of the quadratic (4). The midpoint of the link is $Z_{10} \subset \mathbb{P}(1,1,1,3,5)$, obtained by eliminating $y$ in favour of $w = yz$, and with equation $g_{10} = w^2 + a_5 w + b_7 w$. The map $X \dashrightarrow Z_{10}$ is the Kawamata blowup $Y \to X$ followed by the flopping contraction morphism $Y \to Z$; this is a morphism because $Y$ is the graph of $x_0 : x_1 : x_2$, locally over $P$. As before, $Y \to Z_{10}$ is flopping because quasismooth implies that modulo $z$, the residues $\bar{a}_5, \bar{b}_7 \in k[x_0, x_1, x_2]$ have no common factor.

**Remark 4.7 (starred monomial $*\eta \xi^2$)** In this case, the possibility that $zy^2 \notin F$ does not contradict quasismooth, and what happens is also interesting: we will see later that $P$ is then excluded as a maximal centre (see 7.4.2). The Big Table commemorates this extra generality assumption by starring the key monomial $*zy^2$. The same dichotomy involving a starred monomial $*\eta \xi^2$ happens in 8 other cases, Nos. 12, 13, 20, 25, 31, 33, 38, 58.

**Example 4.8** No. 4, $X_6 \subset \mathbb{P}(1,1,1,2,2)$, centre $P \in P_3 P_4 = 3 \times \frac{1}{3}(1,1,2)$ and No. 6, $X_8 \subset \mathbb{P}(1,1,1,2,4)$, centre $P \in P_3 P_4 = 2 \times \frac{1}{2}(1,1,1)$

The new trick in these cases is simply to arrange that $P$ is one of the coordinate points. In No. 4, $P_3 P_4$ is the $(y_1, y_2)$-line, and the restriction of $F_6$ is a cubic in $(y_1, y_2)$. Since $P$ is one the roots, we can take $P = (0,1)$ and $F = y_1 y_2^2 + \cdots$. (This monomial must occur by quasismoothness at $P$.) Then

$$F_6 = y_1 y_2^2 + a_4 y_2 + b_6,$$

and we proceed as before. Nos. 9, 17, 27 involve the same trick.

In No. 6, $P_3 P_4$ is the line $\mathbb{P}(2,4)$, a badly-formed $\mathbb{P}^1$, with coordinates $y^2, z$. Quasismoothness at $P_4$ implies that $P_4 \notin X$, that is, $z^2 \in F$. Thus $F_8$ restricted to $P_3 P_4$ factors as $(z - c_1 y^2)(z - c_2 y^2)$ with $c_1, c_2 \in k$. A change of coordinates $z \mapsto z' = z - c_1 y^2$ takes $P$ to $P_3 = (0,0,0,1,0)$ (with only $y$ nonzero), and the monomial $zy^2$ is forced by quasismoothness. Thus

$$F_8 = zy^2 + a_6 y + b_8,$$

and we proceed as before. Nos. 15, 30, 41, 42, 68 involve the same trick.

**Theorem 4.9** *Suppose that $X_d \subset \mathbb{P}(1, a_1, a_2, a_3, a_4)$ is given by*

$$F_d = x_{i_3}\xi^2 + a_l\xi + b_d = 0, \tag{5}$$

*where $\xi, x_{i_3}$ are two of the coordinates, and $a_l, b_d$ do not involve $\xi$. Write $P_\xi$ for the coordinate point at which $\xi \neq 0$ and all the other $x_i = 0$. Then interchanging the roots of (5) defines a birational involution $i_{P_\xi}$ of $X_d$. It is a Sarkisov link of Type II in the sense of Definition 3.1.4, and untwists $P_\xi$ as a maximal centre.*

This construction provides all the quadratic involutions marked *Q.I.* in the *Big Table*. The assumption of the theorem includes the presence of the key monomial $*x_j x_i^2$. If the starred monomial is not present, the centre is excluded (see 7.4.2).

The Big Table lists the following information: the key monomial $x_j x_i^2$ (here a $*$ indicates that the presence of the monomial is an extra assumption), and the weighted degrees of $a, b$. For example, No. 33, centre $P_3$, $*ut^2$, 12, 17 means that the presence of the monomial $ut^2$ is assumed, $X_{17} \subset \mathbb{P}(1, 2, 3, 5, 7)$ is defined by

$$F_{17} = ut^2 + a_{12}t + a_{17},$$

and the centre $P_3$ is untwisted by the birational involution $i_{P_3}$.

**Proof** Set $\deg\xi = r$. Write $x_{i_0}, x_{i_1}, x_{i_2}$ for the other 3 variables $x_i$ in addition to $\xi, x_{i_3}$ and $a_{i_j} = \deg x_{i_j}$ for their degrees. Since $P_\xi$ is a singularity of type $\frac{1}{r}(1, a, r-a)$, with local coordinates $x_{i_0}, x_{i_1}, x_{i_2}$, it follows that $a_{i_0}, a_{i_1}, a_{i_2}$ are congruent to $1, a, r-a$ modulo $r$ (in some order). We claim the following little coincidence: the degrees $a_{i_0}, a_{i_1}, a_{i_2}$ are actually *equal* to $1, a, r-a$. Indeed, if $\deg x_{i_3} = s$ then, in view of the term $x_{i_3}\xi^2 \in F_d$, it follows that $d = 2r + s$; on the other hand, $d = \sum a_i + 1 = r + s + \sum_{j=0}^{2} a_{i_j} + 1$, so that $a_{i_0}, a_{i_1}, a_{i_2}$ add to $r + 1$.

Now the Kawamata blowup $Y \to X$ of $P_\xi$ is the $(1, a, r-a)$ weighted blowup, so locally at $P_\xi$ it is the graph of the rational map to $\mathbb{P}(1, a, r-a)$ given by $x_{i_0}, x_{i_1}, x_{i_2}$. It follows easily from Proposition 3.4.6, (2) that $x_{i_0}, x_{i_1}, x_{i_2}$ and $x_{i_3}$ belong to the anticanonical ring of $Y$, so that it is of the form

$$R(Y, B) = k[x_{i_0}, x_{i_1}, x_{i_2}, x_{i_3}, w]/(g),$$

where $w = \xi x_{i_3}$ and $g = w^2 + a_l w + x_{i_3} b_d$.

As in Example 4.5, quasismoothness implies that $x_{i_3} = a_l = b_d = 0$ is a finite number of lines through $P_\xi$. Therefore the anticanonical morphism $Y \to Z$ is a flopping contraction. Q.E.D.

## 4.10    Elliptic involutions

There are eight elliptic involutions, indicated by E.I. in the Big Table. The computation is the same in each case, although considerably more amusing than for the quadratic involutions. It involves *two* different key monomials: the centre we want to untwist is $P = P_\xi$, and the monomial defining the tangent space there is $x_{i_2}\xi^3$; however, the same variable $\xi$ also takes part in a monomial $\xi\zeta^2$ defining the tangent space at another point $Q = Q_\zeta$. Thus

$$F = \xi\zeta^2 + \cdots + x_{i_2}\xi^3 + \cdots .$$

Here $x_{i_0}, x_{i_1}, x_{i_2}, \xi, \zeta$ (in some order) are coordinates on $\mathbb{P}$, and necessarily $\deg \zeta > \deg \xi$. Typically, $\xi = x_3$ and $\zeta = x_4$, so that $P = P_3$ and $Q = P_4$.

Whereas our treatment of the quadratic involution was based on eliminating one variable, for the elliptic involution, we eliminate *both* variables $\xi$ and $\zeta$ at once, replacing them by somewhat complicated terms

$$v = \zeta^2 + x_{i_2}\xi^2 + \cdots \quad \text{and} \quad w = \zeta^3 + x_{i_2}\xi^2\zeta + \cdots .$$

These are designed to be plurianticanonical on $Y$ (that is, vanish enough times on the exceptional divisor $E$ of $Y \to X$), and it turns out that, together with the other coordinates $x_{i_0}, x_{i_1}, x_{i_2}$, they generate the anticanonical ring of $Y$, and satisfy a relation of the form

$$w^2 + Aw = v^3 + Bv^2 + Cv + D \tag{6}$$

with $A, B, C, D \in k[x_{i_0}, x_{i_1}, x_{i_2}]$. This equation defines the midpoint of the link (see 4.1), which is a (weak) Fano hypersurface $Z_{6e} \subset \mathbb{P}(1, b_1, b_2, 2e, 3e)$ having a biregular involution $i_Z$ coming from interchanging the two roots of the quadratic equation (6).

The form of (6) makes clear that the argument depends at some level on the fact that the fibres of the rational map to weighted $\mathbb{P}^2$ given by $x_{i_0}, x_{i_1}, x_{i_2}$ are birationally elliptic curves with a section; we first deduced the existence of a birational involution in terms of these fibres. However, on $X$ itself, these curves are quite singular at $P_\xi$: you can get the flavour by trying to understand why a general curve $E_{25} \subset \mathbb{P}(1, 7, 9)$ is birationally an elliptic curve with a section (the basic reason is that its Newton polygon has the shape of a Weierstrass cubic, compare (7)). Our anticanonical treatment is biregular, and seems to us to be easier and more convincing.

The reader not sharing our addiction to the luscious details of a delicious calculation may fast forward to the next chapter. *Let us play:*

**Example 4.11** No. 61, $X_{25} \subset \mathbb{P}(1, 4, 5, 7, 9)$, centre $P_3 = \frac{1}{7}(1, 5, 2)$

Write $x, y, z, t, u$ for coordinates on $\mathbb{P}$, so that only $t$ is nonzero at $P = P_3$. The tangent monomial at $P$ is necessarily $yt^3$, and $P$ is of type $\frac{1}{7}(1, 5, 2)$.

However, $t$ also appears in the tangent monomial $tu^2$ at the point $Q = P_4 = (0, 0, 0, 0, 1)$ of type $\frac{1}{9}(1, 4, 5)$.

Note that $t$ is $\mathbb{Z}/7$-invariant and nonzero at $P$ and in effect, we set $t = 1$ there. The Kawamata blowup $Y \to X$ of $P$ is the $(1, 5, 2)$-weighted blowup, and locally at $P$ it is the graph of the rational map to $\mathbb{P}(1, 5, 2)$ given by $(x : z : u/t)$; let $E = \mathbb{P}(1, 5, 2)$ be the exceptional surface, so that $-K_Y = B = A - \frac{1}{7}E$. By Proposition 3.4.6, (1), since $x, z, u/t$ are the local coordinates for the Kawamata blowup, $u \in H^0(X, \mathcal{O}_X(9A))$ only vanishes $\frac{2}{7}$ times at $E$, so that

$$u \in H^0(Y, 9A - \frac{2}{7}E) = H^0(Y, 9B + E), \quad \text{but} \quad u \notin H^0(Y, 9B);$$

that is, $u$ is *not* anticanonical. On the other hand,

$$x \in H^0(Y, B), \quad z \in H^0(Y, 5B),$$

and we get $y \in H^0(Y, 4B)$ from the form of the equation $F = yt^3 + \cdots = 0$. To complete the calculation of the anticanonical ring of $Y$, we exhibit two further elements $v, w$ of degree $18, 27$ satisfying a relation of the form $w^2 = v^3 + \cdots$ (7.2 explains how numbers such as $18, 27$ can readily be predicted in this kind of context). For $v$, we have to find an element in $H^0(X, 18A)$ with divisor of zeros $\geq \frac{18}{7}E$.

Start by viewing the equation of $X$ as a quadratic in $u$, obtaining $F = tu^2 + \cdots$; any other terms divisible by $tu$ in $F$ can be killed by a change of variables $u \mapsto u' = u + t\alpha$. Thus we can assume that the coefficient of $u$ does not involve $t$, so that $F$ is of the form

$$F = tu^2 + a_{16}u \qquad \text{with } a, b, c, d \in k[x, y, z]. \qquad (7)$$
$$- yt^3 - b_{11}t^2 - c_{18}t + d_{25}$$

Note that $X$ quasismooth implies that $a, d$ don't have any common factor, that is, $(a = d = 0) \subset \mathbb{P}(1, 4, 5)$ is a finite set (20 points in general).

Filtering off the terms divisible by $t$, we define

$$v = u^2 - yt^2 - b_{11}t - c_{18}, \quad \text{so that} \quad F = vt + a_{16}u + d_{25}. \qquad (8)$$

We claim that the divisor of zeros of $v$ on $Y$ is $\geq \frac{18}{7}E$, so $v \in H^0(Y, 18B)$. The point is that the final form of the relation $F = 0$ in (8) allows us to write $v = -(a_{16}u + d_{25})/t$, with $t$ a unit at $P$; now all terms in $d_{25}$ have divisor of zeros $\geq \frac{25}{7}E$, all terms in $a_{16}$ have divisor of zero $\geq \frac{16}{7}E$, and $u$ has divisor of zeros $\frac{2}{7}E$.

Next, we multiply the final form of $F$ by $u$, and substitute for $u^2$ in terms of $v$. We get

$$uF = uvt + a_{16}u^2 + d_{25}u$$
$$= uvt + a_{16}(v + yt^2 + b_{11}t + c_{18}) + d_{25}u.$$

We again filter off the terms divisible by $t$, and set

$$w = vu + a_{16}yt + a_{16}b_{11}, \tag{9}$$

so that

$$uF = wt + a_{16}(v + c_{18}) + d_{25}u. \tag{10}$$

As before, we use the final relation to write $w = -(a_{16}(v + c_{18}) + d_{25}u)/t$, with $t$ a unit at $P$; all terms in $a_{16}(v + c_{18})$ have divisor of zeros $\geq \frac{34}{7}E$, and all the terms in $d_{25}u$ have divisor of zeros $\geq \frac{27}{7}E$, so that $w \in H^0(Y, 27B)$.

Notice that

$$
\begin{aligned}
- uw + (yt + b)F &= -u^2v - auyt - abu + (yt + b)f \\
&= -(v + yt^2 + bt + c)v - auyt - abu + (yt + b)(vt + au + d),
\end{aligned}
$$

and cancelling gives

$$(yt + b)F = wu + dyt - v(v + c) + db. \tag{11}$$

In order to eliminate $t, u$ in favour of $v, w$, note that we can view (8–11) as inhomogeneous linear relations in $t, u$ with coefficients in $k[x, y, z, v, w]$:

| (8) | $F =$ | $a_{16}u$ | $+$ | $vt$ | $+$ | $d_{25}$ | $= 0,$ |
|---|---|---|---|---|---|---|---|
| (10) | $uF =$ | $d_{25}u$ | $+$ | $wt$ | $+$ | $a_{16}(v + c_{18})$ | $= 0,$ |
| (9) definition of $w$: | | $vu$ | $+$ | $a_{16}yt$ | $+$ | $-w + a_{16}b_{11}$ | $= 0,$ |
| (11) | $(yt + b)F =$ | $wu$ | $+$ | $dyt$ | $+$ | $-v(v + c) + db$ | $= 0.$ |

That is,

$$
\begin{pmatrix}
a & v & d \\
d & w & a(v + c) \\
v & ay & -w + ab \\
w & dy & -v(v + c) + db
\end{pmatrix}
\begin{pmatrix}
u \\
t \\
1
\end{pmatrix}
= 0. \tag{12}
$$

The equation relating $v$ and $w$ is

$$w^2 - abw = v^3 + v^2c - (bd + a^2y)v + (-a^2c + d^2)y. \tag{13}$$

It can be derived from (12) in several different ways: by premultiplying the matrix by $(dy, -ay, w, -v)$, or by calculating any of its $3 \times 3$ minors.

Write $Z_{54} \subset \mathbb{P}(1, 4, 5, 18, 27)$ for the hypersurface given by (13), and $S \subset Z_{54}$ for the locus

$$S : \operatorname{rank} \begin{pmatrix} v & w & ay & dy \\ a & d & v & w \end{pmatrix} \leq 1. \tag{14}$$

Solving (12) clearly gives $t, u$ as rational functions on $Z$, regular outside $S$. This shows that the correspondence $(t, u) \mapsto (v, w)$ is birational.

Write $r = v/a = w/d =$ etc. for the common ratio. It is a rational section of $\mathcal{O}_S(2)$, regular outside $a = d = v = w = 0$ (a finite set). Also, $r^2 = y$. Thus the normalisation of $S$ is $\mathbb{P}(1, 5, 2)$ with coordinates $x, z, r$, and $S$ is the image of $\mathbb{P}(1, 5, 2)$ under the birational inclusion morphism $\mathbb{P}(1, 5, 2) \to \mathbb{P}(1, 4, 5, 18, 27)$ defined by

$$(x, z, r) \mapsto (x, y = r^2, z, v = ar, w = dr).$$

Here $\mathbb{P}(1, 5, 2)$ is the exceptional surface of the blowup $Y \to X$. The map $\mathbb{P}(1, 4, 5, 7, 9) \dashrightarrow \mathbb{P}(1, 4, 5, 18, 27)$ is a "projection" map, that takes $\mathbb{P}(1, 5, 2)$ to the nonnormal surface $S$. It identifies the pairs of points with $\pm r$ in the set $a(x, z, r^2) = d(x, z, r^2) = 0$ (there are 20 pairs).

The equations

$$(a = d = v = 0) \subset X_{25} \subset \mathbb{P}(1, 4, 5, 7, 9)$$

(where, as in (8), $v = u^2 - yt^2 - bt - c$) define a finite number of curves through $P$ (in general 20) with birational transforms contracted by $Y \to Z_{54}$. For general $X$, we prove that these are the *only* contracted curves; as explained in 4.1, they are then exactly the flopping curves of a birational involution of $X$. For $t, u$ are regular outside $S$ by (12); moreover, (8) and (9) show that they are also regular, except possibly at points of $S$ where $y = b = v = 0$, $u^2 = c$ and $au + d = 0$. But $(y = b = 0, d^2 = a^2c)$ defines the empty set in $\mathbb{P}(1, 4, 5)$ in general.

**Example 4.12** No. 7, $X_8 \subset \mathbb{P}(1, 1, 2, 2, 3)$, centre $P \in P_2P_3 = 4 \times \frac{1}{2}(1, 1, 1)$

This case is very similar to Example 4.11, except for a couple of preliminary skirmishes. We choose coordinates $x_0, x_1, y_1, y_2, z$ such that $P = P_3 = (0, 0, 0, 1, 0)$; the tangent monomial at $P$ must be $y_1y_2^3$. The equation of the tangent plane at $P_4$ is $(\alpha y_1 + \beta y_2)z^2$. The general case is $\beta \neq 0$, and then we can change coordinates, to arrange that the tangent monomial at $P_4$ is $y_2z^2$. Thus we write $*y_2z^2 - y_1y_2^3$ in the Big Table, where, as before, the star reminds us that the presence of the monomial involves an extra generality assumption. Assuming $y_2z^2$ is present, any other multiple of $y_2z$ in $F$ can be killed by a coordinate change $z \mapsto z + e_3(x_0, x_1, y_1)$, so that

$$F = y_2z^2 + a_5z \qquad \text{with } a, b, c, d \in k[x_0, x_1, y_1].$$
$$\quad - y_1y_2^3 - b_4y_2^2 - c_6y_2 + d_8$$

From that point onwards, the calculation is identical to that of Example 4.11 on making the substitution

$$y \mapsto y_1, \quad t \mapsto y_2, \quad u \mapsto z.$$

We get $X_8 \dashrightarrow Z_{18} \subset \mathbb{P}(1, 1, 2, 6, 9)$.

**Theorem 4.13** *In the notation of Section 4.10, let $X_d \subset \mathbb{P}(1, a_1, a_2, a_3, a_4)$ be a general hypersurface defined by*

$$F_d = \xi \zeta^2 + a\zeta - x_{i_2}\xi^3 - b\xi^2 - c\xi + d,$$

*where $x_0, x_{i_1}, x_{i_2}, \xi, \zeta$ are the 5 coordinates (in some order), and $a, b, c, d \in k[x_0, x_{i_1}, x_{i_2}]$. Then $P_\xi$ is a quotient singularity of type $\frac{1}{r}(1, a_{i_1}, r - a_{i_1})$, where $r = \deg \xi$, having local coordinates $x_{i_0}, x_{i_1}, \zeta$ (in the sense explained in Proposition 3.4.6, (2)), and $\deg \zeta = 2r - a_{i_1}$.*

*Let $Y \to X$ be the Kawamata blowup of $P_\xi$. Then the anticanonical ring of $Y$ is of the form $R(Y, B) = k[x_0, x_{i_1}, x_{i_2}, v, w]/(G)$, where,*

$$v = \zeta^2 - x_{i_2}\xi^2 - b\xi - c, \quad w = v\zeta + ax_{i_2}\xi + ab, \quad and$$
$$G = -w^2 + abw + v^3 + v^2c - (bd + a^2y)v + (-a^2c + d^2)y.$$

*Write $Z = \operatorname{Proj} R(Y, B)$; then $Y \to Z$ is a morphism and contracts finitely many curves in $Y$ obtained as the birational transform of $(a = d = v = 0)$. The composite $X \dashrightarrow Z$ gives rise to an involution as described in 4.1.*

*This construction provides all the elliptic involutions marked E.I. in the Big Table.*

**Proof**  Write $r = \deg \xi$ and $s = \deg \zeta$. Then

$$d = a_{i_1} + a_{i_2} + r + s = r + 2s = a_{i_2} + 3r,$$

because $d = \sum a_i$ (for all our hypersurfaces), and because of the monomials $\xi \zeta^2, x_{i_2}\xi^3 \in F_d$. This gives $s = a_{i_1} + a_{i_2} = 2r - a_{i_1}$. The rest of the proof is the same as the calculations in Example 4.11, and we omit it.   Q.E.D.

# 5   Excluding

This chapter proves the following theorem, which is just Theorem 3.2 chopped up into cases for ease of digestion (see Definition 3.1.2 for weak and strong maximal singularity):

**Theorem 5.1** *Let $X$ be a general hypersurface of our list.*

**5.1.1** *No curve $\Gamma \subset X$ can be a strong maximal centre.*

**5.1.2** *No nonsingular point $P \in X$ can be a weak maximal centre.*

**5.1.3** *No singular point $P \in X$ with $B^3 \leq 0$ can be a strong maximal centre (see 3.4.5 for the notation $A = -K_X$, $B = -K_Y = A - a_E E$, etc. for a divisorial extraction $f: (E \subset Y) \to (\Gamma \subset X)$).*

**5.1.4** *Other than those untwisted in Chapter 4, no singular point $P \in X$ with $B^3 > 0$ can be a strong maximal centre.*

## Remarks

(1) See 7.8 for a discussion of *general*; in any case, it always means "outside a Zariski closed subset". In particular we always assume that $X$ is quasi-smooth. The central "exclusion" column of the Big Table of Chapter 6 contains data on the method we use to exclude each singularity.

(2) We guess that every curve on $X$ is excluded as a weak maximal centre; but our methods do not prove this, even a posteori (except for cases when $X$ has no links). It might happen, for example, that a linear system $\mathcal{H}$ with very high multiplicity along a curve $C$ has a singular point $P \in C$ as a strong maximal centre, so that untwisting $P$ flops $C$ and dilutes the singularity along it.

(3) In our treatment, we distinguish between singular points $P \in X$ with $B^3 \leq 0$ and $B^3 > 0$, as explained in 5.4–5.5 (see especially Proposition 5.4.1). In the first case, we have the neat statement 5.1.3. Singular points with $B^3 > 0$ fall into two groups: some are untwisted by the quadratic or elliptic involutions described in Chapter 4, and are therefore strong maximal centres. The remainder are excluded by 5.1.4.

The proof divides into 6 sections:

5.2 The test class method, and curves .................................. 204

5.3 Two variations on the test class method, and nonsingular points .. 207

5.4 Determination of $\overline{\mathrm{NE}}\,Y$, and singular points with $B^3 \leq 0$ ......... 212

5.5 Bad links, and singular points with $B^3 > 0$ ...................... 217

5.6 Proof of Theorem 5.3.1: construction of the test class for nonsingular points ......................................................... 219

5.7 Singular points: construction of $T$ .............................. 222

Sections 5.2–5.5 make up the technical and conceptual core of the chapter, introducing the different methods for excluding: the test class method and variations on it, the $\Gamma = S \cap T$ method of determining $\overline{\mathrm{NE}}\,Y$, and the bad link method. At the same time, they prove Theorems 5.1.1–5.1.4, modulo some bulky calculations that we relegate to 5.6 and 5.7 to avoid interrupting the flow of ideas. As in the preceding chapter, rather than reproducing here all of the (several hundred) calculations involved, which would take a disproportionate amount of space, we introduce all the methods and work out detailed examples of each; the numerical data on each case is summarised in the Big Table.

In 7.6 we discuss and compare the different methods of exclusion, and how they are used. As we explain in the final section 7.9, it seems doubtful that a single method could cover all the cases we exclude.

## 5.2    The test class method, and curves

In this section we introduce the test class method in its pure form and use it to prove that no curve on $X$ is a maximal centre (Theorem 5.1.1). The method is based on the following lemma and its corollary.

**Lemma 5.2.1** *Let $\Gamma \subset X$ be a strong maximal centre, and $f\colon (E \subset Y) \to (\Gamma \subset X)$ a maximal extraction. Then the 1-dimensional cycle $B^2 \in N_1 Y$ lies in the interior of the Mori cone of $Y$:*

$$B^2 \in \operatorname{Int} \overline{NE}\, Y$$

*(in fact $\overline{NE}\, Y = NE\, Y$).*

**Proof** The assumption means that there is a linear system $\mathcal{H} \subset |nA|$ with multiplicity $m_E$ along $E$ large compared with the discrepancy of $E$:

$$m_E(\mathcal{H}) > n a_E(K_X).$$

We set $m = m_E(\mathcal{H})$ and $a = a_E(K_X)$. Let $\mathcal{H}' = f_*^{-1}\mathcal{H}$ be the birational transform, and write

$$B = \frac{m - na}{m} A + \frac{a}{m} \mathcal{H}'$$

$$= \alpha A + \beta \mathcal{H}' \quad \text{with } \alpha, \beta > 0.$$

Squaring both sides gives:

$$B^2 = \alpha^2 A^2 + 2\alpha\beta A \cdot \mathcal{H}' + \beta^2 \mathcal{H}'^2 \quad \text{with } \alpha^2, \alpha\beta, \beta^2 > 0.$$

Now rank $N^1 Y = 2$, and the classes $A^2$, $A \cdot \mathcal{H}'$ and $\mathcal{H}'^2$ all lie in $\overline{NE}\, Y$, and are not all proportional, because $A - \varepsilon E$ is ample on $Y$ for some $\varepsilon$. Thus any strictly positive combination of $A^2$, $A \cdot \mathcal{H}'$ and $\mathcal{H}'^2$ is in $\operatorname{Int} \overline{NE}\, Y$, and hence $B^2 \in \operatorname{Int} \overline{NE}\, Y$.    Q.E.D.

**Definition 5.2.2** Let $f\colon (E \subset Y) \to (\Gamma \subset X)$ be an extremal divisorial contraction in the Mori category. A *test class* is a nonzero nef class $M \in N^1 Y$.

**Corollary 5.2.3** *Let $f\colon (E \subset Y) \to (\Gamma \subset X)$ be an extremal divisorial contraction in the Mori category. If $M \cdot B^2 \le 0$ for some test class $M$, then $E$ is not a maximal singularity.*

**Proof**   The contradiction is obvious: if $M \geq 0$ on $\overline{NE}$ and $M \cdot B^2 \leq 0$ then $B^2 \notin \operatorname{Int} \overline{NE}$.   Q.E.D.

We now explain the simplest and most common way of constructing a test class.

**Definition 5.2.4** Let $L$ be a Weil divisor class on a variety $X$ and $\Gamma \subset X$ an irreducible subvariety of codimension $\geq 2$. For an integer $s > 0$, consider the linear system

$$\mathcal{L}_\Gamma^s = |\mathcal{I}_\Gamma^s(sL)|,$$

where $\mathcal{I}_\Gamma$ is the ideal sheaf of $\Gamma$. We say that the class $L$ *isolates* $\Gamma$, or is $\Gamma$-*isolating*, if

(a) $\Gamma \subset \operatorname{Bs} \mathcal{L}_\Gamma^s$ is an isolated component for some positive integer $s$; in other words, in a neighbourhood of $\Gamma$, the base locus of $\mathcal{L}_\Gamma^s$ is contained in $\Gamma$ (as a set). See Lemma 5.6.4 for the point of the argument at which we sometimes need to use $s > 1$.

(b) In the case $\Gamma$ is a curve, the generic point of $\Gamma$ appears in $\operatorname{Bs} \mathcal{L}_\Gamma^s$ with multiplicity 1.

**Lemma 5.2.5** *Suppose that $L$ isolates $\Gamma \subset X$, and let $s$ be as above. Then, for any extremal divisorial contraction*

$$f\colon (E \subset Y) \to (\Gamma \subset X)$$

*the birational transform $M = f_*^{-1}\mathcal{L}_\Gamma^s$ is a test class on $Y$.*

**Proof**   First, $M = f^*sL - mE$ for some $m > 0$, and is obviously nonzero; we must prove that $M$ is nef, that is, $MC \geq 0$ for every irreducible curve $C \subset Y$. There are 4 cases:

(1) $C$ is contracted by $f$. Then $Cf^*L = 0$ and $C(-E) > 0$, so $CM > 0$.

(2) $C$ is disjoint from $E$. Then $CM = Cf^*L = f(C)L > 0$.

(3) $C$ meets $E$ but is not contained in $E$. Then by assumption, $f(C)$ is not contained in the base locus of $\mathcal{L}_\Gamma^s$, so that $C$ is not contained in the base locus of $|M|$. Therefore $CM \geq 0$.

(4) $C$ is contained in $E$ but not contracted. Then by Definition 5.2.4, (ii), $C$ is not in the base locus of $|M|$, hence $C \cdot M > 0$.   Q.E.D.

**The test class method (pure form)**

In a nutshell, the method is this: start with an extremal divisorial contraction $(E \subset Y) \to (\Gamma \subset X)$. To exclude $E$ as a maximal singularity, we find a test class $M$ on $Y$ with $M \cdot B^2 \leq 0$, by constructing a $\Gamma$-isolating system on $X$ or otherwise, and conclude by Corollary 5.2.3.

We use the method in this section to exclude curves, and in 5.4 to exclude some of the singular points. What these two cases have in common is that the centre has a unique extremal extraction, so that by excluding $E$ (on that biregular model), we exclude the point or curve as a strong maximal centre. The method is not applicable as such to a nonsingular point $P \in \mathrm{NonSing}\, X$, because the extremal extractions from $P$ are not yet completely classified (there are infinitely many of them – for a discussion, see [Co2], Section 3.3, which also contains a plausible conjecture as to their classification). The following section 5.3 develops two variations on this theme, which are used to exclude all nonsingular points as maximal centres.

**Proof of Theorem 5.1.1: No curve on $X$ is a strong maximal centre**
The argument follows closely the original approach of Iskovskikh and Manin [IM], and is in two steps. The first uses a crude numerical argument to exclude the great majority of cases. In the second step, we use the test class method to exclude all the remaining cases.

STEP 1    Let $\Gamma \subset X$ be an irreducible curve. If $\Gamma$ is a strong maximal centre, there is a linear system $\mathcal{H} \subset |nA|$ and a maximal extraction

$$f \colon (E \subset Y) \to (\Gamma \subset X)$$

centred in $\Gamma$ whose exceptional divisor $E$ has multiplicity $m > n$. In particular, $f$ is an extremal extraction, and $\Gamma \subset \mathrm{NonSing}\, X$ by Corollary 3.4.3, hence its anticanonical degree is a strictly positive integer:

$$d = \deg \Gamma = A \cdot \Gamma \in \mathbb{N}.$$

Let $s$ be large enough so that $|sA|$ is a very ample Cartier divisor, $S \in |sA|$ a general member, and $H_1, H_2 \in \mathcal{H}$ general members. Then

$$A^3 s n^2 = S \cdot H_1 \cdot H_2 \geq s m^2 \deg \Gamma > s n^2 \deg \Gamma,$$

so

$$A^3 > \deg \Gamma.$$

In particular, $A^3 > 1$, and a quick inspection of the list then reveals that $X$ and $\deg \Gamma$ must be one of the following:

(a) $X_4 \subset \mathbb{P}^4$, $\deg \Gamma = 1, 2$ or $3$;

(b) $X_5 \subset \mathbb{P}(1, 1, 1, 1, 2)$, $\deg \Gamma = 1$ or $2$;

(c) $X_6 \subset \mathbb{P}(1, 1, 1, 1, 3)$, $\deg \Gamma = 1$;

(d) $X_6 \subset \mathbb{P}(1, 1, 1, 2, 2)$, $\deg \Gamma = 1$;

(e) $X_7 \subset \mathbb{P}(1, 1, 1, 2, 3)$, $\deg \Gamma = 1$.

STEP 2    In each of the remaining cases (a–e), we construct a test class $M$ with $M \cdot B^2 \leq 0$. By Corollary 5.2.3, this contradicts the assumption that $E$ is a maximal singularity and thus completes the proof. The method is the same in all cases, so we only do Case (b) as an illustration, leaving the others to the reader.

Consider thus Case (b). If $\deg \Gamma = 1$, we can choose coordinates so that $\Gamma : (y = x_0 = x_1 = 0) \subset \mathbb{P}$. The blowup $f : (E \subset Y) \to (\Gamma \subset X)$ of the ideal sheaf of $\Gamma$ is the only extremal extraction centred at $\Gamma$ and, by what we just said, $2A$ is a $\Gamma$-isolating class and

$$M = f_*^{-1} |\mathcal{I}_\Gamma(2A)| = 2A - E$$

is a test class. An easy calculation, using $A^3 = \frac{5}{2}$, $A^2 \cdot E = 0$ (projection formula), $A \cdot E^2 = -\deg \Gamma = -1$ (also by the projection formula) and $E^3 = -\deg N_\Gamma X = -\deg \Gamma + 2 - 2p_a(\Gamma) = 1$, shows that

$$M \cdot B^2 = (2A - E)(A - E)^2$$
$$= 2A^3 - (4 + 1)A^2 \cdot E + (2 + 2)A \cdot E^2 - E^3 = 5 - 5 \leq 0,$$

and $\Gamma$ is not a maximal centre by Corollary 5.2.3.

If $\deg \Gamma = 2$, then $\Gamma$ is given by either $z = x_3 = x_0 x_1 + x_2^2 = 0$ or $z^2 + a_4(x_0, x_1) = x_3 = x_2 = 0$. In the first case, $M = 2A - E$ is a test class, in the second case $M = 4A - E$ is a test class. In either case, $M \cdot B^2 < 0$ (we leave the numerics as an exercise to the reader) and, as above, $\Gamma$ cannot be a maximal centre.    Q.E.D.

## 5.3    Two variations on the test class method, and nonsingular points

This section introduces two technical results, Theorems 5.3.2 and 5.3.3, that are variations on the test class method, and applies them to prove Theorem 5.1.2. In simple-minded terms, the idea is to reduce to an intersection calculation on a surface to derive a strong numerical conclusion from a maximal singularity at a point $P \in \text{NonSing}\, X$, without any reference to extremal extractions $Y \to X$; but in setting up this calculation, we make knowledgeable use of adjunction theoretic ideas to guide us to the most powerful conclusion.

**Proof of Theorem 5.1.2: No nonsingular point on $X$ is a maximal centre**  The proof involves a single exceptional case. Here the variety is No. 2, $X_5 \subset \mathbb{P}(1,1,1,1,2)$ studied in Example 4.5; in the notation given there, there is a quadratic involution $i_{P_4}$ that untwists $P_4$ by blowing it up, then flopping the birational transform of the lines $L_i$ (15 in general) of $X_5$ of degree $AL_i = \frac{1}{2}$ defined by $x_0 = a_3 = b_5 = 0$. We label this case the *renegade* for future reference:

$$\text{No. 2, } X_5 \subset \mathbb{P}(1,1,1,1,2), \text{ centre } P \in L_i \setminus P_4 \qquad (*)$$

It is thus convenient to divide what we have to prove into

(A) *If $X$ is general then, except possibly for the renegade case (\*), no non-singular point on $X$ can be a weak maximal centre.*

(B) *No nonsingular point in case (\*) can be a weak maximal centre.*

The bulk of the proof of (A) is contained in the following statement, the existence of a suitable $P$-isolating class on $X$ (Definition 5.2.4). Its proof is a lengthy but mechanical calculation (not without its moment of excitement) that we postpone until 5.6 to avoid interrupting the flow of ideas.

**Theorem 5.3.1** *Suppose that $X = X_d \subset \mathbb{P} = \mathbb{P}(1,a_1,a_2,a_3,a_4)$ is general, and let $P \in \text{NonSing}\, X$ be a nonsingular point. Then except in the renegade case (\*), $lA$ is a $P$-isolating class on $X$ (Definition 5.2.4) for some $l \leq 4/A^3$.*

**Proof**  See 5.6.  $\square$

The remainder of this section carries out the additional work needed to deduce Theorem 5.1.2 from Theorem 5.3.1. We need some preparations. To understand these, to say that $P \in X$ is a weak maximal centre means by definition that there is a linear system $\mathcal{H} \subset |nA|$ and $P = C(X, v_E)$ is the centre of a valuation with multiplicity

$$m_E(\mathcal{H}) > na_E(K_X).$$

Another way to say this is that

$$K_X + \frac{1}{n}\mathcal{H}$$

is not canonical at $P$ (Definition 3.1.1). If $S \subset X$ is a surface passing through $P$ then $K_X + S + \frac{1}{n}\mathcal{H}$ is not log canonical (we explain this below). This is the starting point for the following result.

**Theorem 5.3.2** *Let $P \in X$ be the germ of a nonsingular 3-fold, and $\mathcal{H}$ a mobile linear system on $X$. Assume that*

$$K_X + \frac{1}{n}\mathcal{H}$$

*is not canonical at $P$.*

*(1) If $P \in S \subset X$ is a surface and $\mathcal{L} = \mathcal{H}_{|S}$, then*

$$K_S + \frac{1}{n}\mathcal{L}$$

*is not log canonical.*

*(2) If $Z = H_1 \cap H_2$ is the intersection of two general members $H_1$, $H_2$ of $\mathcal{H}$, then*

$$\operatorname{mult}_P Z > 4n^2.$$

**Proof** This is a special case of [Co2], Corollary 3.4. Here (2) follows from (1) and the next Theorem 5.3.3 (the proof is spelled out in [Co2], Section 3.2). We can prove (1) in a single formal sentence by saying that it is an immediate corollary of [FA], 17.7 and 17.4 (results of Shokurov, reworked by Kollár).

Since this is a key technical point, at which the literature is not excessively friendly, we digress briefly to expand on the proof of (1), for the benefit of the reader and two of the coauthors. What we're doing is a special case of Shokurov's *inversion of adjunction*. We want to prove

$$K_X + \frac{1}{n}\mathcal{H} \text{ not canonical at } P \implies K_S + \frac{1}{n}\mathcal{L} \text{ not log canonical.}$$

The obvious approach makes the result seem trivial: just blow up and count discrepancies. Let $(\widetilde{S} \subset \widetilde{X}) \to (S \subset X)$ be a resolution with exceptional locus having simple normal crossings, and $\widetilde{\mathcal{H}}$ free. Then

$$K_{\widetilde{X}} = K_X + \sum a_i E_i$$
$$\widetilde{S} = S - \sum c_i E_i$$
$$\widetilde{\mathcal{H}} = \mathcal{H} - \sum m_i E_i.$$

Now $K_X + S + \frac{1}{n}\mathcal{H}$ not canonical at $P$ means that $a_i - \frac{1}{n}m_i < 0$ for some $E_i$ over $P$, and then $a_i - c_i - \frac{1}{n}m_i < -1$, because $S$ is a Cartier divisor through $P$, so that $c_i \in \mathbb{Z}$, $c_i > 0$. This proves that $K_X + S + \frac{1}{n}\mathcal{H}$ is not log canonical at $P$. Restricting $K_X + S$ gives $S = K_S$ and $K_{\widetilde{X}} + \widetilde{S}$ gives $K_{\widetilde{S}}$; now if $E_i \cap \widetilde{S} \neq \emptyset$, it is an exceptional divisor on $\widetilde{S}$ at which $K_S + \frac{1}{n}\mathcal{H}$ has discrepancy $< -1$, which gives the required implication.

The trouble is that $E_i$ could be disjoint from the birational transform $\widetilde{S}$. This is a traditional difficulty in all discrepancy arguments of this type (compare [C3-f], Theorem 2.6, II and [YPG], Remark 6.2, ii). However, we easily get round it by the *connectedness theorem* of Shokurov and Kollár ([FA], 17.4): the union of $\widetilde{S}$ and the $E_i$ with $a_i - c_i - \frac{1}{n}m_i < -1$ is connected.   Q.E.D.

The inequality for the multiplicity of $Z$ in Theorem 5.3.2, (2) is enough to prove (A). The next result is used in [Co2], Section 3.2 for the proof of (1) implies (2), and we use it in the proof of (B) below.

**Theorem 5.3.3** *Let $P \in \Delta_1 + \Delta_2 \subset S$ be the analytic germ of a normal crossing curve on a nonsingular surface (so that $\Delta_1 + \Delta_2 \subset S$ is isomorphic to the coordinate axes $(xy = 0) \subset \mathbb{C}^2$).*

*Let $\mathcal{L}$ be a mobile linear system on $S$, and write $(\mathcal{L}^2)_P$ for the local intersection multiplicity $(L_1 \cdot L_2)_P$ of two general members $L_1, L_2 \in \mathcal{L}$ at $P$. Fix rational numbers $a_1, a_2 \geq 0$, and assume that*

$$K_S + (1 - a_1)\Delta_1 + (1 - a_2)\Delta_2 + \frac{1}{n}\mathcal{L}$$

*is not log canonical at $P$. Then*

*(1) If either $a_1 \leq 1$ or $a_2 \leq 1$, then*

$$(\mathcal{L}^2)_P > 4a_1 a_2 n^2.$$

*(2) If both $a_i > 1$, then*

$$(\mathcal{L}^2)_P > 4(a_1 + a_2 - 1)n^2.$$

**Proof** Although the statement is fine-tuned for the most powerful result, the proof is quite elementary, and makes a worthwhile exercise. See [Co2], Section 3.1, Theorem 3.1 for the solution.   □

We are now ready to prove (A) and (B) of 5.3.

**Proof of (A)** Let $P \in X$ be a nonsingular point not in the renegade case (∗). To say that $P \in X$ is a weak maximal centre means that $P = C(X, v_E)$ is the centre of a valuation, and there is a linear system $\mathcal{H} \subset |nA|$ with multiplicity

$$m_E(\mathcal{H}) > n a_E(K_X).$$

Letting $lA$ be the $P$-isolating class of Theorem 5.3.1, we use a general $S \in |\mathcal{I}_P^s(slA)|$ as a "test surface". Since $\text{mult}_P S \geq s$, using Theorem 5.3.2, (2) gives

$$4sn^2 < (S \cdot H_1 \cdot H_2)_P.$$

On the other hand $l \leq 4/A^3$ by Theorem 5.3.1, so that

$$S \cdot H_1 \cdot H_2 = sln^2 A^3 \leq 4sn^2.$$

Combining the 2 previous inequalities, we have a contradiction.   Q.E.D.

**Proof of (B)**   Assume that $P \in C \subset X_5 \subset \mathbb{P}(1,1,1,1,2)$, where $C = L_i$ is one of the lines of degree $\frac{1}{2}$ (see Example 4.5). Although $|\mathcal{I}_P(A)|$ does not isolate $P$ (it has base locus $C$), we still choose a general surface

$$S \in |\mathcal{I}_P(A)|,$$

to use as a "test surface". Restricting $\mathcal{H}$ to $S$ we obtain

$$\mathcal{H}_{|S} = cC + \mathcal{L},$$

where $\mathcal{L}$ is a mobile linear system on $S$ and $c = \text{mult}_C \mathcal{H} \geq 0$ the multiplicity of $\mathcal{H}$ along $C$.

The argument now proceeds much as before in 3 straightforward steps: (a) we show that two general members $L_1, L_2 \in \mathcal{L}$ have a large intersection number at $P$; (b) we estimate $\mathcal{L}^2 = L_1 \cdot L_2$ from above; (c) we derive a contradiction from the 2 previous steps.

STEP A   First, $S$ is nonsingular at $P$; indeed if $F = x_0 y^2 + f_3 y + f_5$ is quasismooth at $P$ the projection of $P$ to $\mathbb{P}^3$ must be a nonsingular point of $(f_5 = 0)$. We now assume that $P$ is a maximal centre, and apply Theorem 5.3.3 to $\Delta_1 = C$, $1 - a_1 = \frac{c}{n}$ and $1 - a_2 = 0$. Then Theorem 5.3.2, (1) implies that $K_S + \frac{c}{n}C + \frac{1}{n}\mathcal{L}$ is not log canonical and by Theorem 5.3.3,

$$(L_1 \cdot L_2)_P > 4\left(1 - \frac{c}{n}\right)n^2.$$

STEP B   $S$ has an ordinary double point at $P_4 = (0,0,0,0,1)$ and

$$(K_S \cdot C)_S = (K_X + S \cdot C)_X = 0,$$

so that

$$(K_S + C)_{|C} = K_C + \text{Diff} = K_C + \frac{1}{2}P_4,$$

and

$$(C \cdot C)_S = -2 + \frac{1}{2} = -\frac{3}{2}.$$

From this, we calculate

$$L_1 \cdot L_2 = (nA_{|S} - cC)_S^2 = \frac{5}{2}n^2 - nc - \frac{3}{2}c^2.$$

STEP C    From the 2 previous steps, since $\mathcal{L}$ is mobile, we conclude that

$$4\left(1 - \frac{c}{n}\right)n^2 < (L_1 \cdot L_2)_P \leq L_1 \cdot L_2 = \frac{5}{2}n^2 - nc - \frac{3}{2}c^2,$$

which gives $\frac{3}{2}(n-c)^2 < 0$, a contradiction.    Q.E.D.

## 5.4    Determination of $\overline{NE}\,Y$, and singular points with $B^3 \leq 0$

In Sections 5.4–5.5, we show how to exclude *singular* points $P \in X$ as strong maximal centres. Recall from 3.4 that we write $f \colon (E \subset Y) \to (P \in X)$ for the unique extremal divisorial contraction to $P$, and $B = -K_Y$. We begin with a short summary of the contents, then, before getting down to business, we make a few philosophical remarks. There are 3 parts:

5.4   Determination of $\overline{NE}\,Y$ ......................................... 212

5.4.1  Singular points with $B^3 \leq 0$ ....................................... 216

5.5   Bad links, and singular points with $B^3 > 0$ ...................... 217

This section 5.4 introduces a new idea, the $Q = S \cap T$ method, that determines the Mori cone $\overline{NE}\,Y$ in all cases, again modulo some bulky calculations that we relegate to 5.7. In 5.4.1, we use the exact shape of $\overline{NE}\,Y = R + Q$ to construct test classes that exclude all the points with $B^3 \leq 0$, with the single maverick exception (∗∗) of Lemma 5.4.3. In 5.5, we introduce the bad link method, and exclude points with $B^3 > 0$ and the maverick (∗∗). Together, the results give our main Theorems 5.1.3 and 5.1.4 and complete the proof of Theorem 1.3.

It is natural to try to use the test class method to exclude a centre whenever possible. We first observe that this can only work when $B^3 \leq 0$.

**Proposition 5.4.1** *Let $P \in X$ be a point centre and $f \colon (E \subset Y) \to (P \in X)$ an extremal divisorial contraction. If $B^3 > 0$, then $B^2 \in \operatorname{Int} \overline{NE}\,Y$. In other words, the test class method only has a fighting chance if $B^3 \leq 0$.*

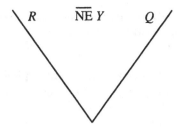

Figure 3: The Mori cone $\overline{NE}\,Y$

**Proof**   Indeed, let $R \subset \overline{NE}\,Y$ be the extremal ray contracted by $f$, and $Q$ the other ray (see Figure 3). If $B \cdot Q > 0$ then $B$ is ample (this never actually happens) and we are done; thus we assume that $B \cdot Q \leq 0$.

Choosing $\Delta \in R$ and $\Gamma \in Q$, we write the 1-dimensional cycle $B^2$ as

$$B^2 = \alpha\Delta + \beta\Gamma;$$

we now show that both $\alpha$ and $\beta$ are $> 0$. First

$$0 < A^3 = A \cdot B^2 = \alpha A \cdot \Delta + \beta A \cdot \Gamma = \beta A \cdot \Gamma,$$

and, since $A \cdot \Gamma > 0$, we get $\beta > 0$. Then

$$0 < B^3 = \alpha B \cdot \Delta + \beta B \cdot \Gamma \leq \alpha B \cdot \Delta,$$

and, since $B \cdot \Delta > 0$, also $\alpha > 0$. Q.E.D.

In fact, we use the test class method in its pure form (5.2) to exclude all singular points with $B^3 \leq 0$ except for the maverick $(**)$, for which $B^2 \in$ Int $\overline{NE}\,Y$ (see Remark 5.4.9 below), so that the test class method cannot be used. In many cases, it is quite easy to find a $P$-isolating system on $X$, satisfying suitable numerical conditions, in the same way as for nonsingular points (and often easier). However, such linear systems quickly become harder and harder to detect. We even found in some examples that the most efficient way to write down the test class was to start by determining $\overline{NE}\,Y$ precisely. Below, we calculate $\overline{NE}\,Y$ in all cases. While, in a few instances (10 in all, listed in 5.7.2), it would have been easier just to write down a convenient isolating class, our presentation has the advantages that

(1) it provides a unified exposition of the material,

(2) it allows us to write down the best possible test class in all cases,

(3) this is precisely the sort of thing that we need in the bad link method, and therefore affords a unified treatment of points with $B^3 \leq 0$ and points with $B^3 > 0$, and

(4) it is the same calculation used to present the anticanonical ring of $Y$ in the case when an untwisting exists, so the method also ties in with the untwisting of Chapter 4.

We do not really have a particularly good explanation as to why some points are excluded and some are not. We discuss this at more length in our concluding remarks 7.7.

### Determination of $\overline{NE}\,Y$

**Notation 5.4.2** If $P \in X = X_d \subset \mathbb{P}(1, a_1, a_2, a_3, a_4)$ is a singular point on $X$ and $f \colon (E \subset Y) \to (P \in X)$ the extremal divisorial contraction (3.4.2), we denote by $S$ the surface

$$S = f_*^{-1}(x_0 = 0).$$

It is always the case that $S \in |B|$ and, if $X$ is general (which we always assume), $S$ is also irreducible. (See the discussion in 7.2.3.)

The main results here are the next two lemmas, whose proof consists of a lengthy but mechanical calculation which we postpone until 5.7 to avoid interrupting the flow of ideas.

**Lemma 5.4.3** *Assume $B^3 \leq 0$. Then, with the sole exception of the following maverick:*

$$\text{No. 82: } X_{36} \subset \mathbb{P}(1, 1, 5, 12, 18), \text{ centre } P_2 = \frac{1}{5}(1, 2, 3), \qquad (**)$$

*there are integers $b, c$ with $b > 0$ and $b/r \geq c \geq 0$ and a surface $T \in |bB + cE|$ such that*

*(1) The intersection $\Gamma = S \cap T$ is an irreducible curve (here and below, we say $S \cap T$ is irreducible to mean that the scheme theoretic complete intersection is a reduced and irreducible curve); and*

*(2) $T \cdot \Gamma \leq 0$.*

*The excluding column of the Big Table lists the class $T \sim bB + cE$ in each case.*

**Proof** See 5.7. □

**Definition 5.4.4** A *special surface* is an irreducible surface $T \in |bB + cE|$, where $b > 0$ and $c < 0$ are integers.

**Lemma 5.4.5** *As usual, let $P \in X$ be a singular point. In all 18 cases labelled "bad link" in the Big Table (this consists of those with $B^3 > 0$, in addition to the maverick (**) of Lemma 5.4.3), there is a special surface $T$ with the properties*

*(1) The scheme theoretic intersection $\Gamma = S \cap T$ is an irreducible curve and*

*(2) $B \cdot \Gamma \leq 0$.*

*The excluding column of the Big Table lists the class $T \sim bB + cE$ in each case.*

**Proof** See 5.7. □

We now discuss some of the implications.

**Corollary 5.4.6** *For $P \in \operatorname{Sing} X$, let $T$ and $\Gamma = S \cap T$ be as in the conclusion of Lemma 5.4.3 or 5.4.5. As usual, write $R$ for the extremal ray of $\overline{\mathrm{NE}}\, Y$ contracted by $f \colon Y \to X$. Then $Q = \mathbb{R}_+[\Gamma]$ is the other ray of $\overline{\mathrm{NE}}\, Y$, so that $\overline{\mathrm{NE}}\, Y = R + Q$. (See Figure 3.)*

**Proof** Clearly $\Gamma \not\subset E$ because $\Gamma \notin R$, and hence $E \cdot \Gamma \geq 0$. Then from the stated inequalities, we read off at once

$$\begin{cases} 0 \geq T \cdot \Gamma \geq bB \cdot \Gamma & \text{in the conclusion of Lemma 5.4.3, and} \\ 0 \geq bB \cdot \Gamma > T \cdot \Gamma & \text{in the conclusion of Lemma 5.4.5.} \end{cases}$$

In the case of Lemma 5.4.3, if $T$ is nef then $T \cdot \Gamma = 0$ and $\Gamma$ is in the boundary of $\overline{\mathrm{NE}}$. Otherwise there exists a curve with $T \cdot C < 0$. Now $C \not\subset E$ (because $T \cdot R \geq 0$), so that necessarily also

$$B \cdot C = \frac{1}{b}(T - cE) \cdot C < 0,$$

and perforce $C \subset S \cap T$. But $S \cap T$ is $\Gamma$, so $C = \Gamma$. The case of Lemma 5.4.5 is similar and left to the reader.   Q.E.D.

**Remark 5.4.7** In fact, the following result also holds, although we don't need it in what follows. Denote by $\overline{\mathrm{NM}}^1 Y \subset N^1 Y$ the closed cone of *effective* divisors. In the case of Lemma 5.4.3, we have

$$\overline{\mathrm{NM}}^1 Y = \mathbb{R}_+[E] + \mathbb{R}_+[B],$$

and, in the case of Lemma 5.4.5,

$$\overline{\mathrm{NM}}^1 Y = \mathbb{R}_+[E] + \mathbb{R}_+[T].$$

**Proof**  Let $D \sim \alpha T + \beta E \geq 0$ be an effective divisor. Assuming that $S$ is not contained in the support of $D$, we show that $\alpha, \beta \geq 0$; this obviously implies the statement. Restricting to $S$, we have

$$D_{|S} = \alpha T_{|S} + \beta E_{|S} = \alpha \Gamma + \beta \Delta \geq 0,$$

where we write $\Delta = E_{|S}$. Note that $\Gamma$ and $\Delta$ are irreducible curves on $S$ with $\Gamma^2 \leq 0$ (by Lemmas 5.4.3 and 5.4.5) and $\Delta^2 < 0$ (because $\Delta$ is contractible).

We now prove that if $C \sim \alpha \Gamma + \beta \Delta \geq 0$ is any effective divisor on $S$ then $\alpha, \beta \geq 0$. We may assume that neither $\Gamma$ nor $\Delta$ is contained in the support of $C$. Note that one of $\alpha$ or $\beta \geq 0$, since otherwise $-C > 0$.

If $\alpha \geq 0$ then

$$0 \leq C \cdot \Gamma = (\alpha \Gamma + \beta \Delta) \cdot \Gamma = \alpha \Gamma^2 + \beta \Delta \cdot \Gamma \leq \beta \Delta \cdot \Gamma.$$

Then since $\Delta \cdot \Gamma > 0$, necessarily $\beta \geq 0$. A similar argument shows that $\beta \geq 0$ implies $\alpha \geq 0$.   Q.E.D.

### 5.4.1   Singular points with $B^3 \leq 0$

We now prove Theorem 5.1.3 with the one exception.

**Theorem 5.4.8**  *Assume that $B^3 \leq 0$. Then, with the sole exception of the maverick case (∗∗) of Lemma 5.4.3, we have $B^2 \notin \operatorname{Int} \overline{NE}\, Y$.*

*In particular, $P$ is not a strong maximal centre by Lemma 5.2.1.*

We stress that the maverick is of course not an exception to $P \in X$ being excluded as a maximal centre (see Theorem 5.5.1 below).

**Proof**  Let $R, Q$ be as before, so that $\overline{NE}\, Y = R + Q$ (see Figure 3). By Corollary 5.4.6,

$$Q = \mathbb{R}_+ \big[ B \cdot (bB + cE) \big]$$

for some integers $b > 0$, $c \geq 0$. This shows that $B^2 \notin \operatorname{Int} \overline{NE}\, Y$.   Q.E.D.

**Remark 5.4.9**  The test class method cannot be used in the case of the maverick (∗∗). Indeed, we know from Lemma 5.4.5 that a "special surface" $T = f_*^{-1}(x_1 = 0) \in |B - E|$ exists such that $T \cdot B^2 < 0$, which shows that $B^2 \in \operatorname{Int} \overline{NE}\, Y$.

## 5.5   Bad links, and singular points with $B^3 > 0$

The next result completes the proof of Theorem 5.1.3 and proves 5.1.4.

**Theorem 5.5.1** *Let $P \in X$ be a singular point and $f\colon (E \subset Y) \to (P \in X)$ the Kawamata blowup. If a special surface $T \in Y$ exists as in the conclusion of Lemma 5.4.5, $P$ is not a strong maximal centre.*

**Proof**   Assume by contradiction that $P$ is a strong maximal centre. We start from the Kawamata blowup $f\colon (E \subset Y) \to (P \in X)$, the unique extremal divisorial extraction centred at $P$ (Theorem 3.4.2), and play a 2-ray game on $Y$ from the starting position given in Corollary 5.4.6: $\overline{\mathrm{NE}}\, Y = R + Q$, where $R$ is the ray contracted by $f$, and $Q = \mathbb{R}_+[\Gamma]$. (Consult [Co2], Chapter 1 for the rules and general conventions of the 2-ray game.)

By the Sarkisov program [Co1], the 2-ray game must exist and terminate in a Sarkisov link; we show that, on the contrary, the game is always aborted, that is, the "link" goes "bad". The crux of the argument is that the 2-ray game must take place entirely in the Mori category: if contracting a ray leads at any stage to a canonical but nonterminal singularity, or to a fibre space other than a Mori fibre space, the rules of the game are violated. This is a contradiction, and proves that $P$ is not a strong maximal centre.

The analysis breaks up in 2 cases: $B \cdot \Gamma = 0$ and $B \cdot \Gamma < 0$.

CASE 1: $B \cdot \Gamma = 0$ (16 instances)   Let $\varphi\colon Y \to W$ be the contraction of the extremal ray $Q = \mathbb{R}_+[\Gamma]$. First, note that $\varphi$ cannot be a flopping contraction: if it were, $\Gamma$ would be a contractible curve in $T$, which is impossible, since $\Gamma = S \cap T$ is a $\mathbb{Q}$-Cartier divisor on $T$ with selfintersection

$$(\Gamma^2)_T = B^2 \cdot T = B \cdot \Gamma = 0.$$

It follows that $\varphi$ is either a divisorial contraction, with the special surface $T$ as exceptional divisor, or is a fibre space. In either case the 2-ray game is aborted. If $\varphi$ is birational, then since $K_Y \cdot \Gamma = -B \cdot \Gamma = 0$, it is crepant, that is, the exceptional divisor $T$ has discrepancy $a(T, K_W) = 0$ over $W$, so that $W$ has a canonical but nonterminal singularity. Also $K_Y \cdot \Gamma = 0$ implies that $\varphi$ is not a Mori fibre space. (If $\dim W < 3$, it is a fibre space of elliptic curves or K3 surfaces.) This concludes Case 1.

CASE 2: $B \cdot \Gamma < 0$   Although we don't actually need this in what follows, we note first that there are only 3 examples, namely

- No. 47, $X_{21} \subset \mathbb{P}(1, 1, 5, 7, 8)$, centre $P = P_2 = \frac{1}{5}(1, 2, 3)$

- No. 62, $X_{26} \subset \mathbb{P}(1, 1, 5, 7, 13)$, centre $P = P_2 = \frac{1}{5}(1, 2, 3)$

- No. 82, $X_{36} \subset \mathbb{P}(1,1,5,12,18)$, centre $P = P_2 = \frac{1}{5}(1,2,3)$; this is the maverick case $(**)$ of Lemma 5.4.3 with $B^3 = 0$, excluded from Theorem 5.4.8.

Write $\varphi \colon Y \to W$ for the contraction of the extremal ray $Q = \mathbb{R}_+[\Gamma]$. Then $-K_Y \cdot \Gamma = B \cdot \Gamma < 0$ means that $\varphi$ is an *inverse flipping* contraction. Set $m_1 = -S \cdot \Gamma$, $m_2 = -T \cdot \Gamma$. Then $m_1, m_2 > 0$, and, locally around $\Gamma$, we obtain the inverse flip $t \colon Y \dashrightarrow Y'$ as the normalisation of the map $t \colon Y \dashrightarrow \mathbb{P}^1 \times W$ given by the linear system $|m_1 T, m_2 S|$; in particular, it exists. If $S', T'$, etc. are the birational transforms, this implies in particular that $S' \cap T' = \emptyset$. We write $-K_{Y'} = B' \sim S'$.

The structure morphism $\varphi' \colon Y' \to W$ is the flipping contraction of an extremal ray $R' = \mathbb{R}_+[\Gamma']$ of $\overline{\mathrm{NE}}\,Y'$; let $Q'$ be the other extremal ray. We show first that $T' \cdot Q' < 0$. The usual flip formalism together with the case assumption $B \cdot \Gamma < 0$ on $Y$ gives $B' \cdot \Gamma' > 0$ on $Y'$; but $B'$ is zero on any curves in $T'$, so that $B' \cdot Q' \le 0$. In the same way, $E \cdot \Gamma > 0$ implies $E' \cdot \Gamma' < 0$, and hence (since an effective divisor is positive on some curves) $E' \cdot Q' > 0$. Therefore

$$T' \cdot Q' = (bB' + cE') \cdot Q' < 0$$

(recall that $c < 0$ in Lemma 5.4.5).

It follows that every curve $C \in Q'$ is contained in $T'$. Thus $S' \cdot Q' = 0$, and hence the contraction

$$\psi \colon Y' \to W'$$

of the ray $Q'$ is given by $|nS'|$ for some $n \gg 0$ (anticanonical model). Because $S' \cap T' = \emptyset$, this contracts $T'$ to a point, and does not contract any curves not in $T'$. Thus $\psi$ is a divisorial contraction with exceptional divisor $T'$, and is crepant over $W'$, so that $W'$ has nonterminal singularities.   Q.E.D.

**Remark 5.5.2** For all these bad links, we can view the 2-ray game entirely in terms of toric modifications of the ambient projective space. As such, they are basically similar to the quadratic involutions of Chapter 4, except that they lead to a contradiction rather than an untwisting. The first troop of 16 all fit into the pattern

No. 21 $X_{14} \subset \mathbb{P}(1,1,2,4,7)$, centre $P_3 = \dfrac{1}{4}(1,1,3)$

$\dashrightarrow Z_{18} \subset \mathbb{P}(1,1,2,6,9)$ given by $(x_0, x_1, y, z, t) \mapsto (x_0, x_1, y, yz, yt)$

$\dots$

No. 63 $X_{26} \subset \mathbb{P}(1,2,3,8,13)$, centre $P_3 = \dfrac{1}{8}(1,3,5)$

$\dashrightarrow Z_{30} \subset \mathbb{P}(1,2,3,10,15)$ given by $(x, y, z, t, u) \mapsto (x, y, z, yt, yu)$

etc. (we give a hint on how to do the calculation in 7.2), whereas the final eccentric trio involving inverse flips go thus:

No. 47, $X_{21} \subset \mathbb{P}(1,1,5,7,8)$, centre $P = P_2 = \frac{1}{5}(1,2,3)$

$\dashrightarrow Z_{24} \subset \mathbb{P}(1,1,6,8,9)$ given by $(x_0, x_1, y, z, t) \mapsto (x_0, x_1, x_1y, x_1z, x_1t)$

No. 62, $X_{26} \subset \mathbb{P}(1,1,5,7,13)$, centre $P = P_2 = \frac{1}{5}(1,2,3)$

$\dashrightarrow Z_{30} \subset \mathbb{P}(1,1,6,8,15)$ given by $(x_0, x_1, y, z, t) \mapsto (x_0, x_1, x_1y, x_1z, x_1^2t)$

No. 82, $X_{36} \subset \mathbb{P}(1,1,5,12,18)$, centre $P = P_2 = \frac{1}{5}(1,2,3)$

$\dashrightarrow Z_{42} \subset \mathbb{P}(1,1,6,14,21)$ given by $(x_0, x_1, y, z, t) \mapsto (x_0, x_1, x_1y, x_1^2z, x_1^3t)$

In each case, the final model $Z$ has $-K_Z$ ample, but has a strictly canonical singularity, so is not a Fano 3-fold in the Mori category.

## 5.6 Proof of Theorem 5.3.1: construction of the test class for nonsingular points

Theorem 5.3.1 is trivial for the quartic $X_4 \subset \mathbb{P}^4$, and we exclude this case here. Fix a general Fano 3-fold hypersurface $X = X_d \subset \mathbb{P} = \mathbb{P}(1, a_1, a_2, a_3, a_4)$. This section discusses how to prove that a suitable class $lA$ on $X$ is $P$-isolating for a nonsingular point $P \in \operatorname{NonSing} X$.

**Notation 5.6.1** We write $x_0, x_1, \ldots, x_4$ (or $x, y, \ldots$ etc.) for homogeneous coordinates on $\mathbb{P}$, and $P = (\xi_0, \xi_1, \xi_2, \xi_3, \xi_4)$ (or $P = (\xi, \eta, \ldots)$) etc. for the coordinates of a point $P$. Let

$$\pi \colon X \dashrightarrow \mathbb{P}(1, a_1, a_2, a_3)$$

be the projection to the first 4 coordinates. Note that $\pi$ is a morphism except possibly at $P_4$; it is generically finite, contracting at most a finite set of curves of $X$, that we write $\operatorname{Exc} \pi$.

**Theorem 5.6.2** *With the exception of the renegade (*) of 5.3 (p. 208), we get $P$-isolating classes $lA$ (Definition 5.2.4) on $X$ as follows:*

*(1) If $\xi_0 \neq 0$ then $a_4A$ isolates $P$; if moreover $P \notin \operatorname{Exc} \pi$, then already $a_3A$ isolates $P$.*

*(2) If $\xi_0 = 0$ and $\xi_1 \neq 0$ then $a_1a_3A$ isolates $P$.*

*(3) If $\xi_0 = \xi_1 = 0$, set*

$$l = \min\Big\{ \{\mathrm{lcm}(a_i, a_j)\} \cup \{a_i + a_j \mid a_k \text{ divides } a_i + a_j\} \Big\},$$

*(taken over all $\{i, j, k\} = \{2, 3, 4\}$); then $lA$ isolates $P$.*

**Proof of Theorem 5.3.1**    The conclusion

$$l \leq 4/A^3.$$

is a straightforward corollary of this claim, together with a direct inspection of the list.    Q.E.D.

We can write the equation of $X$ in one of the forms[2]

$$F = x_4^3 + ax_4 + b \quad \text{or} \quad x_4^2 + b \quad \text{or} \tag{15}$$

$$= x_j x_4^2 + ax_4 + b \quad \text{with } j \neq 4, \tag{16}$$

where $a(x_0, \ldots, x_3)$ and $b(x_0, \ldots, x_3)$ are weighted homogeneous polynomials of appropriate degrees. In case (15), $\pi$ is a finite morphism; in case (16), Exc $\pi$ : $(x_j = a = b = 0)$, which is a finite set of curves, by the argument given in Example 4.5.

**Definition 5.6.3**    Let $\{g_i\}$ be a finite set of homogeneous polynomials of degrees $l_i$ in the variables $x_0, \ldots, x_4$. We say that the set $\{g_i\}$ *isolates $P$* if $P$ is a component of the set

$$X \cap \bigcap_i (g_i = 0).$$

**Lemma 5.6.4**    *If $\{g_i\}$ isolate $P$, then $lA$ isolates $P$, where $l = \max\{l_i\}$.*

**Proof**    This is the point at which we use $s > 1$ in Definition 5.2.4 of isolating: write $d = \mathrm{lcm}\{l_i\}$ for the least common multiple of the degrees of the $g_i$. By replacing each $g_i$ by its power $g_i^{d/l_i}$, we arrange that they all have the same degree $d = sl$, and all belong to $\mathcal{I}_P^s(slA)$.    Q.E.D.

We now show how to find a set of isolating polynomials of reasonably small degree. We work with the case division of Theorem 5.6.2.

---

[2]This can be proved by inspecting the list, or directly by a Newton polygon argument. Namely, by [Fl], Section 8, for each $i$, quasismoothness implies that one of $x_i^{n_i}$ or $x_i^{n_i} x_j$ appears in $F$, and so $\sum_{i=1}^4 \frac{1}{n_i} \geq 1$ by the adjunction formula $d = a_1 + a_2 + a_3 + a_4$ for a K3 $S_d \subset \mathbb{P}(a_1, \ldots, a_4)$. Therefore $\frac{1}{n_4} \geq \frac{1}{4}$, and $1/4$ is only possible for $X_4 \in \mathbb{P}^4$. In all other cases, the $x_4$ of biggest degree appears in a quadratic or cubic monomial.

CASE 1: $\xi_0 \neq 0$   The following is obviously an isolating set of polynomials:

$$\left\{x_1', x_2', x_3', x_4'\right\}, \quad \text{where } x_i' = x_i - \frac{\xi_i}{\xi_0^{a_i}} x_0^{a_i}.$$

We deduce that $a_4 A$ isolates $P$. Moreover, if $P \notin \operatorname{Exc} \pi$, the set

$$\left\{x_1', x_2', x_3'\right\},$$

isolates $P$, and hence so does $a_3 A$. This concludes Case 1.

CASE 2: $\xi_0 = 0$ and $\xi_1 \neq 0$   With some exceptions to be discussed shortly, the following set of polynomials isolates $P$:

$$\left\{x_0, \; \xi_1^{a_2} x_2^{a_1} - \xi_2^{a_1} x_1^{a_2}, \; \xi_1^{a_3} x_3^{a_1} - \xi_3^{a} x_1^{a_3}\right\}, \tag{$*$}$$

which gives Theorem 5.6.2, (2). Since $(*)$ gives us $x_0$, $x_2$ and $x_3$ as algebraic functions of $x_1$, it only remains to restrict $x_4$, and this comes from $F$ unless we are in case (16) and $P \in \operatorname{Exc} \pi$. Thus we are home if $(x_0 = 0) \cap \operatorname{Exc} \pi = \emptyset$. This is equivalent to saying that $a, b$ restricted to the affine line $x_0 = x_j = x_4 = 0$, $x_1 \neq 0$ do not have common zeros. This holds in almost all cases, provided that $X$ is general. Consider for example No. 31, $X_{16} \subset \mathbb{P}(1, 1, 4, 5, 6)$, with equation $F = yt^2 + a_{10}t + b_{16}$; after setting $x_0 = y = t = 0$, we have the $x_1 z$ line, and $a, b$ have no common zeros in general because we could take $a = z^5, b = x_1^{16}$.

However, there is a bizarre quartet of exceptions to this, namely:

- No. 69, $X = X_{28} \subset \mathbb{P}(1, 4, 6, 7, 11)$

- No. 74, $X = X_{30} \subset \mathbb{P}(1, 3, 4, 10, 13)$

- No. 76, $X = X_{30} \subset \mathbb{P}(1, 5, 6, 8, 11)$

- No. 79, $X = X_{33} \subset \mathbb{P}(1, 3, 5, 11, 14)$.

(There are altogether about 40 cases to check, exactly the cases in the Big Table where $P_4$ is associated with a quadratic involution.)

We work out the first of these in detail to show what goes wrong, and to explain how to fix it. In this case the equation is $zu^2 + a_{17}u + b_{28}$, and $\operatorname{Exc} \pi$ is given by

$$z = a_{17}(x, y, z, t) = b_{28}(x, y, z, t) = 0.$$

The peculiar thing here is that $a_{17}(0, y, z, t) = yzt$ has $z$ as a factor, so that $a$ vanishes identically on the affine $yt$ line, and $b_{28}(0, y, 0, t) = 0$ gives a common zero (say, the point $y^7 = t^4$); in other words, (*) fails to isolate $P$. If $P = (0, \eta, 0, \tau, v)$, however, we obtain an isolating set by simply adding the polynomial

$$\eta\tau u - vyt$$

of degree $11 \leq a_1 a_3 = 28$, so it is still true that $a_1 a_3 A$ isolates $P$. The other cases can be cured similarly.

CASE 3: $P = (0, 0, \xi_2, \xi_3, \xi_4)$   We isolate $P$ using $x_0$, $x_1$, and one further polynomial, which we choose to be of smallest degree $\gamma$ among the following

$$\xi_j^{a_i'} x_i^{a_j'} - \xi_i^{a_j'} x_j^{a_i'}, \quad \xi_i \xi_j x_k^{c_k} - \xi_k^{c_k} x_i x_j,$$

where $\operatorname{lcm}(a_i, a_j) = a_i' a_j$ and $a_i + a_j = a_k c_k$ when $a_k \mid a_i + a_j$. For instance, consider No. 13, $P = (0, 0, \eta, \zeta, \tau) \in X_{11} \subset \mathbb{P}(1, 1, 2, 3, 5)$. Then $\gamma = 5$ and the following is a $P$-isolating set

$$\{x_1, x_2, \eta\zeta t - \tau yz\}.$$

In saying this, we use that $\Gamma : (x_0 = x_1 = 0) \cap X$ is an irreducible curve for general $X$, and so does not contain the curve $(x_0 = x_1 = \eta\zeta t - \tau yz = 0)$.   Q.E.D.

## 5.7   Singular points: construction of $T$

We prove Lemmas 5.4.3 and 5.4.5. There are 2 subsections

5.7.1 Hyperplane sections and easy cases

5.7.2 Harder cases

Most of the calculations are fairly trivial, and we treat these in 5.7.1. However, there are 10 harder cases, which we collect and work out in voluptuous detail in 5.7.2; one of these is a real champion that deserves, and gets, an extraordinarily florid treatment.

### 5.7.1   Hyperplane sections and easy cases

In almost all of the cases (to be precise, all but 10, several hundred in all), we obtain $T \in Y$ simply as the birational transform of a suitably chosen hyperplane section: after slightly adjusting the coordinate system if necessary, we can take it to be $(x_i = 0) \subset Y$. In other words, we have

$$T = f_*^{-1}(x_i = 0) \quad \text{for some } i \in \{1, 2, 3, 4\}.$$

All the special surfaces of Lemma 5.4.5 arise thus. Here the calculations are quick and entirely straightforward, but (apart from the special surfaces, which occur predictably, see below) it seems difficult to discern a common pattern. We give several examples illustrating that any coordinate $x_i$ can appear, and in a variety of combinations. For the convenience of the reader and ease of reference, we separate the cases with $B^3 \leq 0$ from the special surfaces.

## $B^3 \leq 0$

**Example 5.7.1** No. 9, $X_9 \subset \mathbb{P}(1, 1, 2, 3, 3)$, centre $P = P_2 = \frac{1}{2}(1, 1, 1)$

Recall (7.2.3) that $S : (x_0 = 0)$. Take $T : (x_1 = 0) \in |B|$, so that $(b, c) = (1, 0)$ in the third column of the Big Table. The curve $\Gamma$ is the birational transform of $C_9 \subset \mathbb{P}(2, 3, 3)$, typically given by $y^3 z_1 + z_1^3 + z_2^3$. This is irreducible and has degree $\frac{1}{2}$ in $\mathbb{P}$, but passes through $P$, and $T \cdot \Gamma$ drops by $\frac{1}{2}$ on blowing up. Alternatively, we calculate $T \cdot \Gamma = T^2 B = B^3 = (A - \frac{1}{2}E)^3 = \frac{1}{2} - \frac{1}{2} = 0$

**Example 5.7.2** No. 18, $X_{12} \subset \mathbb{P}(1, 2, 2, 3, 5)$, centre $P \in P_1 P_2 = 6 \times \frac{1}{2}(1, 1, 1)$

We first change coordinates so that $P = P_1$, the point at which $y_0 = 1$, all the other coordinates $= 0$. We may write the equation of $X$ as

$$F = y_0^5 y_1 + O(\tfrac{2}{2}E),$$

where $O(\frac{2}{2}E)$ means a function that vanishes to order at least $\frac{2}{2}$ along $E$, in the same sense as in Proposition 3.4.6, (2). We then take $T : (y_1 = 0) \in |2A - E| = 2B$.

**Example 5.7.3** No. 43, $X_{20} \subset \mathbb{P}(1, 2, 4, 5, 9)$, centre $P \in P_1 P_2 = 5 \times \frac{1}{2}(1, 1, 1)$

As before, we change coordinates so that $P = P_1$, and write the equation of $X$ as

$$F = y^8 z + O(\tfrac{2}{2}E).$$

Then set $T : (z = 0) \in |4A - E| = |4B + E|$. A small calculation shows

$$(4B + E)^2 B = (4A - E)^2 (A - \frac{1}{2}E) = \frac{16}{18} - 2 < 0.$$

**Example 5.7.4** No. 23, $X_{14} \subset \mathbb{P}(1, 2, 3, 4, 5)$, centre $P \in P_1 P_3 = 3 \times \frac{1}{2}(1, 1, 1)$

We may assume that $P = P_1$, and write the equation of $X$ as

$$F = y^5 t + O(\tfrac{2}{2}E),$$

set $T : (t = 0) \in |4B + E|$, and check that $(4B + E)^2 \cdot B \leq 0$.

**Example 5.7.5** No. 33, $X_{17} \subset \mathbb{P}(1,2,3,5,7)$, centre $P = P_1 = \frac{1}{2}(1,1,1)$

We assume by generality that the monomial $y^5u$ appears in $F$. Then all terms divisible by $y^5$ in $F$ can be collected together as $y^5u'$ by setting $u' = u + yt + y^2t + y^3x$. Write the equation of $X$ as

$$F = y^5u' + O(\tfrac{3}{2}E).$$

Then set $T : (u' = 0) \in |7A - \frac{3}{2}E| = |7B + 2E|$. We have

$$(7B + 2E)^2B = (7A - \frac{3}{2}E)^2(A - \frac{1}{2}E) = 49 \times \frac{17}{210} - \frac{9}{2} < 0.$$

**Special surfaces (Lemma 5.4.5)**

There are 18 special surfaces in all. We give $3 + 1$ examples, then briefly comment on the general shape of special surfaces.

**Example 5.7.6** No. 10, $X_{10} \subset \mathbb{P}(1,1,1,3,5)$, centre $P = P_3 = \frac{1}{3}(1,1,2)$

Choosing suitable coordinates $x_0, x_1, x_2, y, z$ on $\mathbb{P}$, the equation of $X$ is

$$F = y^3(x_0 + x_1 + x_2) + O(\tfrac{4}{3}E).$$

The special surface is then given by

$$T : (x_1 = 0) \in |A - \frac{4}{3}E| = |B - E|;$$

this vanishes along $E$ to unexpectedly high order. We easily check that

$$T \cdot B^2 = (A - \frac{4}{3}E)(A - \frac{1}{3}E)^2 = \frac{2}{3} - \frac{4}{6} = 0.$$

**Example 5.7.7** No. 21, $X_{14} \subset \mathbb{P}(1,1,2,4,7)$, centre $P = P_3 = \frac{1}{4}(1,1,3)$

Again, in suitable coordinates, we may write the equation of $X$ as

$$z^3y + O(\tfrac{6}{4}E),$$

and $T : (y = 0) \in |2B - E|$. We easily calculate

$$T \cdot B^2 = (2A - \frac{6}{4}E)(A - \frac{1}{4}E)^2 = \frac{2}{4} - \frac{6}{12} = 0.$$

**Example 5.7.8** No. 55, $X_{24} \subset \mathbb{P}(1,2,3,7,12)$, centre $P = P_3 = \frac{1}{7}(1,2,5)$

We choose coordinates so that the equation of $X$ is

$$F = t^3z + O(\tfrac{10}{7}E),$$

set $T : (z = 0) \in |3B - E|$, and calculate $T \cdot B^2 = 0$.

**Example 5.7.9** No. 82, $X_{36} \subset \mathbb{P}(1,1,5,12,18)$, centre $P = P_2 = \frac{1}{5}(1,2,3)$

This is slightly different from the above. Choose coordinates so that the equation of $X$ is

$$F = y^7 x_1 + O(\tfrac{6}{5}E),$$

set $T : (x_1 = 0) \in |B - E|$, calculate

$$T \cdot B^2 = (A - \frac{6}{5}E)(A - \frac{1}{5}E)^2 = \frac{1}{30} - \frac{6}{30} < 0.$$

**Remark 5.7.10** Special surfaces $T$ follow a nice general pattern, namely,

$$\begin{cases} \text{either} & P = P_3, \quad T : (z = 0), \quad T \cdot B^2 = 0; \\ \text{or} & P = P_2, \quad T : (y = 0), \quad T \cdot B^2 < 0. \end{cases}$$

### 5.7.2  Harder cases

There remain 10 cases to discuss, honoured with a star in the Big Table in recognition of their distinguished difficulty. In each of these harder cases, as we shall soon see, it is easy to locate a $P$-isolating linear system $\mathcal{L}$ on $X$ that gives a test class $M$ on $Y$ with $M \cdot B^2 \le 0$, which shows, of course, that $B^2 \notin \overline{NE}\,Y$ and immediately excludes $P \in X$ as a strong maximal centre, just as in 5.2. Thus, for the proof of Theorem 1.3, we do not actually need to do all this work to determine $\overline{NE}\,Y$ precisely, and the impatient reader may fast forward through this section.

Taking into account affinities of behaviour, these 10 cases divide naturally into three groups, containing 6, 3, and 1 examples.

**First group**

**Example 5.7.11** No. 19, $X_{12} \subset \mathbb{P}(1,2,3,3,4)$, centre $P \in P_1 P_4 = 3 \times \frac{1}{2}(1,1,1)$

We prove that $M = 3B + E$ is a test class, and $T \in |6B + E|$.

Changing coordinates, we may assume that $P = P_1$. We write everything in the coordinate patch $(y = 1)$ of the weighted projective space, which is an affine quotient of type $\frac{1}{2}(1,1,1,0)$. Abusing notation slightly, we still denote the coordinate functions by $x, z_0, z_1, t$, of weights $1, 1, 1, 2$ locally around $P$.

Note first that $\{x, z_0, z_1, t\}$ is a $P$-isolating set of polynomials, giving rise to the test class

$$M = 3A - \frac{1}{2}E = 3B + E$$

on $Y$. A small calculation gives

$$M \cdot B^2 = (3A - \frac{1}{2}E)(A - \frac{1}{2}E)^2 = \frac{3}{6} - \frac{1}{2} = 0,$$

which shows that $B^2 \notin \text{Int} \overline{NE}\, Y$, and excludes $P$ as a maximal centre. The reader only interested in the proof of the rigidity theorem may stop reading now and move on to the next example, while we proceed to construct the surface $T$.

We expand the equation $F$ of $X$ as a sum of polynomials which are weighted homogeneous with respect to the local weighting

$$F = F_2 + F_4 + \cdots$$
$$= (t + z^2 + zx + x^2) + F_4 + \cdots,$$

where, for brevity, $z^2$ stands for an arbitrary quadratic polynomial in $z_0, z_1$, etc., and we simply omit coefficients in front of monomials. We only write out the first term of the expansion, which is all that we need. Indeed, it is clear that the function

$$(t + z^2 + zx + x^2)_{|X}$$

vanishes to order at least $\frac{4}{2}$ along $E$, in other words, reinstating $y$,

$$u = z^2 + y(t + zx) + y^2 x^2 \in |6A - \frac{4}{2}E| = |6B + E|.$$

Let $T : (u = 0)$. Using Bertini, it is easy to check that, if $X$ is general, the curve $\Gamma = T \cap S$ is irreducible, and a small calculation gives

$$T \cdot \Gamma = (6A - \frac{4}{2}E)^2 \cdot (A - \frac{1}{2}E) = \frac{36}{6} - \frac{16}{2} < 0.$$

**Remark 5.7.12** There are 4 other cases that behave almost exactly the same. The list, including the case just discussed, is:

- No. 19, $X_{12} \subset \mathbb{P}(1, 2, 3, 3, 4)$, centre $P \in P_1 P_4 = 3 \times \frac{1}{2}(1, 1, 1)$

- No. 38, $X_{18} \subset \mathbb{P}(1, 2, 5, 5, 8)$, centre $P \in P_1 P_4 = 2 \times \frac{1}{2}(1, 1, 1)$

- No. 66, $X_{27} \subset \mathbb{P}(1, 5, 6, 7, 9)$, centre $P = P_1 = \frac{1}{5}(1, 1, 4)$

- No. 75, $X_{30} \subset \mathbb{P}(1, 4, 5, 6, 15)$, centre $P = P_1 = \frac{1}{4}(1, 1, 3)$

- No. 77, $X_{32} \subset \mathbb{P}(1, 2, 5, 9, 16)$, centre $P \in P_1 P_4 = 2 \times \frac{1}{2}(1, 1, 1)$

In all cases, after a coordinate change, $P = P_1$, $\{x, z, t, u\}$ is a $P$-isolating set of polynomials producing a test class $M$ with $M \cdot B^2 \leq 0$, and

$$T = \text{either } \{z^2 + y(t+?)\} \text{ or } \{t^2 + y(z+?)\},$$

(depending on the example). The calculations are straightforward and left to the reader.

The final example in this group differs only slightly from those just discussed.

**Example 5.7.13** No. 86, $X_{38} \subset \mathbb{P}(1, 5, 6, 8, 19)$, centre $P = P_1 = \frac{1}{5}(1, 1, 4)$
  We will show that $M = 6B + E$ is a test class, and $T \in |18B + 2E|$.
  First, the $P$-isolating set $\{x, z, t, u\}$ gives the test class $M = 6A - \frac{1}{5}E$, and one checks that $M \cdot B^2 \leq 0$, which excludes $P$ as a maximal centre. The reader only interested in the proof of the rigidity theorem may pass on to the next case.
  We write the equation of $X$ as follows

$$F = y^4\left(z^3 + yz^2x + y^2(zx^2 + t) + y^3x^3\right) + O(\tfrac{8}{5}E),$$

where $O(\frac{8}{5}E)$ denotes a function which vanishes $\frac{8}{5}$ times along $E$. Because $y$ is a unit at $P_1$, this says that

$$v = z^3 + yz^2x + y^2(zx^2 + t) + y^3x^3 \in |18A - \frac{8}{5}E| = |18B + 2E|.$$

This is all very nice, since

$$(18B + 2E)^2B = 4(9A - \frac{4}{5}E)^2(A - \frac{1}{5}E) = 4\left(\frac{81}{120} - \frac{16}{20}\right) < 0.$$

As before we set $T = (v = 0)$, and it is easy to check that $\Gamma = S \cap T$ is irreducible.

**Second group**

**Example 5.7.14** No. 81, $X_{34} \subset \mathbb{P}(1, 4, 6, 7, 17)$, centre $P \in P_1P_2 = 2 \times \frac{1}{2}(1, 1, 1)$
  We will show that $M = 17B + 8E$ is a test class, and $T \in |12B + E|$.
  Indeed, $\{x, t, u\}$ is a $P$-isolating set, the corresponding test class is $M = 17A - \frac{1}{2}E = 17B + 8E$ and

$$M \cdot B^2 = (17A - \frac{1}{2}E)(A - \frac{1}{2}E)^2 = \frac{17}{84} - \frac{1}{2} < 0;$$

hence, $P$ is not a maximal centre.

It is also easy to determine $T$. Indeed

$$F(0, y, z, 0, 0) = yz(y^6 + y^3 z^2 + z^4) = yz(z^2 - \alpha^2 y^3)(z^2 - \beta^2 y^3),$$

where we may assume that $P = (0, 1, \alpha, 0, 0)$. We simply need to observe that

$$v = z^2 - \alpha^2 y^3$$

vanishes along $E$, that is, $v \in |12A - E| = |12B + 5E|$, and

$$(12B + 5E)^2 (A - \frac{1}{2}E) = 4(6A - \frac{1}{2}E)^2 (A - \frac{1}{2}E) = 4\left(\frac{36}{84} - \frac{1}{2}\right) < 0.$$

Then set $T = (v = 0)$, etc.

The next example is similar but harder

**Example 5.7.15** No. 60, $X_{24} \subset \mathbb{P}(1, 4, 5, 6, 9)$, centre $P \in P_1 P_3 = 2 \times \frac{1}{2}(1, 1, 1)$

We will show that $M = 9B + 4E$ is a test class, and $T \in |18B + 7E|$.

$\{x, z, u\}$ isolates $P$, the corresponding test class is $M = 9A - \frac{1}{2}E = 9B + 4E$ and $M \cdot B^2 = \frac{9}{45} - \frac{1}{2} < 0$, so $P$ is not a maximal centre.

To find $T$, we first try to proceed as before

$$F(0, y, 0, t, 0) = (t^2 - \alpha^2 y^3)(t^2 - \beta^2 y^3)$$

and, assuming that $P = (0, 1, 0, \alpha, 0)$,

$$v = t^2 - \alpha^2 y^3 \in |12A - E| = |12B + 5E|.$$

Unfortunately

$$(12B + 5E)^2 B = 4(6A - \frac{1}{2}E)^2 (A - \frac{1}{2}E) = 4\left(\frac{36}{45} - \frac{1}{2}\right) > 0.$$

Since this doesn't work, there must be a function vanishing to higher order along $E$. Let us write the equation of $X$ as

$$F(x, y, z, t, u) = (t^2 - \alpha^2 y^3)(t^2 - \beta^2 y^3) + $$
$$+ t\big(y(t^2 + y^3)x^2 + y^2 z^2 + u^2 + (t^2 + y^3)xz + y^2 xu + yzu\big)$$
$$+ O_{x,z,u}(4).$$

Now simply take

$$v = \left(\frac{\beta^2 - \alpha^2}{\alpha^2}\right) t(t^2 - \alpha^2 y^3) + $$
$$y(t^2 + y^3)x^2 + y^2 z^2 + u^2 + (t^2 + y^3)xz + y^2 xu + yzu,$$

so that $F = tv + \frac{\beta^2}{\alpha^2}(t^2 - \alpha^2 y^3)^2 + O_{x,z,u}(4)$, and $v$ vanishes along $\frac{4}{2}E$, hence $T = (v = 0) \in |18A - 2E| = |18B + 7E|$, etc.

**Remark 5.7.16** The final example in this group is entirely analogous to that just shown, and we will not do it in detail:

No. 69, $X_{28} \subset \mathbb{P}(1,4,6,7,11)$, centre $P \in P_1P_2 = 2 \times \frac{1}{2}(1,1,1)$.

Here $M = 11B + 5E$ is a test class, and $T \in |22B + 9E|$.

### 5.7.3 The amazing example

The final case, because of its crazy complexity, deserves a place of its own. We speak of

No. 39, $X_{18} \subset \mathbb{P}(1,3,4,5,6)$, centre $P \in P_1P_4 = 3 \times \frac{1}{3}(1,1,2)$.

We will show that $M = 5B + E$ is a test class, and $T \in |20B + 3E|$.

To begin with, note that $u \in |6A - \frac{3}{3}E| = |6B + E|$ (unfortunately, you can check that $(6B+E)^2 \cdot B > 0$, so we are not done yet) and $t \in |5A - \frac{2}{3}E| = |5B + E|$, so that $\{x, t, u\}$ is a $P$-isolating set of polynomials, giving rise to the test class $M = 5B + E$. A small calculation yields

$$M \cdot B^2 = (5B + E)B^2 = (5A - \frac{2}{3}E)(A - \frac{1}{3}E)^2 = \frac{5}{20} - \frac{2}{6} < 0,$$

which shows that $B^2 \notin \overline{NE}\,Y$ and rules out $P$ as a maximal centre.

We now construct $T$ and prove that it satisfies Lemma 5.4.3. The argument takes us 4 intricate steps, so you might again prefer to press fast forward (or maybe better still, just go out for a drink).

Before we plunge into the details of this calculation, we point out that it is not difficult to guess that we should be looking for a curve $\Gamma \in |D|$, with $D = 20B + 3E_{|S}$ (for a more systematic discussion, compare 7.2). Indeed, the Riemann–Roch formula of [YPG] on $S$ immediately gives

$$h^0(D) - h^1(D) = 2 + \frac{1}{2}D^2 - \frac{1}{2}\left(\frac{1}{2} + 2 \times \frac{2}{3}\right) = 1,$$

where we used that $D^2 = -\frac{1}{6}$. The problem is, we have no way to guarantee that, *for $X$ general*, the section $\Gamma \in |D|$ is *irreducible*. In fact, $\Gamma$ is almost certain to break up as $X$ specialises. This is why we still need to find an explicit equation for $\Gamma$ and show that it is irreducible.

STEP 1 To simplify the notation, we ignore the $x$-coordinate, in effect working with the K3 surface

$$S = S_{18} \subset \mathbb{P}(3,4,5,6)$$

with coordinates $y, z, t, u$ of weights $3, 4, 5, 6$. With the single allowed coordinate change, we may arrange that $P = P_1 = \frac{1}{3}(1, 2) \in S$.

We write the equation of $S$ as follows

$$
\begin{aligned}
F = \; & y^2 z^3 + a_1 y^3 zt + a_2 y^4 u + \\
& + (a_3 z^3 u + a_4 z^2 t^2) + y(a_5 ztu + a_6 t^3) + a_7 y^2 u^2 + \\
& + a_8 u^3,
\end{aligned}
\tag{1}
$$

where terms on the first, second and third line, vanish along $E$ to order $\frac{3}{3}$, $\frac{6}{3}$, $\frac{9}{3}$.

To make life simpler now and later, it is convenient to normalise $t$ and $u$ so that $a_5 = a_1 a_3 + 1$ and $a_6 = a_1 a_4 + 1$, and we assume that this has already been done, although it will not come up for quite some time. As a first step, it is easy to find a function

$$
v = z^3 + a_1 yzt + a_2 y^2 u \in |12A - \tfrac{6}{3}E|.
\tag{2}
$$

We will often use the formula

$$
z^3 = -a_1 yzt - a_2 y^2 u + v.
\tag{3}
$$

STEP 2    The main problem, which makes it hard to write down functions with high order of vanishing at $P$, is the term $a_3 z^3 u + a_4 z^2 t^2$ in $F$. In this step we use equation (3) to get rid of this term. This allows us to write down a function $w \in |19A - \frac{10}{3}E| = |19B + 3E|$. Unfortunately, you can check that $(19B + 3E)^2 > 0$, so we need to work even harder.

First, substitute (3) in the expression (1) for $F$, to obtain

$$
\begin{aligned}
F = \; & a_4 z^2 t^2 + y\big((a_5 - a_1 a_3) ztu + a_6 t^3\big) + y^2 \big((a_7 - a_2 a_3) u^2 + v\big) + \\
& + a_8 u^3 + a_3 uv,
\end{aligned}
\tag{4}
$$

where the first and second lines contain terms that vanish to order $\frac{6}{3}$ and $\frac{9}{3}$ along $E$. This eliminates the unwanted term $a_3 z^3 u$. To eliminate the remaining $a_4 z^2 t^2$, we first multiply (4) by $z$:

$$
\begin{aligned}
zF = \; & a_4 z^3 t^2 + y\big((a_5 - a_1 a_3) z^2 tu + a_6 zt^3\big) + y^2 \big((a_7 - a_2 a_3) zu^2 + zv\big) + \\
& + a_8 zu^3 + a_3 zuv
\end{aligned}
$$

and then again substitute for $z^3$ using (3)

$$
\begin{aligned}
zF = \; & y\big((a_5 - a_1 a_3) z^2 tu + (a_6 - a_1 a_4) zt^3\big) + \\
& + y^2 \big(-a_2 a_4 t^2 u + (a_7 - a_2 a_3) zu^2 + zv\big) + \\
& + a_8 zu^3 + a_3 zuv + a_4 t^2 v.
\end{aligned}
\tag{5}
$$

Grouping together the multiples of $y$, we can rewrite (5) as

$$zF = yw + (a_8 zu^3 + a_3 zuv + a_4 t^2 v),$$

where

$$
\begin{aligned}
w &= (a_5 - a_1 a_3) z^2 tu + (a_6 - a_1 a_4) zt^3 + y\big(-a_2 a_4 t^2 u + (a_7 - a_2 a_3) zu^2 + zv\big) \\
&= z^2 tu + zt^3 + y\big(-a_2 a_4 t^2 u + (a_7 - a_2 a_3) zu^2 + zv\big) \\
&\in |19A - \tfrac{10}{3} E| = |19B + 3E|,
\end{aligned}
\tag{6}
$$

where we made use of our initial choice $a_5 = a_1 a_3 + 1$, $a_6 = a_1 a_4 + 1$. We will use (6) in the form

$$zt^3 = -z^2 tu - y\big(-a_2 a_4 t^2 u + (a_7 - a_2 a_3) zu^2 + zv\big) + w. \tag{7}$$

STEP 3    In this step, we write down our final element $\omega \in |20A - \tfrac{11}{3} E| = |20B + 3E|$. We begin multiplying $F$, as given in (4), by $t$

$$
\begin{aligned}
tF = a_4 z^2 t^3 &+ y\big(zt^2 u + a_6 t^4\big) + y^2\big((a_7 - a_2 a_3) tu^2 + tv\big) + \\
&+ a_8 tu^3 + a_3 tuv.
\end{aligned}
\tag{8}
$$

The plan is to eliminate from (8) the unwanted monomial $a_4 z^2 t^3$. We use (7) and then (3) again to get

$$
\begin{aligned}
z^2 t^3 &= -z^3 tu + y\big(a_2 a_4 zt^2 u + (a_2 a_3 - a_7) z^2 u^2 - z^2 v\big) + zw = \\
&= y\big((a_1 + a_2 a_4) zt^2 u + (a_2 a_3 - a_7) z^2 u^2 - z^2 v\big) + a_2 y^2 t u^2 - \\
&\quad - tuv + zw,
\end{aligned}
\tag{9}
$$

where we isolated in the last line terms which vanish $\tfrac{11}{3}$ times along $E$. Finally, we substitute (9) into (8) to get

$$
\begin{aligned}
tF = y\big((1 &+ a_1 a_4 + a_2 a_4^2) zt^2 u + a_4(a_2 a_3 - a_7) z^2 u^2 + a_6 t^4 - a_4 z^2 v\big) + \\
&+ y^2\big((a_7 - a_2 a_3 + a_2 a_4) tu^2 + tv\big) + \\
&+ a_8 tu^3 + a_3 tuv - a_4 tuv + a_4 zw,
\end{aligned}
\tag{10}
$$

where the last line contains terms vanishing $\tfrac{11}{3}$ times along $E$. We can now rewrite (10) as

$$tF = y\omega + a_8 tu^3 + a_3 tuv - a_4 tuv + a_4 zw, \tag{11}$$

where

$$
\begin{aligned}
\omega = (1 &+ a_1 + a_2 a_4) zt^2 u + a_4(a_2 a_3 - a_7) z^2 u^2 + a_6 t^4 - a_4 z^2 v + \\
&+ y\big((a_7 - a_2 a_3 + a_2 a_4) tu^2 + tv\big) \in |20A - \tfrac{11}{3} E| = |20B + 3E|.
\end{aligned}
\tag{12}
$$

STEP 4  We leave to the reader the entirely straightforward task to use Bertini's theorem and the explicit equations (1) and (12) to verify that the curve $\Gamma = (F = \omega = 0)$ is irreducible (it is not hard).

This completes the proof of all the currently outstanding steps, lemmas and theorems.  Q.E.D.

# 6  The Big Table

We tabulate the singular points of our hypersurfaces

$$X = X_d \subset \mathbb{P} = \mathbb{P}(1, a_1, a_2, a_3, a_4),$$

and, as promised in Theorem 3.2, we indicate for each point either how to exclude it as a maximal centre, or how to untwist it by a quadratic or elliptic involution.

The famous 95 are ordered as in Fletcher [Fl]. We choose coordinates $x, y, z, t, u$ or $x, y_1, y_2, z, t$ etc. in order of ascending degree; for example, in No. 12, $X_{10} \subset \mathbb{P}(1, 1, 2, 3, 4)$, the coordinates are $x_0, x_1, y, z, t$. In all cases $x$ or $x_0, x_1$ etc. are of degree 1, and $(x_0 = 0) : S \subset X$ is the general elephant, which we assume to be a quasismooth K3 surface, as discussed in 7.2.3.

The first column tabulates the types of singularities. For example, in No. 12, $X_{10} \subset \mathbb{P}(1, 1, 2, 3, 4)$, the entry $P_3 = \frac{1}{3}(1, 1, 2)$ means that the only nonzero coordinate $z$ at $P_3 = (0, 0, 0, 1, 0)$ has weight 3, and the local eigencoordinates at $P_3$ are $x_0, x_1, y$ (compare Proposition 3.4.6).

The second column of the table documents the argument we use for exclusion. As in Notation 3.4.5, we write $A = -K_X$ and $B = -K_Y = A - \frac{1}{r}E$, where $Y \to X$ is the Kawamata blowup. Each entry first distinguishes $B^3 \leq 0$ versus $B^3 > 0$, as discussed in Remark 4 at the start of Chapter 5. Next, we indicate the class $T = bB + cE$ used in either the numerical exclusion of 5.4 or the bad link method 5.5. As we explained in Remark 5.4.7, this information completely determines $\overline{NE}\,Y$, as well as the cone of effective divisors $\overline{NM}^1 Y$. Finally, (†) indicates the 19 bad links of 5.5 and (∗) the 10 hard cases of the computation of $\overline{NE}\,Y$ of 5.7.2.

The right-hand column tells us how to untwist the centre. The quadratic involutions are treated in 4.4 and Theorem 4.9. For example, in No. 65, $X_{27} \subset \mathbb{P}(1, 2, 5, 9, 11)$, centre $P_4 = \frac{1}{11}(1, 2, 9)$, the entry Q.I. $zu^2, 16, 27$ says $F = zu^2 + a_{16}(x, y, z, t)u + b_{27}$; the degrees of $a$ and $b$ are trivial to figure out. The star in front of a monomial such as $*ut^2$ indicates an extra generality assumption (compare 7.4.2).

Refer to 4.10 and Theorem 4.13 for the elliptic involutions. In Example 4.11, No. 61, $X_{25} \subset \mathbb{P}(1, 4, 5, 7, 9)$, centre $P_3 = \frac{1}{7}(1, 5, 2)$, the table says E.I., $tu^2 - yt^3, 16, 25$. This means that the equation of $X$ is $F =$

$tu^2 + ua_{16}(x, y, z) - yt^3 - \cdots + d_{25}(x, y, z, t)$. A star again indicates an extra generality assumption.

No. 1: $X_4 \subset \mathbb{P}(1,1,1,1,1)$, degree $A^3 = 4$

| smooth | N/A | none |
|---|---|---|

No. 2: $X_5 \subset \mathbb{P}(1,1,1,1,2)$, degree $A^3 = 5/2$

| $P_4 = \frac{1}{2}(1,1,1)$ | | Q.I. $x_1 z^2, 3, 5$ |
|---|---|---|

No. 3: $X_6 \subset \mathbb{P}(1,1,1,1,3)$, degree $A^3 = 2$

| smooth | N/A | none |
|---|---|---|

No. 4: $X_6 \subset \mathbb{P}(1,1,1,2,2)$, degree $A^3 = 3/2$

| $P_3 P_4 = 3 \times \frac{1}{2}(1,1,1)$ | | Q.I. $y_1 y_2^2, 4, 6$ |
|---|---|---|

No. 5: $X_7 \subset \mathbb{P}(1,1,1,2,3)$, degree $A^3 = 7/6$

| $P_4 = \frac{1}{3}(1,1,2)$ | | Q.I. $x_1 z^2, 4, 7$ |
|---|---|---|
| $P_3 = \frac{1}{2}(1,1,1)$ | | Q.I. $*zy^2, 5, 7$ |

No. 6: $X_8 \subset \mathbb{P}(1,1,1,2,4)$, degree $A^3 = 1$

| $P_3 P_4 = 2 \times \frac{1}{2}(1,1,1)$ | | Q.I. $zy^2, 6, 8$ |
|---|---|---|

No. 7: $X_8 \subset \mathbb{P}(1,1,2,2,3)$, degree $A^3 = 2/3$

| $P_4 = \frac{1}{3}(1,1,2)$ | | Q.I. $y_2 z^2, 6, 8$ |
|---|---|---|
| $P_2 P_3 = 4 \times \frac{1}{2}(1,1,1)$ | | E.I. $*y_2 z^2 - y_1 y_2^3, 5, 8$ |

No. 8: $X_9 \subset \mathbb{P}(1,1,1,3,4)$, degree $A^3 = 3/4$

| $P_4 = \frac{1}{4}(1,1,3)$ | | Q.I. $x_1 z^2, 5, 9$ |
|---|---|---|

No. 9: $X_9 \subset \mathbb{P}(1,1,2,3,3)$, degree $A^3 = 1/2$

| $P_2 = \frac{1}{2}(1,1,1)$ | $B^3 = 0, B$ | |
|---|---|---|
| $P_3 P_4 = 3 \times (1,1,2)$ | | Q.I. $z_1 z_2^2, 6, 9$ |

No. 10: $X_{10} \subset \mathbb{P}(1,1,1,3,5)$, degree $A^3 = 2/3$

| $P_3 = \frac{1}{3}(1,1,2)$ | $B^3 > 0, B - E$  (†) | none |
|---|---|---|

No. 11: $X_{10} \subset \mathbb{P}(1,1,2,2,5)$, degree $A^3 = 1/2$

| $P_2 P_3 = 5 \times \frac{1}{2}(1,1,1)$ | $B^3 = 0, B$ | none |
|---|---|---|

No. 12: $X_{10} \subset \mathbb{P}(1,1,2,3,4)$, degree $A^3 = 5/12$

| $P_4 = \frac{1}{4}(1,1,3)$ | | Q.I. $yt^2, 6, 10$ |
|---|---|---|
| $P_3 = \frac{1}{3}(1,1,2)$ | | Q.I. $*z^2 t, 7, 10$ |
| $P_2 P_4 = 2 \times \frac{1}{2}(1,1,1)$ | $B^3 < 0, B$ | |

No. 13: $X_{11} \subset \mathbb{P}(1,1,2,3,5)$, degree $A^3 = 11/30$

| | | |
|---|---|---|
| $P_4 = \frac{1}{5}(1,2,3)$ | | Q.I. $x_1t^2, 6, 11$ |
| $P_3 = \frac{1}{3}(1,1,2)$ | | Q.I. $*z^2t, 8, 11$ |
| $P_2 = \frac{1}{2}(1,1,1)$ | $B^3 < 0, B$ | |

No. 14: $X_{12} \subset \mathbb{P}(1,1,1,4,6)$, degree $A^3 = 1/2$

| | | |
|---|---|---|
| $P_3P_4 = 1 \times \frac{1}{2}(1,1,1)$ | $B^3 = 0, B$ | none |

No. 15: $X_{12} \subset \mathbb{P}(1,1,2,3,6)$, degree $A^3 = 1/3$

| | | |
|---|---|---|
| $P_3P_4 = 2 \times \frac{1}{3}(1,1,2)$ | | Q.I. $tz^2, 9, 12$ |
| $P_2P_4 = 2 \times \frac{1}{2}(1,1,1)$ | $B^3 < 0, B$ | |

No. 16: $X_{12} \subset \mathbb{P}(1,1,2,4,5)$, degree $A^3 = 3/10$

| | | |
|---|---|---|
| $P_4 = \frac{1}{5}(1,1,4)$ | | Q.I. $yt^2, 8, 12$ |
| $P_2P_3 = 3 \times \frac{1}{2}(1,1,1)$ | $B^3 < 0, B$ | |

No. 17: $X_{12} \subset \mathbb{P}(1,1,3,4,4)$, degree $A^3 = 1/4$

| | | |
|---|---|---|
| $P_3P_4 = 3 \times \frac{1}{4}(1,1,3)$ | | Q.I. $z_1z_2^2, 8, 12$ |

No. 18: $X_{12} \subset \mathbb{P}(1,2,2,3,5)$, degree $A^3 = 1/5$

| | | |
|---|---|---|
| $P_4 = \frac{1}{5}(1,2,3)$ | | Q.I. $y_1t^2, 7, 12$ |
| $P_1P_2 = 6 \times \frac{1}{2}(1,1,1)$ | $B^3 < 0, 2B$ | |

No. 19: $X_{12} \subset \mathbb{P}(1,2,3,3,4)$, degree $A^3 = 1/6$

| | | |
|---|---|---|
| $P_1P_4 = 3 \times \frac{1}{2}(1,1,1)$ | $B^3 < 0, 6B + E \ (*)$ | none |
| $P_2P_3 = 4 \times \frac{1}{3}(1,2,1)$ | $B^3 = 0, 2B$ | |

No. 20: $X_{13} \subset \mathbb{P}(1,1,3,4,5)$, degree $A^3 = 13/60$

| | | |
|---|---|---|
| $P_4 = \frac{1}{5}(1,1,4)$ | | Q.I. $yt^2, 8, 13$ |
| $P_3 = \frac{1}{4}(1,1,3)$ | | Q.I. $*z^2t, 9, 13$ |
| $P_2 = \frac{1}{3}(1,1,2)$ | | E.I. $yt^2 - *zy^3, 8, 13$ |

No. 21: $X_{14} \subset \mathbb{P}(1,1,2,4,7)$, degree $A^3 = 1/4$

| | | |
|---|---|---|
| $P_3 = \frac{1}{4}(1,1,3)$ | $B^3 > 0, 2B - E \quad (\dagger)$ | none |
| $P_2P_3 = 3 \times \frac{1}{2}(1,1,1)$ | $B^3 < 0, B$ | |

No. 22: $X_{14} \subset \mathbb{P}(1,2,2,3,7)$, degree $A^3 = 1/6$

| | | |
|---|---|---|
| $P_3 = \frac{1}{3}(1,2,1)$ | $B^3 = 0, 2B$ | none |
| $P_1P_2 = 7 \times \frac{1}{2}(1,1,1)$ | $B^3 < 0, 2B$ | |

No. 23: $X_{14} \subset \mathbb{P}(1,2,3,4,5)$, degree $A^3 = 7/60$

| | | |
|---|---|---|
| $P_4 = \frac{1}{5}(1,2,3)$ | | Q.I. $tu^2, 9, 14$ |
| $P_3 = \frac{1}{4}(1,3,1)$ | | E.I. $tu^2 - yt^3, 9, 14$ |
| $P_2 = \frac{1}{3}(1,2,1)$ | $B^3 < 0, 2B$ | |
| $P_1P_3 = 3 \times \frac{1}{2}(1,1,1)$ | $B^3 < 0, 4B + E$ | |

No. 24: $X_{15} \subset \mathbb{P}(1,1,2,5,7)$, degree $A^3 = 3/14$

| | | |
|---|---|---|
| $P_4 = \frac{1}{7}(1,2,5)$ | | Q.I. $x_1t^2, 8, 15$ |
| $P_2 = \frac{1}{2}(1,1,1)$ | $B^3 < 0, B$ | |

No. 25: $X_{15} \subset \mathbb{P}(1,1,3,4,7)$, degree $A^3 = 5/28$

| | | |
|---|---|---|
| $P_4 = \frac{1}{7}(1,3,4)$ | | Q.I. $x_1t^2, 8, 15$ |
| $P_3 = \frac{1}{4}(1,1,3)$ | | Q.I. $*tz^2, 11, 15$ |

No. 26: $X_{15} \subset \mathbb{P}(1,1,3,5,6)$, degree $A^3 = 1/6$

| | | |
|---|---|---|
| $P_4 = \frac{1}{6}(1,1,5)$ | | Q.I. $yt^2, 9, 15$ |
| $P_3P_4 = 2 \times \frac{1}{3}(1,1,2)$ | $B^3 = 0, B$ | |

No. 27: $X_{15} \subset \mathbb{P}(1,2,3,5,5)$, degree $A^3 = 1/10$

| | | |
|---|---|---|
| $P_3P_4 = 3 \times \frac{1}{5}(1,2,3)$ | | Q.I. $t_1t_2^2, 10, 15$ |
| $P_1 = \frac{1}{2}(1,1,1)$ | $B^3 < 0, 5B + E$ | |

No. 28: $X_{15} \subset \mathbb{P}(1,3,3,4,5)$, degree $A^3 = 1/12$

| | | |
|---|---|---|
| $P_3 = \frac{1}{4}(1,3,1)$ | $B^3 = 0, 3B$ | none |
| $P_1P_2 = 5 \times \frac{1}{3}(1,1,2)$ | $B^3 < 0, 3B$ | |

No. 29: $X_{16} \subset \mathbb{P}(1,1,2,5,8)$, degree $A^3 = 1/5$

| | | |
|---|---|---|
| $P_3 = \frac{1}{5}(1,2,3)$ | $B^3 > 0, B - E$   (†) | none |
| $P_2P_4 = 2 \times \frac{1}{2}(1,1,1)$ | $B^3 < 0, B$ | |

No. 30: $X_{16} \subset \mathbb{P}(1,1,3,4,8)$, degree $A^3 = 1/6$

| | | |
|---|---|---|
| $P_3P_4 = 2 \times \frac{1}{4}(1,1,3)$ | | Q.I. $tz^2, 12, 16$ |
| $P_2 = \frac{1}{3}(1,1,2)$ | $B^3 = 0, B$ | |

No. 31: $X_{16} \subset \mathbb{P}(1,1,4,5,6)$, degree $A^3 = 2/15$

| | | |
|---|---|---|
| $P_4 = \frac{1}{6}(1,1,5)$ | | Q.I. $yt^2, 10, 16$ |
| $P_3 = \frac{1}{5}(1,1,4)$ | | Q.I. $*tz^2, 11, 16$ |
| $P_2P_4 = 1 \times \frac{1}{2}(1,1,1)$ | $B^3 < 0, B$ | |

No. 32: $X_{16} \subset \mathbb{P}(1,2,3,4,7)$, degree $A^3 = 2/21$

| | | |
|---|---|---|
| $P_4 = \frac{1}{7}(1,3,4)$ | | Q.I. $yu^2, 9, 16$ |
| $P_2 = \frac{1}{3}(1,2,1)$ | $B^3 < 0, 2B$ | |
| $P_1P_3 = 4 \times \frac{1}{2}(1,1,1)$ | $B^3 < 0, 4B + E$ | |

No. 33: $X_{17} \subset \mathbb{P}(1,2,3,5,7)$, degree $A^3 = 17/210$

| | | |
|---|---|---|
| $P_4 = \frac{1}{7}(1,2,5)$ | | Q.I. $zu^2, 10, 17$ |
| $P_3 = \frac{1}{5}(1,2,3)$ | | Q.I. $*ut^2, 12, 17$ |
| $P_2 = \frac{1}{3}(1,2,1)$ | $B^3 < 0, 2B$ | |
| $P_1 = \frac{1}{2}(1,1,1)$ | $B^3 < 0, 7B + 2E$ | |

No. 34: $X_{18} \subset \mathbb{P}(1,1,2,6,9)$, degree $A^3 = 1/6$

| | | |
|---|---|---|
| $P_3 P_4 = 1 \times \frac{1}{3}(1,1,2)$ | $B^3 = 0, B$ | none |
| $P_2 P_3 = 3 \times \frac{1}{2}(1,1,1)$ | $B^3 < 0, B$ | |

No. 35: $X_{18} \subset \mathbb{P}(1,1,3,5,9)$, degree $A^3 = 2/15$

| | | |
|---|---|---|
| $P_3 = \frac{1}{5}(1,1,4)$ | $B^3 > 0, 3B - E$  (†) | none |
| $P_2 P_4 = 2 \times \frac{1}{3}(1,1,2)$ | $B^3 < 0, B$ | |

No. 36: $X_{18} \subset \mathbb{P}(1,1,4,6,7)$, degree $A^3 = 3/28$

| | | |
|---|---|---|
| $P_4 = \frac{1}{7}(1,1,6)$ | | Q.I. $yt^2, 11, 18$ |
| $P_2 = \frac{1}{4}(1,1,3)$ | | E.I. $yt^2 - zy^3, 11, 18$ |
| $P_2 P_3 = 1 \times \frac{1}{2}(1,1,1)$ | $B^3 < 0, B$ | |

No. 37: $X_{18} \subset \mathbb{P}(1,2,3,4,9)$, degree $A^3 = 1/12$

| | | |
|---|---|---|
| $P_3 = \frac{1}{4}(1,3,1)$ | $B^3 = 0, 2B$ | none |
| $P_2 P_4 = 2 \times \frac{1}{3}(1,2,1)$ | $B^3 < 0, 2B$ | |
| $P_1 P_3 = 4 \times \frac{1}{2}(1,1,1)$ | $B^3 < 0, 4B + E$ | |

No. 38: $X_{18} \subset \mathbb{P}(1,2,3,5,8)$, degree $A^3 = 3/40$

| | | |
|---|---|---|
| $P_4 = \frac{1}{8}(1,3,5)$ | | Q.I. $yu^2, 10, 18$ |
| $P_3 = \frac{1}{5}(1,2,3)$ | | Q.I. $*ut^2, 13, 18$ |
| $P_1 P_4 = 2 \times \frac{1}{2}(1,1,1)$ | $B^3 < 0, 10B + 3E$ (∗) | |

No. 39: $X_{18} \subset \mathbb{P}(1,3,4,5,6)$, degree $A^3 = 1/20$

| | | |
|---|---|---|
| $P_3 = \frac{1}{5}(1,4,1)$ | $B^3 = 0, 3B$ | none |
| $P_2 = \frac{1}{4}(1,3,1)$ | $B^3 < 0, 3B$ | |
| $P_2 P_4 = \frac{1}{2}(1,1,1)$ | $B^3 < 0, 3B + E$ | |
| $P_1 P_4 = 3 \times \frac{1}{3}(1,1,2)$ | $B^3 < 0, 20B + 3E$ (∗) | |

No. 40: $X_{19} \subset \mathbb{P}(1,3,4,5,7)$, degree $A^3 = 19/420$

| | | |
|---|---|---|
| $P_4 = \frac{1}{7}(1,3,4)$ | | Q.I. $tu^2, 12, 19$ |
| $P_3 = \frac{1}{5}(1,3,2)$ | | E.I. $tu^2 - zt^3, 12, 19$ |
| $P_2 = \frac{1}{4}(1,3,1)$ | $B^3 < 0, 3B$ | |
| $P_1 = \frac{1}{3}(1,1,2)$ | $B^3 < 0, 7B + E$ | |

No. 41: $X_{20} \subset \mathbb{P}(1,1,4,5,10)$, degree $A^3 = 1/10$

| $P_3P_4 = 2 \times \frac{1}{5}(1,1,4)$ | | Q.I. $tz^2, 15, 20$ |
|---|---|---|
| $P_2P_4 = \frac{1}{2}(1,1,1)$ | $B^3 < 0, B$ | |

No. 42: $X_{20} \subset \mathbb{P}(1,2,3,5,10)$, degree $A^3 = 1/15$

| $P_2 = \frac{1}{3}(1,2,1)$ | $B^3 < 0, 2B$ | |
|---|---|---|
| $P_3P_4 = 2 \times \frac{1}{5}(1,2,3)$ | | Q.I. $tz^2, 15, 20$ |
| $P_1P_4 = 2 \times \frac{1}{2}(1,1,1)$ | $B^3 < 0, 10B + 3E$ | |

No. 43: $X_{20} \subset \mathbb{P}(1,2,4,5,9)$, degree $A^3 = 1/18$

| $P_4 = \frac{1}{9}(1,4,5)$ | | Q.I. $yu^2, 11, 20$ |
|---|---|---|
| $P_1P_2 = 5 \times \frac{1}{2}(1,1,1)$ | $B^3 < 0, 4B + E$ | |

No. 44: $X_{20} \subset \mathbb{P}(1,2,5,6,7)$, degree $A^3 = 1/21$

| $P_4 = \frac{1}{7}(1,2,5)$ | | Q.I. $tu^2, 13, 20$ |
|---|---|---|
| $P_3 = \frac{1}{6}(1,5,1)$ | | E.I. $tu^2 - yt^3, 13, 20$ |
| $P_1P_3 = 3 \times \frac{1}{2}(1,1,1)$ | $B^3 < 0, 6B + 2E$ | |

No. 45: $X_{20} \subset \mathbb{P}(1,3,4,5,8)$, degree $A^3 = 1/24$

| $P_4 = \frac{1}{8}(1,3,5)$ | | Q.I. $zu^2, 12, 20$ |
|---|---|---|
| $P_1 = \frac{1}{3}(1,1,2)$ | $B^3 < 0, 8B + E$ | |
| $P_2P_4 = 2 \times \frac{1}{4}(1,3,1)$ | $B^3 < 0, 3B$ | |

No. 46: $X_{21} \subset \mathbb{P}(1,1,3,7,10)$, degree $A^3 = 1/10$

| $P_4 = \frac{1}{10}(1,3,7)$ | | Q.I. $xt^2, 11, 21$ |
|---|---|---|

No. 47: $X_{21} \subset \mathbb{P}(1,1,5,7,8)$, degree $A^3 = 3/40$

| $P_4 = \frac{1}{8}(1,1,7)$ | | Q.I. $yt^2, 13, 21$ |
|---|---|---|
| $P_2 = \frac{1}{5}(1,2,3)$ | $B^3 > 0, B - E$  (†) | |

No. 48: $X_{21} \subset \mathbb{P}(1,2,3,7,9)$, degree $A^3 = 1/18$

| $P_4 = \frac{1}{9}(1,2,7)$ | | Q.I. $zu^2, 12, 21$ |
|---|---|---|
| $P_1 = \frac{1}{2}(1,1,1)$ | $B^3 < 0, 3B + E$ | |
| $P_2P_4 = 2 \times \frac{1}{3}(1,2,1)$ | $B^3 < 0, 2B$ | |

No. 49: $X_{21} \subset \mathbb{P}(1,3,5,6,7)$, degree $A^3 = 1/30$

| $P_3 = \frac{1}{6}(1,5,1)$ | $B^3 = 0, 3B$ | none |
|---|---|---|
| $P_2 = \frac{1}{5}(1,3,2)$ | $B^3 = 0, 3B$ | |
| $P_1P_3 = 3 \times \frac{1}{3}(1,2,1)$ | $B^3 < 0, 7B + E$ | |

No. 50: $X_{22} \subset \mathbb{P}(1,1,3,7,11)$, degree $A^3 = 2/21$

| $P_3 = \frac{1}{7}(1,3,4)$ | $B^3 > 0, B - E$  (†) | none |
|---|---|---|
| $P_2 = \frac{1}{3}(1,1,2)$ | $B^3 < 0, B$ | |

No. 51: $X_{22} \subset \mathbb{P}(1,1,4,6,11)$, degree $A^3 = 1/12$

| $P_3 = \frac{1}{6}(1,1,5)$ | $B^3 > 0, 4B - E$   (†) | none |
|---|---|---|
| $P_2 = \frac{1}{4}(1,1,3)$ | $B^3 = 0, B$ | |
| $P_2P_3 = \frac{1}{2}(1,1,1)$ | $B^3 < 0, B$ | |

No. 52: $X_{22} \subset \mathbb{P}(1,2,4,5,11)$, degree $A^3 = 1/20$

| $P_3 = \frac{1}{5}(1,4,1)$ | $B^3 = 0, 2B$ | none |
|---|---|---|
| $P_2 = \frac{1}{4}(1,1,3)$ | $B^3 < 0, 2B$ | |
| $P_1P_2 = 5 \times \frac{1}{2}(1,1,1)$ | $B^3 < 0, 4B + E$ | |

No. 53: $X_{24} \subset \mathbb{P}(1,1,3,8,12)$, degree $A^3 = 1/12$

| $P_3P_4 = 1 \times \frac{1}{4}(1,1,3)$ | $B^3 = 0, B$ | none |
|---|---|---|
| $P_2P_4 = 2 \times \frac{1}{3}(1,1,2)$ | $B^3 < 0, B$ | |

No. 54: $X_{24} \subset \mathbb{P}(1,1,6,8,9)$, degree $A^3 = 1/18$

| $P_4 = \frac{1}{9}(1,1,8)$ | | Q.I. $yt^2, 15, 24$ |
|---|---|---|
| $P_2P_4 = 1 \times \frac{1}{3}(1,1,2)$ | $B^3 < 0, B$ | |
| $P_2P_3 = \frac{1}{2}(1,1,1)$ | $B^3 < 0, B$ | |

No. 55: $X_{24} \subset \mathbb{P}(1,2,3,7,12)$, degree $A^3 = 1/21$

| $P_3 = \frac{1}{7}(1,2,5)$ | $B^3 > 0, 3B - E$   (†) | none |
|---|---|---|
| $P_2P_4 = 2 \times \frac{1}{3}(1,2,1)$ | $B^3 < 0, 2B$ | |
| $P_1P_4 = 2 \times \frac{1}{2}(1,1,1)$ | $B^3 < 0, 3B + E$ | |

No. 56: $X_{24} \subset \mathbb{P}(1,2,3,8,11)$, degree $A^3 = 1/22$

| $P_4 = \frac{1}{11}(1,3,8)$ | | Q.I. $yu^2, 13, 24$ |
|---|---|---|
| $P_1P_3 = 3 \times \frac{1}{2}(1,1,1)$ | $B^3 < 0, 3B + E$ | |

No. 57: $X_{24} \subset \mathbb{P}(1,3,4,5,12)$, degree $A^3 = 1/30$

| $P_3 = \frac{1}{5}(1,3,2)$ | $B^3 = 0, 3B$ | none |
|---|---|---|
| $P_2P_4 = 2 \times \frac{1}{4}(1,3,1)$ | $B^3 < 0, 3B$ | |
| $P_1P_4 = 2 \times \frac{1}{3}(1,1,2)$ | $B^3 < 0, 12B + 2E$ | |

No. 58: $X_{24} \subset \mathbb{P}(1,3,4,7,10)$, degree $A^3 = 1/35$

| $P_4 = \frac{1}{10}(1,3,7)$ | | Q.I. $zu^2, 14, 24$ |
|---|---|---|
| $P_3 = \frac{1}{7}(1,3,4)$ | | Q.I. $*ut^2, 17, 24$ |
| $P_2P_4 = \frac{1}{2}(1,1,1)$ | $B^3 < 0, 3B + E$ | |

No. 59: $X_{24} \subset \mathbb{P}(1,3,6,7,8)$, degree $A^3 = 1/42$

| $P_3 = \frac{1}{7}(1,6,1)$ | $B^3 = 0, 3B$ | none |
|---|---|---|
| $P_2P_4 = \frac{1}{2}(1,1,1)$ | $B^3 < 0, 3B + E$ | |
| $P_1P_2 = 4 \times \frac{1}{3}(1,1,2)$ | $B^3 < 0, 6B + E$ | |

No. 60: $X_{24} \subset \mathbb{P}(1,4,5,6,9)$, degree $A^3 = 1/45$

| | | |
|---|---|---|
| $P_4 = \frac{1}{9}(1,4,5)$ | | Q.I. $tu^2$, 15, 24 |
| $P_2 = \frac{1}{5}(1,4,1)$ | $B^3 < 0, 4B$ | |
| $P_3P_4 = 1 \times \frac{1}{3}(1,1,2)$ | $B^3 < 0, 5B + E$ | |
| $P_1P_3 = 2 \times \frac{1}{2}(1,1,1)$ | $B^3 < 0, 18B + 7E$ (*) | |

No. 61: $X_{25} \subset \mathbb{P}(1,4,5,7,9)$, degree $A^3 = 5/252$

| | | |
|---|---|---|
| $P_4 = \frac{1}{9}(1,4,5)$ | | Q.I. $tu^2$, 16, 25 |
| $P_3 = \frac{1}{7}(1,5,2)$ | | E.I. $tu^2 - yt^3$, 16, 25 |
| $P_1 = \frac{1}{4}(1,3,1)$ | $B^3 < 0, 9B + E$ | |

No. 62: $X_{26} \subset \mathbb{P}(1,1,5,7,13)$, degree $A^3 = 2/35$

| | | |
|---|---|---|
| $P_3 = \frac{1}{7}(1,1,6)$ | $B^3 > 0, 5B - E$ (†) | none |
| $P_2 = \frac{1}{5}(1,2,3)$ | $B^3 > 0, B - E$ (†) | |

No. 63: $X_{26} \subset \mathbb{P}(1,2,3,8,13)$, degree $A^3 = 1/24$

| | | |
|---|---|---|
| $P_3 = \frac{1}{8}(1,3,5)$ | $B^3 > 0, 2B - E$ (†) | none |
| $P_2 = \frac{1}{3}(1,2,1)$ | $B^3 < 0, 2B$ | |
| $P_1P_3 = 3 \times \frac{1}{2}(1,1,1)$ | $B^3 < 0, 3B + E$ | |

No. 64: $X_{26} \subset \mathbb{P}(1,2,5,6,13)$, degree $A^3 = 1/30$

| | | |
|---|---|---|
| $P_3 = \frac{1}{6}(1,5,1)$ | $B^3 = 0, 2B$ | none |
| $P_2 = \frac{1}{5}(1,2,3)$ | $B^3 = 0, 2B$ | |
| $P_1P_3 = 4 \times \frac{1}{2}(1,1,1)$ | $B^3 < 0, 6B + 2E$ | |

No. 65: $X_{27} \subset \mathbb{P}(1,2,5,9,11)$, degree $A^3 = 3/110$

| | | |
|---|---|---|
| $P_4 = \frac{1}{11}(1,2,9)$ | | Q.I. $zu^2$, 16, 27 |
| $P_2 = \frac{1}{5}(1,4,1)$ | $B^3 < 0, 2B$ | |
| $P_1 = \frac{1}{2}(1,1,1)$ | $B^3 < 0, 11B + 4E$ | |

No. 66: $X_{27} \subset \mathbb{P}(1,5,6,7,9)$, degree $A^3 = 1/70$

| | | |
|---|---|---|
| $P_3 = \frac{1}{7}(1,5,2)$ | $B^3 = 0, 5B$ | none |
| $P_2 = \frac{1}{6}(1,5,1)$ | $B^3 < 0, 5B$ | |
| $P_1 = \frac{1}{5}(1,1,4)$ | $B^3 < 0, 12B + E$ (*) | |
| $P_2P_4 = 1 \times \frac{1}{3}(1,2,1)$ | $B^3 < 0, 5B + E$ | |

No. 67: $X_{28} \subset \mathbb{P}(1,1,4,9,14)$, degree $A^3 = 1/18$

| | | |
|---|---|---|
| $P_3 = \frac{1}{9}(1,4,5)$ | $B^3 > 0, B - E$ (†) | none |
| $P_2P_4 = \frac{1}{2}(1,1,1)$ | $B^3 < 0, B$ | |

No. 68: $X_{28} \subset \mathbb{P}(1,3,4,7,14)$, degree $A^3 = 1/42$

| $P_1 = \frac{1}{3}(1,1,2)$ | $B^3 < 0, 7B + E$ | |
|---|---|---|
| $P_3P_4 = 2 \times \frac{1}{7}(1,3,4)$ | | Q.I. $ut^2, 21, 28$ |
| $P_2P_4 = \frac{1}{2}(1,1,1)$ | $B^3 < 0, 3B + E$ | |

No. 69: $X_{28} \subset \mathbb{P}(1,4,6,7,11)$, degree $A^3 = 1/66$

| $P_4 = \frac{1}{11}(1,4,7)$ | | Q.I. $zu^2, 17, 28$ |
|---|---|---|
| $P_2 = \frac{1}{6}(1,1,5)$ | $B^3 < 0, 4B$ | |
| $P_1P_2 = 2 \times \frac{1}{2}(1,1,1)$ | $B^3 < 0, 22B + 9E$ (∗) | |

No. 70: $X_{30} \subset \mathbb{P}(1,1,4,10,15)$, degree $A^3 = 1/20$

| $P_2 = \frac{1}{4}(1,1,3)$ | $B^3 < 0, B$ | none |
|---|---|---|
| $P_3P_4 = \frac{1}{5}(1,1,4)$ | $B^3 = 0, B$ | |
| $P_2P_3 = 1 \times \frac{1}{2}(1,1,1)$ | $B^3 < 0, B$ | |

No. 71: $X_{30} \subset \mathbb{P}(1,1,6,8,15)$, degree $A^3 = 1/24$

| $P_3 = \frac{1}{8}(1,1,7)$ | $B^3 > 0, 6B - E$  (†) | none |
|---|---|---|
| $P_2P_4 = 1 \times \frac{1}{3}(1,1,2)$ | $B^3 < 0, B$ | |
| $P_2P_3 = 1 \times \frac{1}{2}(1,1,1)$ | $B^3 < 0, B$ | |

No. 72: $X_{30} \subset \mathbb{P}(1,2,3,10,15)$, degree $A^3 = 1/30$

| $P_3P_4 = 1 \times \frac{1}{5}(1,2,3)$ | $B^3 = 0, 2B$ | none |
|---|---|---|
| $P_2P_4 = 2 \times \frac{1}{3}(1,2,1)$ | $B^3 < 0, 2B$ | |
| $P_1P_3 = 3 \times \frac{1}{2}(1,1,1)$ | $B^3 < 0, 3B + E$ | |

No. 73: $X_{30} \subset \mathbb{P}(1,2,6,7,15)$, degree $A^3 = 1/42$

| $P_3 = \frac{1}{7}(1,6,1)$ | $B^3 = 0, 2B$ | none |
|---|---|---|
| $P_2P_4 = 1 \times \frac{1}{3}(1,2,1)$ | $B^3 < 0, 2B$ | |
| $P_1P_2 = 5 \times \frac{1}{2}(1,1,1)$ | $B^3 < 0, 6B + 2E$ | |

No. 74: $X_{30} \subset \mathbb{P}(1,3,4,10,13)$, degree $A^3 = 1/52$

| $P_4 = \frac{1}{13}(1,3,10)$ | | Q.I. $zu^2, 17, 30$ |
|---|---|---|
| $P_2 = \frac{1}{4}(1,3,1)$ | $B^3 < 0, 3B$ | |
| $P_2P_3 = \frac{1}{2}(1,1,1)$ | $B^3 < 0, 3B + E$ | |

No. 75: $X_{30} \subset \mathbb{P}(1,4,5,6,15)$, degree $A^3 = 1/60$

| $P_1 = \frac{1}{4}(1,1,3)$ | $B^3 < 0, 10B + E$ (∗) | none |
|---|---|---|
| $P_3P_4 = 1 \times \frac{1}{3}(1,1,2)$ | $B^3 < 0, 5B + E$ | |
| $P_2P_4 = 2 \times \frac{1}{5}(1,4,1)$ | $B^3 < 0, 4B$ | |
| $P_1P_3 = 2 \times \frac{1}{2}(1,1,1)$ | $B^3 < 0, 5B + 2E$ | |

No. 76: $X_{30} \subset \mathbb{P}(1,5,6,8,11)$, degree $A^3 = 1/88$

| $P_4 = \frac{1}{11}(1,5,6)$ | | Q.I. $tu^2$, 19, 30 |
|---|---|---|
| $P_3 = \frac{1}{8}(1,5,3)$ | | E.I. $tu^2 - zt^3$, 22, 30 |
| $P_2P_3 = 1 \times \frac{1}{2}(1,1,1)$ | $B^3 < 0, 5B + 2E$ | |

No. 77: $X_{32} \subset \mathbb{P}(1,2,5,9,16)$, degree $A^3 = 1/45$

| $P_3 = \frac{1}{9}(1,2,7)$ | $B^3 > 0, 5B - E$ (†) | none |
|---|---|---|
| $P_2 = \frac{1}{5}(1,4,1)$ | $B^3 < 0, 2B$ | |
| $P_1P_4 = 2 \times \frac{1}{2}(1,1,1)$ | $B^3 < 0, 18B + 7E$ (*) | |

No. 78: $X_{32} \subset \mathbb{P}(1,4,5,7,16)$, degree $A^3 = 1/70$

| $P_3 = \frac{1}{7}(1,5,2)$ | $B^3 = 0, 4B$ | none |
|---|---|---|
| $P_2 = \frac{1}{5}(1,4,1)$ | $B^3 < 0, 4B$ | |
| $P_1P_4 = 2 \times \frac{1}{4}(1,1,3)$ | $B^3 < 0, 7B + E$ | |

No. 79: $X_{33} \subset \mathbb{P}(1,3,5,11,14)$, degree $A^3 = 1/70$

| $P_4 = \frac{1}{14}(1,3,11)$ | | Q.I. $zu^2$, 19, 33 |
|---|---|---|
| $P_2 = \frac{1}{5}(1,1,4)$ | $B^3 < 0, 3B$ | |

No. 80: $X_{34} \subset \mathbb{P}(1,3,4,10,17)$, degree $A^3 = 1/60$

| $P_3 = \frac{1}{10}(1,3,7)$ | $B^3 > 0, 4B - E$ (†) | none |
|---|---|---|
| $P_2 = \frac{1}{4}(1,3,1)$ | $B^3 < 0, 3B$ | |
| $P_1 = \frac{1}{3}(1,1,2)$ | $B^3 < 0, 10B + 2E$ | |
| $P_2P_3 = 1 \times \frac{1}{2}(1,1,1)$ | $B^3 < 0, 3B + E$ | |

No. 81: $X_{34} \subset \mathbb{P}(1,4,6,7,17)$, degree $A^3 = 1/84$

| $P_3 = \frac{1}{7}(1,4,3)$ | $B^3 = 0, 6B$ | none |
|---|---|---|
| $P_2 = \frac{1}{6}(1,1,5)$ | $B^3 < 0, 4B$ | |
| $P_1 = \frac{1}{4}(1,3,1)$ | $B^3 < 0, 7B + E$ | |
| $P_1P_2 = 2 \times \frac{1}{2}(1,1,1)$ | $B^3 < 0, 12B + 5E$ (*) | |

No. 82: $X_{36} \subset \mathbb{P}(1,1,5,12,18)$, degree $A^3 = 1/30$

| $P_2 = \frac{1}{5}(1,2,3)$ | $B^3 = 0, B - E$ (†) | none |
|---|---|---|
| $P_3P_4 = 1 \times \frac{1}{6}(1,1,5)$ | $B^3 = 0, B$ | |

No. 83: $X_{36} \subset \mathbb{P}(1,3,4,11,18)$, degree $A^3 = 1/66$

| $P_3 = \frac{1}{11}(1,4,7)$ | $B^3 > 0, 3B - E$ (†) | none |
|---|---|---|
| $P_2P_4 = 1 \times \frac{1}{2}(1,1,1)$ | $B^3 < 0, 3B + E$ | |
| $P_1P_4 = 2 \times \frac{1}{3}(1,1,2)$ | $B^3 < 0, 18E + 6B$ | |

No. 84: $X_{36} \subset \mathbb{P}(1,7,8,9,12)$, degree $A^3 = 1/168$

| | | |
|---|---|---|
| $P_2 = \frac{1}{8}(1,7,1)$ | $B^3 < 0,7B$ | none |
| $P_1 = \frac{1}{7}(1,2,5)$ | $B^3 < 0,8B$ | |
| $P_3P_4 = 1 \times \frac{1}{3}(1,1,2)$ | $B^3 < 0,8B + 2E$ | |
| $P_2P_4 = 1 \times \frac{1}{4}(1,3,1)$ | $B^3 < 0,7B + E$ | |

No. 85: $X_{38} \subset \mathbb{P}(1,3,5,11,19)$, degree $A^3 = 2/165$

| | | |
|---|---|---|
| $P_3 = \frac{1}{11}(1,3,8)$ | $B^3 > 0,5B - E$  (†) | none |
| $P_2 = \frac{1}{5}(1,1,4)$ | $B^3 < 0,3B$ | |
| $P_1 = \frac{1}{3}(1,2,1)$ | $B^3 < 0,5B + E$ | |

No. 86: $X_{38} \subset \mathbb{P}(1,5,6,8,19)$, degree $A^3 = 1/120$

| | | |
|---|---|---|
| $P_3 = \frac{1}{8}(1,5,3)$ | $B^3 = 0,5B$ | none |
| $P_2 = \frac{1}{6}(1,5,1)$ | $B^3 < 0,5B$ | |
| $P_1 = \frac{1}{5}(1,1,4)$ | $B^3 < 0,18B + 2E$ (∗) | |
| $P_2P_3 = 1 \times \frac{1}{2}(1,1,1)$ | $B^3 < 0,5B + 2E$ | |

No. 87: $X_{40} \subset \mathbb{P}(1,5,7,8,20)$, degree $A^3 = 1/140$

| | | |
|---|---|---|
| $P_2 = \frac{1}{7}(1,1,6)$ | $B^3 < 0,5B$ | none |
| $P_3P_4 = 1 \times \frac{1}{4}(1,1,3)$ | $B^3 < 0,7B + E$ | |
| $P_1P_4 = 2 \times \frac{1}{5}(1,2,3)$ | $B^3 < 0,20B + 3E$ | |

No. 88: $X_{42} \subset \mathbb{P}(1,1,6,14,21)$, degree $A^3 = 1/42$

| | | |
|---|---|---|
| $P_3P_4 = 1 \times \frac{1}{7}(1,1,6)$ | $B^3 = 0,B$ | none |
| $P_2P_4 = 1 \times \frac{1}{3}(1,1,2)$ | $B^3 < 0,B$ | |
| $P_2P_3 = 1 \times \frac{1}{2}(1,1,1)$ | $B^3 < 0,B$ | |

No. 89: $X_{42} \subset \mathbb{P}(1,2,5,14,21)$, degree $A^3 = 1/70$

| | | |
|---|---|---|
| $P_2 = \frac{1}{5}(1,4,1)$ | $B^3 < 0,2B$ | none |
| $P_3P_4 = 1 \times \frac{1}{7}(1,2,5)$ | $B^3 = 0,2B$ | |
| $P_1P_3 = 3 \times \frac{1}{2}(1,1,1)$ | $B^3 < 0,5B + 2E$ | |

No. 90: $X_{42} \subset \mathbb{P}(1,3,4,14,21)$, degree $A^3 = 1/84$

| | | |
|---|---|---|
| $P_2 = \frac{1}{4}(1,3,1)$ | $B^3 < 0,3B$ | none |
| $P_3P_4 = \frac{1}{7}(1,3,4)$ | $B^3 = 0,3B$ | |
| $P_1P_4 = 2 \times \frac{1}{3}(1,1,2)$ | $B^3 < 0,21B + 5E$ | |
| $P_2P_3 = \frac{1}{2}(1,1,1)$ | $B^3 < 0,3B + E$ | |

No. 91: $X_{44} \subset \mathbb{P}(1,4,5,13,22)$, degree $A^3 = 1/130$

| $P_3 = \frac{1}{13}(1,4,9)$ | $B^3 > 0, 5B - E$   (†) | none |
|---|---|---|
| $P_2 = \frac{1}{5}(1,3,2)$ | $B^3 < 0, 4B$ | |
| $P_1 P_4 = 1 \times \frac{1}{2}(1,1,1)$ | $B^3 < 0, 5B + 2E$ | |

No. 92: $X_{48} \subset \mathbb{P}(1,3,5,16,24)$, degree $A^3 = 1/120$

| $P_2 = \frac{1}{5}(1,1,4)$ | $B^3 < 0, 3B$ | none |
|---|---|---|
| $P_3 P_4 = 1 \times \frac{1}{8}(1,3,5)$ | $B^3 = 0, 3B$ | |
| $P_1 P_4 = 2 \times \frac{1}{3}(1,2,1)$ | $B^3 < 0, 5B + E$ | |

No. 93: $X_{50} \subset \mathbb{P}(1,7,8,10,25)$, degree $A^3 = 1/280$

| $P_2 = \frac{1}{8}(1,7,1)$ | $B^3 < 0, 7B$ | none |
|---|---|---|
| $P_1 = \frac{1}{7}(1,3,4)$ | $B^3 < 0, 8B$ | |
| $P_3 P_4 = 1 \times \frac{1}{5}(1,2,3)$ | $B^3 < 0, 8B + E$ | |
| $P_2 P_3 = 1 \times \frac{1}{2}(1,1,1)$ | $B^3 < 0, 7B + 3E$ | |

No. 94: $X_{54} \subset \mathbb{P}(1,4,5,18,27)$, degree $A^3 = 1/180$

| $P_2 = \frac{1}{5}(1,3,2)$ | $B^3 < 0, 4B$ | none |
|---|---|---|
| $P_1 = \frac{1}{4}(1,1,3)$ | $B^3 < 0, 18B + 3E$ | |
| $P_3 P_4 = 1 \times \frac{1}{9}(1,4,5)$ | $B^3 = 0, 4B$ | |
| $P_1 P_3 = 1 \times \frac{1}{2}(1,1,1)$ | $B^3 < 0, 5B + 2E$ | |

No. 95: $X_{66} \subset \mathbb{P}(1,5,6,22,33)$, degree $A^3 = 1/330$

| $P_1 = \frac{1}{5}(1,2,3)$ | $B^3 < 0, 6B$ | none |
|---|---|---|
| $P_3 P_4 = 1 \times \frac{1}{11}(1,5,6)$ | $B^3 = 0, 5B$ | |
| $P_2 P_4 = \frac{1}{3}(1,2,1)$ | $B^3 < 0, 5B + E$ | |
| $P_2 P_3 = 1 \times \frac{1}{2}(1,1,1)$ | $B^3 < 0, 5B + 2E$ | |

# 7  Final remarks

Although our proof of Theorem 1.3 is now complete, many mysteries remain. Some properties that we observe to hold for our general hypersurfaces certainly fail for general Fano 3-folds. Some features of our proof, both methods of exclusion and constructions of untwisting, occur over a whole range of cases, including many not treated here, and can sometimes even be stated as general properties of all Fanos, although we have no conceptual a priori proofs. As we discuss in the final 7.9, the many individual cases of special behaviour that have kept us company throughout this long paper possibly indicate that no such simple logic is applicable.

## 7.1   Iskovskikh–Manin by our methods

The classical result of Iskovskikh–Manin on the nonsingular quartic 3-fold [IM] becomes quite straightforward by our methods. See 5.2 for the method for excluding curves as maximal centres, where we took their argument as a model. For excluding any point as a maximal centre, our method is very short and simple: it uses only the statement of the log surface method in Theorem 5.3.2 and the obvious test class $A$ with $A^3 = 4$. Compare also [Pu1] and [Co2].

## 7.2   Numerics

This is a short tutorial on some nice methods of calculation that we have found useful on many occasions.

### 7.2.1   Predicting the degrees

The invariants $d, a_1, \ldots, a_4$ and $A^3 = (-K_X)^3$ of our anticanonical hypersurfaces $X_d \subset \mathbb{P}(1, a_1, \ldots, a_4)$ satisfy the two relations

$$\sum a_i = d \quad \text{and} \quad A^3 = \frac{d}{\prod a_i}.$$

Note that by Proposition 3.4.6, (1), for a Kawamata blowup $Y \to X$

$$-K_Y = B = A - \frac{1}{r}E \quad \text{so that} \quad B^3 = A^3 - \frac{1}{ra(r-a)}.$$

Thus if we suspect that the anticanonical ring of $Y$ is a hypersurface, we can usually predict the degrees of the generators and relations without effort; this is particularly effective when used in conjunction with the lists of Fletcher [Fl]. It applies to all the involutions constructed in Chapter 4: when $Y \to X$ leads to a quadratic involution, all but one of the generators of $X$ survive, so that if the anticanonical model of $Y$ is $Z_{d'} \subset \mathbb{P}(1, a_1, \ldots, a'_i, \ldots, a_4)$, the values of $a'_i$ and $d'$ are determined. (The calculation we left implicit for the first troop of 16 bad links in Remark 5.5.2 is exactly the same.) Similarly, if $Y \to X$ leads to an elliptic involution, we eliminate two of the generators $x_i, x_j$ in favour of two new generators $x'_i, x'_j$ that should appear in the Weierstrass equation (6) and thus have degrees $a'_i, a'_j = 2e, 3e$; for example, in Example 4.11,

$$(2e)(3e) = 9d \quad \text{and} \quad 2e + 3e = d - 9,$$

which has only one solution $e = 9$ and $d = 54$ in integers.

### 7.2.2  Hilbert function

A slightly more sophisticated form of this argument involves the Hilbert function (or Hilbert polynomial) of graded rings as in Fletcher [Fl], II.8, and developed more systematically in Altınok [A]. Write $P_n(X) = \chi(X, nA)$ and define the Hilbert polynomial by $P_X(t) = \sum_{n \geq 0} P_n t^n$; in our present case, since $A$ is ample, $\chi(X, nA) = h^0(X, nA)$ for each $n$ by vanishing, and the Hilbert polynomial equals the Hilbert function of $R(X, A)$. If $X_d \subset \mathbb{P}(1, a_1, \ldots, a_4)$ then obviously

$$P_X(t) = \frac{1 - t^d}{\prod_{i=0}^4 (1 - t^{a_i})}. \tag{17}$$

We can define the relative Hilbert polynomial of a morphism $f \colon Y \to X$ by $P_f = P_X - P_Y$. Here we use this only for $f$ the Kawamata blowup of a point of type $\frac{1}{r}(1, a, r - a)$; a calculation using the plurigenus formula of [YPG, §10] gives

$$P_f = P_X - P_Y = \frac{t^r}{(1 - t)(1 - t^a)(1 - t^{r-a})(1 - t^r)}. \tag{18}$$

It is clear that $\chi(Y, nB) = h^0(Y, nB)$ for all $n \geq 0$ if and only if we are in the "projectively normal" case when $R(Y, nB) \to R(E, nB_{|E})$ is surjective for all $n \geq 0$. It is a fun exercise to verify $P_f = P_X - P_Y$ in all the cases of Theorem 4.9, where we take $P_X$ and $P_Y$ to be given by (17) and $P_f$ by the right hand side of (18).

The method can be modified to work when $R(Y, B) \to R(E, B_{|E})$ is not surjective, which happens in the elliptic involution cases, and even when inverse flips intervene as in the final eccentric trio of Remark 5.5.2.

### 7.2.3  Restriction from Fano to K3

By the generality assumption, we can assume that $S : (x_0 = 0)$ is a quasismooth K3 surface, and $x_0$ is the local coordinate of degree 1 at every point of type $\frac{1}{r}(1, a, r - a)$ in the sense of Proposition 3.4.6, (2), so that $S \subset X$ is the general elephant.

We write $A_S = A_{|S} = -K_{X|S}$. The restriction map $H^0(X, nA) \to H^0(S, nA_S)$ is surjective for all $n \geq 0$, so that $R(S, A_S) = R(X, A)/(x_0)$. For many purposes, there is not much difference in working with linear systems on $X$ or their restrictions to $S$, other than technical convenience. The convenience, however, may sometimes be quite substantial, because linear systems on $S$ can be treated in terms of $\mathbb{Q}$-divisor on the minimal resolution, and we have pretty good intuition for how linear systems on K3s behave. For example, let $Y \to X$ be the Kawamata blowup of a $\frac{1}{r}(1, a, r - a)$ point $P \in X$, and $T \subset Y$ the birational transform of $S$. Then $P \in S$ of type $\frac{1}{r}(a, r - a)$ is an $A_{r-1}$ Du Val, and the $(a, r - a)$ blowup $T \to S$ extracts the $a$th curve of its resolution graph, leaving two cyclic singularities of index $a$ and $r - a$.

We explain the calculation used in the amazing example of 5.7.3 in these terms. There $S_{18} \subset \mathbb{P}(3, 4, 5, 6)$ is a K3 surface with $A^2 = \frac{1}{20}$, and we extract a curve $E$ by making the $(1, 2)$ blowup of the quotient singularity $\frac{1}{3}(1, 2)$ at $P_1 = (1, 0, 0, 0)$. Now $B = A - \frac{1}{3}E$, and $E^2 = -\frac{3}{2}$, and we look at the class $20B + 3E = 20A - \frac{11}{3}E$, obtaining $(20B + 3E)^2 = -\frac{1}{6}$, and $\chi(20B + 3E) = 1$. Write $y, z, t, u$ for the coordinates of $\mathbb{P}(3, 4, 5, 6)$ so that $z, t$ are local coordinates at $P_1$; they vanish along $E$ with multiplicity $\frac{1}{3}, \frac{2}{3}$ respectively. It is not hard to write out a monomial basis of the 11-dimensional space $H^0(20A)$. For an element to vanish $\frac{11}{3}$ times along $E$, we have to write it in terms of the local coordinates $z, t$, then set to zero the 10 coefficients of the monomials

$$
\begin{array}{lllll}
z^2 & t & & & \\
z^5 & z^3t & zt^2 & & \\
z^8 & z^t & z^4t^2 & z^2t^3 & t^4
\end{array}
$$

Translating to the local coordinates $z, t$ using the equation $F_{18}$ is the tricky part, and what makes the calculations of 5.7.3 so amazingly hard.

## 7.3 Involutions and projections

Our construction of quadratic and elliptic involutions in Chapter 4 consisted of calculating the subring $R(Y, B) \subset R(X, A)$ obtained by imposing zeros of order $\frac{d}{r}$ along $E$ as "antiadjunction" conditions on $H^0(X, \mathcal{O}_X(dA))$, which amounts algebraically to eliminating one or more of the variables. We explain briefly how to view this geometrically as a *projection*.

The case of quadratic involution is straightforward: in the notation of Theorem 4.9, we eliminate $\xi$ and introduce $w = \xi\eta$ in its place. Geometrically, $X \dashrightarrow Z$ projects away from the $\xi$ point, and the exceptional locus maps to a normal embedding of $\mathbb{P}(1, a, r - a)$ as the codimension 2 subvariety $(\eta = w + a = 0)$. The image $Z$ has equation $g = w(w + a) + \eta b = 0$; the inverse map $Z \dashrightarrow X$ consists of setting $\xi = w/\eta = b/(w + a)$. This fits into a *generic* pattern: write $\Pi : (x = y = 0)$ for a codimension 2 complete intersection and $Z : (Ay - Bx = 0)$ for a hypersurface containing $\Pi$; here $x, y, A, B$ could be independent indeterminates, or a regular sequence in a suitable ambient space. Setting $z = A/x = B/y$ defines a rational map (an inverse projection or *unprojection*) $Z \dashrightarrow X$ to the codimension 2 complete intersection $X : (zx = A, zy = B)$. The familiar unprojection of a cubic surface $S_3 \subset \mathbb{P}^3$ to $S_4 \subset \mathbb{P}^4$ fits into this generic case. Whereas in general, we expect the codimension to increase by 1 under this unprojection, in each case of Theorem 4.9, $A = w + a$ involved the variable $w$ linearly, so that $X$ happens to remain a hypersurface.

The elliptic involution is more subtle, and we only mention briefly the meaning of the curious equations around (12) of Example 4.11. If we treat $a, d, v, w, y$ as independent indeterminates, it is not hard to see that 4.11, (14) defines the nonnormal toric variety $\Sigma$ obtained as the image of the normalisation

$$\mathbb{C}^3 \to \Sigma \subset \mathbb{C}^5 \quad \text{given by} \quad (a, d, x) \mapsto (a, d, v = ax, w = dx, y = x^2),$$

which folds the $x$-axis in half, identifying $(0, 0, \pm x)$. Write $Z$ for the generic hypersurface containing $\Sigma$, given by

$$A(vd - aw) + B(v^2 - a^2 y) + C(vw - ady) + D(w^2 - dy^2) = 0.$$

Treating $a, d, v, w, y, A, B, C, D$ as independent indeterminates, $\Sigma$ can still be "unprojected" by introducing rational functions $u, t$ satisfying

$$\begin{pmatrix} a & v & aC - dD \\ d & w & aB \\ v & ay & wD + aA \\ w & dy & dA - vB + wC \end{pmatrix} \begin{pmatrix} u \\ t \\ 1 \end{pmatrix} = 0.$$

These equations are the first 4 Pfaffians of the $5 \times 5$ skewsymmetric matrix

$$\begin{pmatrix} a & d & v & w \\ & t & D & -u \\ & & u + C & B \\ -\text{sym} & & & A + yt \end{pmatrix}.$$

(The fifth Pfaffian is not linear in $u, t$.) In the generic context, $u, t$ define a birational map $Z \dashrightarrow X$ to the Gorenstein codimension 3 variety $X$ defined by the five Pfaffians. Whereas in general this type of unprojection should increase the codimension by 2, in the cases of Theorem 4.13, $D = 1$, so that the equations (12) involve $w$ and $a$ linearly, and $X$ happens to be a hypersurface.

There are many other types of unprojection: for example, let $\Gamma \subset \mathbb{P}^3$ be a nonnormal twisted quartic, and $S = S_4 \subset \mathbb{P}^3$ a nonsingular quartic surface containing it. Then $\Gamma$ can be contracted to a node by a map $S \to \overline{S}$, and it is easy to check that $\overline{S}$ is a codimension 5 K3 surface $\overline{S} \subset \mathbb{P}(1^7, 2)$. This case is essentially the hyperplane section of the double projection of the genus 6 Fano 3-fold $X \subset \mathbb{P}^7$, which has image a quartic of $\mathbb{P}^4$ containing the nonsingular projection of the Veronese.

We confidently expect that the unprojection idea will in the fullness of time serve as the basis for a *classification of all Fano 3-folds*. In other words, for the Fano 3-folds whose anticanonical ring are of codimension $\geq 5$, a first guess would be that the great majority of them have a singular point $\frac{1}{r}(1, a, r - a)$ for which three of the generators of the canonical ring have degrees equal to $1, a, r - a$. Then they project down to a graded ring of smaller codimension containing a projectively normal $\mathbb{P}(1, a, r - a)$. We have in mind here an analogy with Fano's study of nonsingular Fanos by projection and double projection and the experience of Selma Altınok's thesis [A] on Fanos with anticanonical ring of codimension 3 and 4.

## 7.4    Comments on the results

### 7.4.1    Mini-definition of rigidity

We could define rigidity via the *mini-definition*: every Sarkisov link $X \dashrightarrow Y$ has $Y \cong X$. In other words, we could take a statement like Theorem 3.2 as the definition of rigid, although, of course, the equivalence of the two definitions depends on knowing the Sarkisov program.

### 7.4.2    The key monomials

When a key monomial $*y^2z$ is absent (compare Example 4.6), we expect the point $P_t$ to be excluded as a maximal centre by a bad link. We haven't studied all these cases systematically, but some of them are easy enough, and what happens supports Conjecture 1.3 in a rather interesting way. Consider No. 5, $X_7 \subset \mathbb{P}(1, 1, 1, 2, 3)$, centre $P = P_3 = \frac{1}{2}(1, 1, 1)$, Example 4.6. An interesting degeneration takes place here: in general, the projection $X \dashrightarrow \mathbb{P}^2$ given by $\{x_0 : x_1 : x_2\}$ has fibres that are nonsingular elliptic curves $E_7 \subset \mathbb{P}(1, 2, 3)$ passing through the singularities $P_3$ and $P_4$. If the monomial $y^2z$ is not present, $X$ contains the $(y, z)$-line $\Gamma = P_3P_4$, and these fibres $E_7$ degenerate to nonsingular elliptic curves $E_6' \subset \mathbb{P}(1, 2, 3)$ plus $\Gamma$, where $E_6'$ meets $\Gamma$ transversally at one point; $\Gamma$ would be a $-1$-curve on a nonsingular model of $\mathbb{P}(1, 2, 3)$.

Assume that the tangent monomial at $P_y$ is $x_2y^3$. If $Y \to X$ is the Kawamata blowup then $B \cdot \Gamma < 0$, and we get a bad link involving an inverse flip as in 5.5. In fact the anticanonical model of $Y$ is $X \dashrightarrow Z_9 \subset \mathbb{P}(1, 1, 1, 3, 4)$, given by

$$(x_0, x_1, x_2, y, z) \mapsto (x_0, x_1, x_2, x_2y, x_2z),$$

exactly as at the end of 5.5.

### 7.4.3    Comments on $\operatorname{Bir} X$

(i) *Why on earth does it always have to be involutions?* For our Fano hypersurfaces $X$, we have proved, among other things, that $\operatorname{Bir} X$ is generated by birational involutions; this seems to hold also for every Fano variety for which rigidity has been established up to now. It may also be a general feature of badly nonrigid varieties, e.g., Max Noether's case of $\mathbb{P}^2$, which, although familiar, lacks any convincing explanation.

(ii) There are many cases when, even though $\operatorname{Bir} X$ is an infinite group, $X$ still has an invariant elliptic fibration or an invariant pencil of K3s. Then $\operatorname{Bir} X$ can be interpreted as the group of biregular automorphisms of the generic fibre. For example, in the case just discussed, $X_7 \subset \mathbb{P}(1,1,1,2,3)$ has in general two involutions $i_y$ and $i_z$, both of which leave invariant the projection to $\mathbb{P}^2$, and generate an infinite reflection group of biregular automorphisms of the generic fibre $E_7$; in fact $i_z$ interchanges $y$ and $z$, and $i_y$ is the reflection in $z$. There seems to be an analogy here with results on automorphisms of K3s, and even with the currently popular finiteness theorems and conjectures for the Kähler cones of Calabi–Yau 3-folds and their automorphism groups.

(iii) When the coefficient of the term $z^2t$ in a standard quadratic involution degenerates to zero as in 7.4.2, the birational involution of $X$ remains present, but degenerates to an involution no longer untwisting $P_z$. Recall the example of the projection from an Eckardt point in 2.11, where a birational involution *degenerated to a biregular involution*. The example of 7.4.2 is more interesting: in a varying family, as $c \to 0$, the involutions $i_y$ *degenerates to coincide with* $i_z$. (This is easily proved: the bad link argument given in 7.4.2 exclude $P_y$ as a maximal centre, so that, together with the main argument of Chapter 5, it shows that the variety $X_7$ with $c = 0$ has only the single involution $i_z$. But $i_y$ acted by $-1$ on the generic fibre $E_7$, and that can't change into $+1$.) Thus we see the group $\operatorname{Bir} X$ jumping down from an infinite reflection group to $\mathbb{Z}/2$. As far as the generic fibre $E_7$ is concerned, it just means that the points $P_y$ and $P_z$ come together.

### 7.4.4    Inverse flips

In the construction of a Sarkisov link, if flips and inverse flips occur, then we necessarily do inverse flips first, then flops, and flips last. This is just an automatic consequence of the way the 2-ray game works: on the extraction $Y \to X$, it is a MMP directed by $\mathcal{H}$.

If we do an extremal extraction $Y \to X$ then an inverse flip means that the maps $R(X, mA) \to R(E, mA_{|E})$ fails to be surjective, with the cokernel growing like $m^3$.

Inverse flips certainly exist in Sarkisov links on Fano 3-folds. Consider, for example, the case treated by Iskovskikh, and later by him and Pukhlikov, of the involution centred in a conic $q$ on $X = X_{2,3} \subset \mathbb{P}^5$ which is *special* in the sense that its plane also contains a line $l$ of $X$. This involution is a Sarkisov link of Type II (Definition 3.1.4) conveniently derived using the method of Chapter 4 based on the anticanonical ring. It is obtained by first making the blowup $Y \to X$ of $q$, followed by the inverse flip $Y \dashrightarrow Y'$ in the birational transform of $l$, which is a $(-1, -2)$-line on $Y$. Then $Y'$ has anticanonical model $Z_{12} \subset \mathbb{P}(1, 1, 1, 4, 6)$ which has a biregular involution $i_Z$ as explained in 4.1.

Why do none of these appear on our 95 hypersurfaces? Equivalently, why is $B = -K_Y$ nef? Maybe we're just not looking deep enough into singular varieties.

## 7.5    The Sarkisov category and the rigid boundary

Our main theorem is really very surprising: a priori, it is really hard to imagine any particular reason why a Sarkisov link $X \dashrightarrow X'$ cannot exist between two Fano 3-folds $X$ and $X'$ in different families (in fact we know that this does occur in other contexts); or between $X$ and $X'$ in the same family but with different moduli. Of course, the intermediate Jacobian says something, but we don't expect it always to give a blanket contradiction.

The following simple example shows how terminal singularities (even quite mild ones) may drastically affect the boundary line between rigid and nonrigid varieties.

**Example 7.5.1** A quartic 3-fold $X \subset \mathbb{P}^4$ may have a terminal singularity of local type $x^2 + y^2 + z^4 + t^4$, so that every surface in the pencil $\lambda x + \mu y = 0$ has an essential singularity. Then $X$ is birational to a pencil of rational surfaces, and so is certainly not rigid.

For a long time, in the absence of examples, we entertained the conjecture that a Fano must either be rigid, or have a strict Mori fibre space structure. Now Corti and Mella have a few "birigid" examples having *exactly two* Fano minimal models and no Mori fibre space; Grinenko and Pukhlikov have proposed related examples. It seems there will many more phenomena of this nature, intermediate between rigid and wildly nonrigid cases. (In fact this area is developing rapidly, and what we write here may soon be out of date.)

At one point in our studies, we hoped that rigid versus nonrigid should be a firm dichotomy in 3-folds; examples such as Example 7.5.1 show that this is too optimistic. Exactly as with questions of rationality or unirationality, it seems that the boundary is mysterious and possibly chaotic. Probably a lot of things suddenly explode together at the boundary: we get links to completely different Mori fibre space models, so that untwisting by selfmaps becomes out of the question, the group Bir $X$ becomes wildly infinite dimensional, etc.

## 7.6   The spectrum: all the arguments of the rainbow

Logically speaking, there cannot be any intersection between the cases when our methods of untwisting and of excluding apply, so it might seem to be a stroke of luck that they cover precisely every case. In fact, there really is a big overlap: our methods based on the anticanonical ring work more or less uniformly across a whole range of cases, sometimes giving an untwisting, and sometimes ending up in an anticanonical contraction other than a Mori fibre space, and thus a bad link. In this context, it is instructive to compare the calculations of Chapter 4 that lead to an untwisting with those of Chapter 5 that lead to a bad link (see for example Remark 5.5.2).

We can view the different excluding or untwisting arguments used in this paper as forming a *spectrum* ranging from entirely numerical arguments (obtaining a contradiction from an intersection calculation) to entirely anticanonical Mori theoretic arguments. This point was already mentioned in 2.11. The attentive reader will surely have noticed items of several different colours:

(i)   Simple numerical arguments such as intersection number on a surface; see for example Lemma 2.5, p. 180, and Step 1 of the proof of Theorem 5.1.1 on p. 206.

(ii)   The pure test class method 5.2; here we use intersection number on extremal extraction $Y \to X$, but still no extra clever choices.

(iii)   Variations on the test class method 5.3; note that Theorems 5.3.2 and 5.3.3 make more sophisticated use of the minimal model program and log MMP to set up a more refined intersection number calculation.

(iv)   The calculation of $\overline{\mathrm{NE}}$ in 5.4, which we first used in 5.4.1 to help find a test class, again feeding into a numerical calculation.

(v)   The bad link method discussed in 2.11 and 5.5; this used the knowledge of $\overline{\mathrm{NE}}$, together with facts about the Sarkisov program.

(vi)   Anticanonical model; when $|-K_Y|$ (or a multiple) is big on an extremal extraction $Y \to X$ we can in many practical cases follow through a 2-ray game all the way to an anticanonical model. This may lead to a contradiction, as we have seen in 5.5, or it may give a Sarkisov link that untwists our centre as in Chapter 4.

The *test class method* is at the numerical end of this spectrum; our version of it simplifies and strengthens the original approach used by Iskovskikh and Manin to treat the nonsingular quartic 3-fold. It requires little detailed knowledge of the geometry of $X$, and in several cases it implies that $\Gamma$ is not even a weak maximal centre. At the biregular end we have the *bad link method* (see 5.5), which rests on the inner logic of the Sarkisov program and needs a finer understanding of some of the geometry of $X$. Our treatment has followed this organisation roughly.

It would be nice to consolidate our methods, and do everything by the bad link method: after making an extraction $Y \to X$, we have $\overline{\mathrm{NE}}\,Y = R + Q$, and for a link to exist, it must be possible to detect $Q$, and to contract it (or inverse flip it) in the Mori category. In almost all cases, $Q$ can be thought of a curve (or limit of curves) on the K3 ($x_0 = 0$), usually rational. We know that when this fails, we get a contradiction of "bad link" type. However, in practice, this method is hard to use systematically.

As a simple illustration, suppose that we are trying to exclude curves on the quartic $C \subset X_4 \subset \mathbb{P}^4$ as part of the proof of the Manin–Iskovskikh theorem (and we tie our hands behind our backs, not allowing ourselves to use intersection numbers). If $C$ is a line, we blow up $Y \to X$; the anticanonical system $|-K_Y|$ of $Y$ gives a fibre space $\varphi \colon Y \to \mathbb{P}^2$ of plane cubics. Therefore $\overline{\mathrm{NE}}\,Y = R + Q$ where $Q$ is the fibre of $\varphi$. Since this is not a Mori fibre space, the 2-ray game stops, and we have a contradiction to $C$ a maximal centre.

If $C$ is a conic, we blowup $Y \to X$, but not the residual conic $C'$; then $|-K_Y|$ is a pencil of K3s having $C'$ as its fixed curve, but is not nef. If you want to continue the two ray game, the inverse flip $Y \dashrightarrow Y'$ of $C'$ exists, and contracts $C'$ to a node on every surface $S \in |-K_Y|$. Then $Y'$ has a strictly canonical singularity.

If $C \subset X_4$ is a twisted cubic, a similar kind of argument works: look at the quadrics through $C$, getting a nef system $|B + E|$ that defines a fibration of $Y$ which is not a Mori fibre space. But we have no idea how to extend this kind of thing to curves of higher degree. The paradox is that we only seem to be able to make the bad link method work for exclusion when there is a reasonably substantial anticanonical system, so that we can at least start out on a 2-ray game (in other words, if our centre is reasonably close to being a maximal singularity).

Note that, as we said in 5.2, we cannot use bad link or other biregular methods to exclude nonsingular points as maximal centres, because we don't have a classification of extremal extractions $Y \to X$; and even if we did, it might be too complicated to use. Thus the test class stuff and the clever

Theorems 5.3.2 and 5.3.3 are the only arguments we know. (Although it would be reasonable to hope that nonsingular points could be excluded for some very coarse reasons.)

## 7.7   Remarks relating to the proof of exclusion

The key question is whether there is a fundamental reason for some centres to be excluded and others not. Let $f\colon (E \subset Y) \to (\Gamma \subset X)$ be an extremal divisorial contraction.

In a vague sense, it is natural to expect that, if $B^3 \le 0$, $E$ is not a maximal singularity. Indeed, if the 2-ray game originating from $f$ is to produce a Sarkisov link, $B$ must at least be big (that is, $\kappa(X, B) = 3$, but not necessarily satisfying $B$ nef or $B^3 > 0$). This explanation works in many cases, but one must not be deceived. Inverse flips increase $B^3$, so it is still possible for $B$ to be big even if $B^3 \le 0$ (this happens, for instance, in the maverick exception No. 82, $X_{36} \subset \mathbb{P}(1, 1, 5, 12, 18)$, centre $P_2 = \frac{1}{5}(1, 2, 3)$, but in many more cases – not touched upon in this paper – when $X$ is not general but still quasismooth). Also, in all bad links, $B$ is big but the point in question is excluded anyway. It should also be noted (see 7.4.4 above) that inverse flips do sometimes occur in links of Fano 3-folds, although never in this paper.

We find it interesting that, although the test class method and the bad link method are fundamentally different in spirit, both rest on the existence of a surface $T \in |bB + cE|$ satisfying suitable conditions. This seems to suggest that there should be a unified method of excluding. At the end of all our calculations, we can prove the following

**Theorem 7.7.1** *Let $X$ be general, and $f\colon (E \subset Y) \to (\Gamma \subset X)$ an arbitrary extremal divisorial contraction. Then $E$ is a maximal singularity if and only if*

$$B^2 \in \operatorname{Int} \overline{\mathrm{NM}}_1 Y,$$

*where $\overline{\mathrm{NM}}_1 Y \subset N_1 Y$ is the cone of nef curves.*

**Proof** If $P$ is a maximal centre, we know from the classification of links in Chapter 4 that $B$ is nef and big. More than that, we know that the anticanonical map

$$Y \to \overline{Y} = \operatorname{Proj} \bigoplus_{n \ge 0} H^0(-nK_Y)$$

only contracts a finite number of curves on $Y$. Then $B^2 \cdot D > 0$ for all effective surfaces $D \subset Y$ and $B^2 \in \operatorname{Int} \overline{\mathrm{NM}}_1 Y$. Assume now that $\Gamma \subset X$

is not a maximal centre. If the reason for excluding it was the test class method, then for all extremal divisorial contractions $E \subset Y \to \Gamma \subset X$, $B^2 \notin \text{Int} \overline{\text{NE}} Y$ and, in particular, $B^2 \notin \text{Int} \overline{\text{NM}}_1 Y$. In all other cases, by Lemma 5.4.5, there is a special surface $T$ on $Y$ with $B^2 \cdot T \leq 0$, in other words $B^2 \notin \text{Int} \overline{\text{NM}}_1 Y$. Q.E.D.

**Remark 7.7.2**    (i) It is important to understand that we only know how to prove Theorem 7.7.1 *after* completing the whole analysis of untwisting and exclusion in this paper.

(ii) If a direct proof of Theorem 7.7.1 could be found, it would greatly simplify the task of classifying maximal centres. For instance, if $B^3 \leq 0$, it is immediate that $B^2 \notin \text{Int} \overline{\text{NM}}_1 Y$. In the case of a singular point with $B^3 > 0$, the existence of a special surface $T$ with $B^2 \cdot T \leq 0$ (even without the additional conditions of Lemma 5.4.5, which are needed in the bad link method) immediately implies that $B^2 \notin \text{Int} \overline{\text{NM}}_1 Y$.

(iii) Theorem 7.7.1 should be contrasted with the backbone of the test class method 5.2: $E$ maximal implies $B^2 \in \text{Int} \overline{\text{NE}} Y$. Because $\overline{\text{NM}}_1 \subset \overline{\text{NE}}$, as already observed, the statement contains the test class method as a special case.

(iv) We do not know any counterexamples to Theorem 7.7.1, even if $X$ is not general, or not a hypersurface.

(v) In some sense, the test class method

$$E \text{ maximal} \implies B^2 \in \text{Int} \overline{\text{NE}} Y$$

is analogous to Pukhlikov's condition on $K^2$ in [Pu2]. Pukhlikov proves that if $X \to S$ is a 3-fold Mori fibration in del Pezzo surfaces (satisfying technical conditions including $X$ nonsingular) of degree $d \leq 3$ and $K_X^2 \notin \text{Int} \overline{\text{NE}} X$, then $X \to S$ is birationally rigid – see [Pu2] and [Co2], Chapter 5 for treatments of this result. Although easily verified in many natural examples, the $K^2$ condition is not very well understood. Could it be that $K_X^2 \notin \text{Int} \overline{\text{NM}}_1 X$ implies $X \to S$ birationally rigid?

## 7.8    Generality

Together with the Main Theorem 1.3, we stated Conjecture 1.3 that the statement holds for any quasismooth $X$; this still looks pretty good to us. We list here the places in the paper where generality assumptions on $X$ beyond quasismoothness are involved.

(i) In 9 of the quadratic involutions and 2 of the elliptic involutions, we assumed the presence of a key monomial such as $*z^2t \in f$ as an extra generality assumption in order to construct the untwisting; as we discuss in 7.4.2, if the monomial has zero coefficient, we expect that the centre $P$ is excluded by a bad link, and the involution degenerates to a birational map that no longer untwists $P$.

(ii) In the proof of Theorem 5.6.2, Case 2: to find convenient polynomials isolating $P$, we required that $a, b$ have no common zeros on a certain affine line; with the bizarre quartet of exceptions discussed there, this holds by Bertini.

(iii) Similarly in Theorem 5.6.2, Case 3, we asked $(x_0 = x_1 = 0) \cap X$ to be an irreducible curve; this again holds by Bertini.

(iv) In the calculation of $\overline{NE}\, Y$ in 5.7, we needed the curve $\Gamma = S \cap T$ to be irreducible. This is also OK by Bertini, but checking it requires a case-by-case inspection; the shape of $T$ does not follow any specially regular pattern.

In summary, generality always mean "outside a Zariski closed set", but the extent to which the open condition can be made explicit varies with the requirement. For the involutions, we only require the presence of a specific monomial (and it is even easy enough to analyse what happens if it does not appear). In contrast, in the proof of Theorem 5.6.2, the condition that $\Gamma = S \cap T$ is irreducible is much less clear-cut: in many cases when $\Gamma$ has fairly low degree, classifying the various exceptions is fun, but $\Gamma$ occasionally has higher degree, for example in the "amazing example" treated in 5.7.3.

Many of these generality assumptions could be removed without much trouble, but in some instances this would be far from straightforward. Relaxing generality allows all sorts of curious and delightful monsters to crawl out from their underground hiding places. Although we have never found any that seriously threaten the truth of our conjecture, it is clear that doing away with generality assumptions altogether would require a volume of calculations that is quite beyond us.

## 7.9 Eccentricity

At a number of points in this paper there are statements that have exceptions or arguments that break down due to antisocial behaviour on the part of a small and unrepresentative group of disruptive individuals. At the top of our list of trouble-makers are

(i) The renegade ($*$) of 5.3 (p. 208). This is an exception to Theorem 5.3.1: if $P \in X_5 \subset \mathbb{P}(1,1,1,1,2)$ lies on an exceptional line, even though the point is excluded, no $P$-isolating system with the required numerical properties exists.

(ii) The maverick exception

$$\text{No. 82:} \quad X_{36} \subset \mathbb{P}(1,1,5,12,18), \text{ centre } P_2 = \frac{1}{5}(1,2,3)$$

to Lemma 5.4.3. This is the only point with $B^3 \le 0$ for which no test class exists; the point is excluded using the bad link method.

(iii) The eccentric trio of Remark 5.5.2 (which includes in its ranks the maverick) Nos. 47, 62 and 82: the bad link starts with an inverse flip.

(iv) The amazing example of 5.7.3, where the calculation of $\overline{NE}\,Y$ becomes fantastically intricate.

(v) The bizarre quartet in the proof of Theorem 5.6.2, Case 2, where an otherwise harmless genericity assumption becomes impossible.

These are just the most striking examples of an already idiosyncratic proof: the tricksy 10 cases of quadratic involutions, the 10 hard cases of calculation of $\overline{NE}\,Y$ and the errant nature of the several hundred easy cases, the unpredictability of singular points $P$ with $B^3 > 0$ – some of these are excluded, some are not – are similar expressions of an unruly population.

There is an amusing and always surprising interplay between generality and eccentricity; relaxing generality encourages eccentricity. For example:

• in 7.4.2, if its starred monomial $x_j x_i^2$ is absent, a quadratic involution degenerates to coincide with some different involution no longer untwisting the centre; or

• the bizarre quartet, stubbornly resisting a perfectly innocuous generality assumption; or

• the 18 special surfaces of Definition 5.4.4, whose very existence is eccentric, but which all line up in a neat predictable pattern.

In the face of this extravagant exuberance of behaviour, two extreme conclusions present themselves. One is that we haven't understood what's really going on; we have cobbled together some ad hoc techniques, but failed to uncover the more powerful methods that are capable of imposing order. The alternative point of view is that our arguments are necessarily complicated, tailored as they must be to encompass the special circumstances of a population of individualists resenting overenthusiastic attempts to legislate.

# References

[A]    S. Altınok, Graded rings corresponding to polarised K3 surfaces and Q-Fano 3-folds, Univ. of Warwick Ph.D. thesis, Sep. 1998, 93 + vii pp.

[CL]   F. Call and G. Lyubeznik, A simple proof of Grothendieck's theorem on the parafactoriality of local rings, in Commutative algebra: syzygies, multiplicities, and birational algebra (South Hadley, 1992), Contemp. Math. **159**, AMS (1994) pp. 15–18

[CKM]  H. Clemens, J. Kollár and S. Mori, Higher dimensional complex geometry, Astérisque **166** Soc. Math. de France (1988)

[Co1]  A. Corti, Factoring birational maps of 3-folds after Sarkisov, J. Alg. Geom. **4** (1995), 223–254

[Co2]  A. Corti, Singularities of linear systems and 3-fold birational geometry, this volume, 259–312

[Fl]   A. Fletcher, Working with weighted complete intersections, this volume, 73–173

[SGA2] A. Grothendieck and others, Séminaire de géométrie algébrique 1962, Cohomologie locale des faisceaux cohérents et théorèmes de Lefschetz locaux et globaux, IHES 1965

[FA]   J. Kollár and others, Flips and abundance for algebraic 3-folds, Astérisque **211**, Soc. Math. de France 1992

[KM]   J. Kollár and S. Mori, Birational geometry of algebraic varieties, C.U.P., 1998

[Isk]  V. A. Iskovskikh, Rational surfaces with a pencil of rational curves and positive square of the canonical class, Mat. Sbornik **83** (1970), 90–119. = Math. USSR Sb. **12** (1970)

[IM]   V. A. Iskovskikh and Yu. I. Manin, Three-dimensional quartics and counterexamples to the Lüroth problem, Math. USSR Sb. **15** (1971), 141–166

[Ka]   Y. Kawamata, Divisorial contractions to 3-dimensional terminal quotient singularities, in Higher-dimensional complex varieties (Trento, 1994), 241–246, de Gruyter, Berlin, 1996

[Ko]   J. Kollár, The structure of algebraic threefolds: an introduction to Mori's program, Bull. Amer. Math. Soc. **17** (1987), 211–273

[C3-f] M. Reid, Canonical 3-folds, in Journées de géométrie algébrique d'Angers, ed. A. Beauville, Sijthoff and Noordhoff, Alphen 1980, 273–310

[YPG] M. Reid, Young person's guide to canonical singularities, in Algebraic Geometry, Bowdoin 1985, ed. S. Bloch, Proc. of Symposia in Pure Math. **46**, A.M.S. (1987), vol. 1, 345–414

[R]    M. Reid, Tendencious survey of 3-folds, in Algebraic Geometry, Bowdoin 1985, ed. S. Bloch, Proc. of Symposia in Pure Math. **46**, A.M.S. (1987), vol. 1, 333–344

[Pu1] A. V. Pukhlikov, Essentials of the method of maximal singularities, this volume, 73–100

[Pu2] A. V. Pukhlikov, Birational automorphisms of three-dimensional algebraic varieties with a pencil of del Pezzo surfaces, Izv. Ross. Akad. Nauk Ser. Mat. **62** (1998), 123–164. = Izv. Math. **62** (1998), 115–155

[S]    B. Segre, The non-singular cubic surfaces, Oxford 1942

[Sa1] V. G. Sarkisov, Birational automorphisms of conic bundles, Izv. Akad. Nauk SSSR Ser. Mat. **44** (1980), 918–945. = Math. USSR Izv. **62** (1998)

[Sa2] V. G. Sarkisov, On conic bundle structures, Izv. Akad. Nauk SSSR Ser. Mat. **46** (1982), 371–408. = Math. USSR Izv. **46** (1982)

Alessio Corti
DPMMS, University of Cambridge,
Centre for Mathematical Sciences,
Wilberforce Road, Cambridge CB3 0WB, U.K.
e-mail: a.corti@dpmms.cam.ac.uk

Alexandr V. Pukhlikov,
Number Theory Section, Steklov Mathematics Institute,
Gubkina, 8,
117966 Moscow, Russia
e-mail: dost@dost.mccme.rssi.ru and pukh@mi.ras.ru

Miles Reid,
Math Inst., Univ. of Warwick,
Coventry CV4 7AL, England
e-mail: miles@maths.warwick.ac.uk
web: www.maths.warwick.ac.uk/~miles

# Singularities of linear systems and 3-fold birational geometry

### Alessio Corti

## Contents

**1 Introduction**     **260**

**2 The Sarkisov program**     **267**
2.1 The Sarkisov degree . . . . . . . . . . . . . . . . . . . 267
2.2 Links and the 2-ray game . . . . . . . . . . . . . . . . 269
2.3 Construction of the untwisting link . . . . . . . . . . . 272

**3 Singularities of linear systems**     **275**
3.1 Linear systems on surfaces . . . . . . . . . . . . . . . . 275
3.2 Shokurov connectedness and its implications . . . . . . . . . 277
3.3 Divisorial contractions . . . . . . . . . . . . . . . . . . 281
3.4 Linear systems on 3-folds . . . . . . . . . . . . . . . . . 283

**4 Conic bundles**     **286**
4.1 Rigid Conic bundles . . . . . . . . . . . . . . . . . . . 286
4.2 Open questions . . . . . . . . . . . . . . . . . . . . . . 293

**5 Del Pezzo fibrations**     **295**
5.1 Rigid Del Pezzo fibrations . . . . . . . . . . . . . . . . 295
5.2 Semistable models . . . . . . . . . . . . . . . . . . . . 298
5.3 Open questions . . . . . . . . . . . . . . . . . . . . . . 299

**6 Fano 3-folds**     **300**
6.1 Rigid Fano 3-fold hypersurfaces . . . . . . . . . . . . . . 300
6.2 Rigid Fano 3-fold complete intersections . . . . . . . . . . 304
6.3 Failure of rigidity . . . . . . . . . . . . . . . . . . . . 309

**References**     **310**

# 1    Introduction

In this paper I explain some new techniques for studying singularities of linear systems, with applications to birational maps between 3-fold Mori fibre spaces, and especially the property of *birational rigidity*, see Definition 1.3 below. These techniques are closely related and all have something to do with *Shokurov's connectedness principle* (see Section 3.2). Though they do not as yet form a coherent "method", they are intended to replace the combinatorial study of the resolution graph, started by Iskovskikh and Manin [IM].

My goal is to provide concise but complete proofs of the known criteria for birational rigidity of 3-fold Mori fibre spaces. I do not give the results in their most general or sharpest form, leaving considerable scope for improvement in various places. For instance, I do not discuss fibre spaces of cubic surfaces, the most intricate parts of [CPR], or *relations* between generators of Bir $X$. My desire is rather to introduce the main ideas in the simplest context in which they appear. I do, however, prove in Section 6.2 the rigidity of a general smooth 3-fold complete intersection of type $2, 3$. I include this argument here because the result is important and the method of proof is an excellent illustration of the power of the new methods. A proof in the spirit of [IM] is outlined in Iskovskikh and Pukhlikov [IP].

I have tried to state many open questions, problems and conjectures, with the aim of understanding how and why rigidity fails. On the other hand, I have made a conscious decision to say nothing about dimension $\geq 4$, even though there have been some interesting recent developments (work of Pukhlikov and Cheltsov).

The subject is classical. It was revived by the work of Iskovskikh and Manin on the nonsingular quartic 3-fold in the early 1970s, and it is practised today by many followers of the Moscow school of birational geometry. Mori theory came roughly a decade later, initially a close but seemingly unrelated development, but now increasingly relevant both as a foundational basis for classification and a source of new techniques:

**(1)**    Mori fibre spaces are one possible outcome of the minimal model program and are hence basic objects of the classification of 3-folds. They provide a significant extension of the classical world: for instance, there are 16 deformation families of smooth and primitive Fano 3-folds, while there are several hundred known deformation families of Fano 3-folds with terminal singularities and Picard number 1 (the total number of families is known to be bounded).

**(2)**    Sarkisov realised that the method going back to Noether, Castelnuovo and Fano of *untwisting* or *factoring* birational maps as a composite of *links*, which is the basis of all the results of the Moscow school, can be extended to the category of Mori fibre spaces as an abstract framework, the *Sarkisov program*.

**(3)**  The Sarkisov program has now been worked out in detail [CPR] for a significant class of singular Fano 3-folds, the 95 weighted hypersurfaces $X_d \subset \mathbb{P}(1, a_1, \ldots, a_4)$ with $-K_X = \mathcal{O}_X(1)$.

**(4)**  The techniques introduced in this paper to quantify and study singularities of linear systems are based on the log category and Shokurov connectedness, important technical tools of Mori theory with applications in other areas of classification theory. These techniques are more powerful than the method of Iskovskikh and Manin, and easier to use. Pukhlikov's article [P4] contains an excellent and accessible exposition of the method of Iskovskikh and Manin, which works well in many cases, and is in some sense more elementary (for example, it does not depend on Kodaira vanishing).

**(5)**  The links of the Sarkisov program are themselves made up of the divisorial contractions and flips associated to the extremal rays of Mori theory. This point of view has suggested some new ways of describing the links themselves, based on methods and techniques of graded rings. This description is one of the crucial innovations of [CPR]. On the other hand, the classification of 3-fold divisorial contractions (Problem 3.8) is a beautiful open problem in Mori theory whose solution would greatly enhance the applicability of the Sarkisov program. The couple of known cases of this problem are the basis of [CPR] and of the modern treatment of the rigidity of a quartic 3-fold with a node, see Theorem 1.9.

In the remainder of the introduction, I give the most important definitions, state the main results, and indicate briefly the contents of the various chapters.

The Mori category is the category of projective varieties with $\mathbb{Q}$-factorial terminal singularities. We always work in the Mori category. Recall

**Definition 1.1**  A *Mori fibre space* is an extremal contraction $f \colon X \to S$ of fibre type. In other words $f_*\mathcal{O}_X = \mathcal{O}_S$ and:

(1)  $-K_X$ is relatively ample for $f$,

(2)  $\operatorname{rank} N^1 X = \operatorname{rank} N^1 S + 1$,

(3)  $\dim S < \dim X$.

If $\dim X = 3$ there are 3 main cases

$$\begin{cases} \dim S = 0 : & X \text{ is a Fano 3-fold,} \\ \dim S = 1 : & X \text{ is a Del Pezzo fibration,} \\ \dim S = 2 : & X \text{ is a conic bundle.} \end{cases}$$

Note that this terminology is only used classically for nonsingular varieties.

**Definition 1.2**  (1) The *Sarkisov category* is the category whose objects are Mori fibre spaces and morphisms are birational maps (regardless of the fibre structure).

(2) Let $X \to S$ and $X' \to S'$ be Mori fibre spaces. A morphism in the Sarkisov category, that is, a birational map $f \colon X \dashrightarrow X'$, is *square* if it fits into a commutative square

$$\begin{array}{ccc} X & \overset{f}{\dashrightarrow} & X' \\ \downarrow & & \downarrow \\ S & \underset{g}{\dashrightarrow} & S' \end{array}$$

where $g$ is a birational map (which thus identifies the function field $L$ of $S$ with that of $S'$) and if, in addition, the induced birational map of generic fibres $f_L \colon X_L \dashrightarrow X'_L$ is biregular. In this case, we say that $X \to S$ and $X' \to S'$ are *square birational*.

(3) A *Sarkisov isomorphism* is a birational map $f \colon X \dashrightarrow X'$ which is biregular and square.

**Definition 1.3** A Mori fibre space $X \to S$ is *birationally rigid* if, given any birational map $\varphi \colon X \dashrightarrow X'$ to another Mori fibre space $X' \to S'$, there exists a birational selfmap $\alpha \colon X \dashrightarrow X$ such that the composite $\varphi \circ \alpha \colon X \dashrightarrow X'$ is square. In other words, $X$ birational to $X'$ implies $X$ square birational to $X'$. I say that $X \to S$ is *birationally rigid over* $S$ if a map $\alpha$ as in the definition can be taken to be defined over $S$.

Note that the definition does not say that every birational map $X \dashrightarrow X'$ is square. The main point is, of course, that rigid implies nonrational in a strong sense. I believe the notion is reasonably well behaved in families (see Chapter 6 for more remarks and examples):

**Conjecture 1.4** *Birational rigidity is open in moduli. In other words, given any scheme $T$, and a flat family of Mori fibre spaces parametrised by $T$*

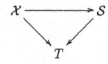

*the set of all* $t \in T$ *such that the corresponding fibre* $\mathcal{X}_t \to \mathcal{S}_t$ *is birationally rigid is open in* $T$ *(possibly empty).*

This paper contains proofs of the following results.

**Theorem 1.5 (Sarkisov [Sa1], [Sa2])** *Let* $X \to S$ *be a 3-fold conic bundle. Assume that the morphism* $X \to S$ *is extremal; that is, it is a Mori fibre space, and*

*(1) X is smooth (this implies that S is also smooth), and*

*(2) the divisor* $4K_S + \Delta$ *is effective, where* $\Delta \subset S$ *denotes the discriminant of the conic bundle.*

*Then X is birationally rigid over S.*

In Chapter 4, besides giving several proofs of this result, I explain why the requirement that $X$ be smooth is not really a restriction, and also indicate how one might hope to weaken the second assumption.

**Theorem 1.6 (Pukhlikov [P5])** *Let* $X \to S$ *be a 3-fold Del Pezzo fibration of degree* $d \le 2$. *Assume that the morphism* $X \to S$ *is extremal; that is, it is a Mori fibre space, and*

*(1) X is smooth, and*

*(2) the 1-cycle* $K_X^2 \in N_1 X$ *does not lie in the interior* $\operatorname{Int} \overline{NE} X$ *of the Mori cone.*

*Then X is birationally rigid over S.*

It is certainly too restrictive in this case to insist that $X$ be smooth, and I think it is fair to say that the second assumption, although in some sense analogous to the corresponding requirement for conic bundles, and easily verified in many natural examples, is poorly understood. $K^2$ seems to also control the rigidity of Fano 3-folds, see [CPR], Theorem 7.7.1 and Remark 7.7.2, especially (v). Chapter 5 gives a proof of this result based on the same idea as the original argument of Pukhlikov [P5], but technically simpler. A similar result for Del Pezzo fibrations of degree 3 [P6] still seems rather hard (but should follow from the methods of this paper).

Anthony Iano-Fletcher [IF] wrote a list of 95 families of Fano 3-fold hypersurfaces $X = X_d \subset \mathbb{P}_w^4$ in weighed projective $\mathbb{P}_w^4 = \mathbb{P}(1, a_1, a_2, a_3, a_4)$, with $-K_X = \mathcal{O}_X(1)$. Chapter 6 opens with an introduction to the following theorem:

**Theorem 1.7 (Corti, Pukhlikov, Reid [CPR])** *Assume that X is a general member of any of the 95 families. Then*

*(1) X is birationally rigid,*

*(2) Bir $X/\operatorname{Aut} X$ is generated by a finite number of explicit rational involutions.*

Chapter 6 treats the special case $X = X_5 \subset \mathbb{P}(1,1,1,1,2)$ of the theorem. We conjecture that the same conclusions hold if $X$ is quasismooth. The rest of Chapter 6 is devoted to some extensions, generalisations, and related results.

Iano-Fletcher also constructed 86 families of Fano 3-fold codimension 2 weighted complete intersections, the first example of which is $X = X_{2,3} \subset \mathbb{P}^5$, the complete intersection of a quadric and a cubic in $\mathbb{P}^5$. We have not as yet made any systematic study of complete intersections, but I prove the following result:

**Theorem 1.8 (Iskovskikh, Pukhlikov [I1], [IP])** *A general $X = X_{2,3} \subset \mathbb{P}^5$ is birationally rigid. Moreover, Bir $X/\operatorname{Aut} X$ is generated by rational involutions centred on the lines and conics of $X$.*

The proof in [IP] is long and complicated; that given here, while not entirely painless, is quite a bit simpler, and is a nice application of Shokurov connectedness.

Selma Altınok [A] compiled a list of 70 families of codimension 3 Pfaffian weighted K3 surfaces, and 138 in codimension 4 (of which 115 are firmly established, and most of the remaining 23 are conjectured to exist), which are anticanonical sections of Fano 3-folds. She also produced 3 candidates for Fano 3-folds of codimension $\geq 4$ with $|-K| = \emptyset$ that do not correspond to any K3 surfaces. It should be possible, in principle, to study all these varieties by our methods.

Finally, I consider some cases where $X$ is not quasismooth, but acquires the simplest kind of singularity.

**Theorem 1.9 (Pukhlikov [P3])** *Let $X = X_4 \subset \mathbb{P}^4$ be a quartic 3-fold with a single ordinary double point $P \in X$. Then $X$ is birationally rigid. Moreover, Bir $X/\operatorname{Aut} X$ is generated by the rational involution centred at the singular point, and those centred at the lines of $X$ passing through the singular point.*

I prove this result as a consequence of a classification of divisorial contractions $f\colon E \subset Y \to P \in X$ for $P \in X$ an ordinary 3-fold double point $(xy + zt = 0) \subset \mathbb{C}^4$ (more precisely, the germ of an ordinary double point); in fact, in Theorem 3.10, I prove that $f$ must be the blowup of the maximal ideal of $P$. The proof of Theorem 3.10 is quite short, and the resulting proof of Theorem 1.9 is much simpler and more local in nature than the original [P3].

# Overview of the methods

Before closing the Introduction, I spend a few words describing the methods in very general terms. The definitions and properties of the Sarkisov program are recalled in Chapter 2. Suppose that we wish to prove that a given Mori fibre space $X \to S$ is rigid, or otherwise restrict the possible models of $X$ as a Mori fibre space.

Let $X' \to S'$ be another Mori fibre space, $X \dashrightarrow X'$ a birational map. Choose a very ample complete linear system $\mathcal{H}'$ on $X'$, and denote its transform on $X$ by $\mathcal{H}$. Roughly speaking, the method is in 3 stages:

STEP 1: SET UP Necessarily $\mathcal{H} \subset |-\mu K_X + A|$, where $\mu > 0$ is a rational number and $A$ a divisor pulled back from $X$. Suitable biregular assumptions on $X$, such as the $K^2$ condition of Theorem 1.6 or the assumptions on the structure of the anticanonical ring of Theorem 1.7, together with the yoga of the Noether–Fano–Iskovskikh inequalities (Theorem 2.4), imply that "$\mathcal{H}$ has a base point with big multiplicity". At first sight it may not be obvious how to make this into a useful quantitative statement, but our experience of working in Mori theory suggests the following:

$$K_X + \frac{1}{\mu}\mathcal{H} \quad \text{is not canonical.}$$

This means that there exists a valuation $E$ with centre $C_E(X)$ on $X$ along which $\mathcal{H}$ has multiplicity $m_E$ that is big compared to the degree $\mu$ and the discrepancy; or more precisely,

$$m_E(\mathcal{H}) > \mu a_E(K_X),$$

where $a_E(K_X)$ is the discrepancy of $E$ with respect to the canonical class of $X$. Alternatively, we can use a condition that $K_X - \sum \lambda_i F_i + \frac{1}{\mu}\mathcal{H}$ is not canonical; see for example the proof of the rigidity theorem for Del Pezzo fibrations, Theorems 1.6 and 5.1.

STEP 2: EXCLUSION In the next stage, we try to use the information that $K + \frac{1}{\mu}\mathcal{H}$ is not canonical to deduce restrictions on the possible centres $\Gamma = C_E(X)$ of the valuation $E$, in some cases even ruling out the possibility that any such centres exist. This is clear in principle, but in practice various approaches come to mind, and I don't know how to decide *a priori* which works best, other than by experience with many calculations on a large number of examples. Chapter 3 contains several variations, refinements, small improvements and proofs of the following key result (see Theorem 3.1, Corollary 3.4, Corollary 3.5, Theorem 3.12).

**Lemma 1.10** *Let $X$ be a 3-fold, $\mathcal{H}$ a mobile linear system on $X$, and suppose that the closed point $X \ni P = C_E(X)$ is the centre of a valuation $E$ with*

$$m_E(\mathcal{H}) > \mu a_E(K_X).$$

*Write $Z = H_1 \cdot H_2$ for the cycle theoretic intersection of two general members of $\mathcal{H}$. Then*

$$\operatorname{mult}_P Z > 4\mu^2.$$

For instance, if $X$ is a smooth quartic 3-fold then

$$4\mu^2 = \deg Z \geq \operatorname{mult}_P Z > 4\mu^2$$

gives a contradiction; this is the hard case in the proof of [IM]. Traditionally, Lemma 1.10 is proved via the combinatorial analysis of the resolution graph of $E$ (the Iskovskikh–Manin graph, see [P4]). Although there are now 3 or 4 different ways of proving Lemma 1.10, it seems to me that we still don't know a truly compelling reason why it holds.

STEP 3: UNTWISTING    Even when $X$ is rigid, birational maps $X \dashrightarrow X$ often exist. In this case Step 2 only gives restrictions on what can happen, typically saying that no curve of large enough degree and no nonsingular point can be a maximal centre, and perhaps also excluding some of the singular points. Step 3 then consists of classifying all possible links $X \dashrightarrow X'$, where $X' \to S'$ is (*a priori*) any Mori fibre space. In fairly simple cases a little ingenuity is sufficient to guess the correct answer but in more complicated situations, for instance in [CPR], the simple-minded approach becomes computationally intractable. In practice, it is often useful to calculate the anticanonical ring $R(Y, -K_Y)$ for a large number of extremal divisorial contractions $E \subset Y \to P \in X$. I give a very few simple examples in this paper and refer the interested reader to [CPR] for a fuller treatment.

Chapter 2 is an exposition of the Sarkisov program, mostly without proofs. Chapter 3 is a study of singularities of linear systems, based on Shokurov connectedness and inversion of adjunction, and represents the technical core of the whole paper. Chapters 4, 5 and 6 give applications to conic bundles, Del Pezzo fibrations and Fano 3-folds. I have tried to cover all the 3-dimensional results of [IP], including some of the more recent material.

These notes grew out of a lecture I gave at the Warwick Algebraic Geometry special year in November 1995. I then taught the material as "Five lectures on 3-fold birational geometry" at RIMS in June 1997. I thank both institutions for their warm hospitality.

It is a great pleasure to acknowledge the influence that Miles Reid has had in shaping my view of the world. For the dozens of hours of mathematical conversations, ideas and criticism, this paper is as much his work as it is mine. Almost all of the sharper results were originally proved by Alexandr Pukhlikov, who is also coauthor of [CPR], and everything here grew out of an attempt to understand his work. Finally, I would like to thank Massimiliano Mella for checking many of the calculations in Chapters 3 and 6, and pointing out various mistakes in several of the earlier versions, and Ivan Cheltsov for detecting an imbecility in one of the later ones.

# 2   The Sarkisov program

In this chapter, I explain the general structure of the Sarkisov program, referring to Corti [C2] and Bruno and Matsuki [BM] for detailed proofs. Apart from the Noether–Fano–Iskovskikh inequalities, the material here is not used until Chapter 6.

Let $X \to S$, $X' \to S'$ be Mori fibre spaces and

$$\varphi\colon X \dashrightarrow X'$$

a birational map. The Sarkisov program factors $\varphi$ as a *chain of links*: a link is an elementary birational map of one of four types discussed and defined below. In favourable cases they can be classified and the method proves that $X$ is birationally rigid, and gives explicit generators of $\operatorname{Bir} X / \operatorname{Aut} X$.

The factorisation program begins by assigning $\varphi$ a *Sarkisov degree* $\deg \varphi$ (morally a discrete invariant, if not in actual fact) and then *untwisting* $\varphi$ by a link $\psi_1$ so that $\deg \varphi\psi_1^{-1} < \deg \varphi$. In other words,

$$\varphi = \varphi_1\psi_1,$$

where $\varphi_1 = \varphi\psi_1^{-1}$ has degree smaller than $\varphi$; we then continue inductively with $\varphi_1$ in place of $\varphi$.

In practice it is only possible to carry out the Sarkisov program explicitly, in the original form discussed here, under suitable strong restrictive assumptions on $X$—for example if $X$ is one of the 95 Fano weighted hypersurfaces (cf. [CPR] and Chapter 6) or one of a handful of the Fano weighted complete intersections. The difficulty with *strict Mori fibrations*, that is, conic bundles and Del Pezzo fibrations, is that they always involve infinitely many links, which are very difficult to control. Applications to these varieties therefore involve circumventing this difficulty.

## 2.1   The Sarkisov degree

Let $X$ be a variety and $\mathcal{H}$ a mobile linear system (that is, without fixed part or free in codimension 1).

A discrete valuation $v$ of $k(X)$ is *geometric* if the residue field $A_v/m_v$ has transcendence degree $\dim X - 1$ or, equivalently, if $X$ has a normal birational model $Z \to X$ containing a prime divisor $E$ for which $v = v_E$ is the valuation along $E$. In the language of Zariski, the model $E \subset Z \to X$ just discussed is a *uniformisation* of $v$. We only consider geometric discrete valuations. I often abuse notation and use $E$ for both the divisor itself and the valuation $v_E$. Write $a_E = a_E(K_X)$ for the discrepancy and $m_E = m_E(\mathcal{H})$ for the multiplicity of $\mathcal{H}$ along $E$; in other words, if $E$ appears as a prime divisor in $Z$, then $a_E$ and $m_E$ are defined by

$$K_Z = K_X + a_E E \quad \text{and} \quad \mathcal{H}_Z = \mathcal{H} - m_E E,$$

locally near the generic point of $E$. (We simplify the notation throughout by suppressing the pullback $f^*$ of $\mathbb{Q}$-divisors.)

**Definition 2.1** Let $t \in \mathbb{Q}_+$ be a positive rational number. The pair $X, t\mathcal{H}$ has *canonical singularities* (respectively *terminal singularities*)—equivalently, I say that the $\mathbb{Q}$-divisor $K_X + t\mathcal{H}$ is *canonical* (or *terminal*)—if

$$m_E(\mathcal{H}) \leq \frac{1}{t} a_E(K_X) \quad \text{for all } E,$$

(respectively, $<$ for all $E$ exceptional over $X$).

**Exercise 2.2**  (1) For $X$ a surface, $X, t\mathcal{H}$ is terminal if and only if $X$ is nonsingular and

$$\mathrm{mult}_P \mathcal{H} < \frac{1}{t} \quad \text{for all points } P \in X.$$

(2) For $X$ a 3-fold, $K + 2\mathcal{H}$ is terminal if and only if $X$ is terminal and $\mathcal{H}$ is free from base points. $K + \mathcal{H}$ is terminal if and only if $X$ is terminal and the scheme theoretic base locus of $\mathcal{H}$ is a finite set of nonsingular points.

Now fix Mori fibre spaces $X \to S$, $X' \to S'$, and a birational map

$$\varphi \colon X \dashrightarrow X'.$$

Our aim in this section is to define the Sarkisov degree $\deg \varphi$ of $\varphi$. For this, choose a very ample complete linear system $\mathcal{H}' = |-\mu'K' + A'|$ on $X'$, where $A'$ is the pullback of a divisor ample on the base $S'$. This choice is made once and for all at the beginning of the factorisation process and $\mathcal{H}'$ remains unchanged throughout the chain. This is of course possible because the target Mori fibre space $X' \to S'$ remains unchanged throughout the chain. The definition of degree depends upon this choice.

**Definition 2.3** The *Sarkisov degree* $\deg \varphi$ is the triple $(\mu, c, e)$ defined as follows:

(1) If $\mathcal{H}$ is the birational transform of $\mathcal{H}'$ on $X$, then because $X \to S$ is a Mori fibre space, we can write

$$\mathcal{H} \subset |-\mu K + A|,$$

where $\mu$ is a positive rational number with bounded denominator and $A$ is the pullback of a divisor (not necessarily nef, or effective) on the base $S$.

(2) $c \in \mathbb{Q}_+$ is the canonical threshold of the pair $X, \mathcal{H}$, that is,

$$c = \max\{t \mid K + t\mathcal{H} \text{ is canonical}\}.$$

(3) $e \in \mathbb{N}$ is the (finite) number of crepant valuations of $K + c\mathcal{H}$, that is,

$$e = \#\left\{E \;\middle|\; m_E(\mathcal{H}) = \frac{1}{c}a_E(K_X)\right\}.$$

We *order* such triples as follows: $(\mu_1, c_1, e_1) > (\mu_2, c_2, e_2)$ if either

(a) $\mu_1 > \mu_2$, or

(b) $\mu_1 = \mu_2$ and $c_1 < c_2$ (no misprint here), or

(c) $\mu_1 = \mu_2$, $c_1 = c_2$ and $e_1 > e_2$.

## 2.2   Links and the 2-ray game

The references for this section are Kollár [Ko2] and [C2]. The 2-ray game is an inductive sequence of forced moves and configurations, starting from a given configuration; it can be played in various categories—terminal (Mori), log terminal, etc. We only play it in the Mori category. I want to emphasise that while the moves are uniquely determined by the initial configuration, there is no guarantee at any point that the next move exists.

Each configuration is a commutative diagram

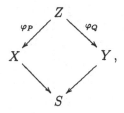

consisting of 4 varieties and projective morphisms, in which the variety on top swaps 2 extremal contractions $\varphi_P$ and $\varphi_Q$ along (pseudo-) extremal rays labelled $P$ and $Q$. The bottom variety $S$ is fixed throughout the game.

To ensure that the game is uniquely determined by the initial configuration we always make the following assumptions:

(1) $Z$ is projective over $S$ and rank $N^1(Z/S) = 2$.

This assumption implies that the Mori cone $\overline{NE}(Z/S)$ is a 2-dimensional closed cone in $N^1(Z/S) \cong \mathbb{R}^2$, so that there can be at most 2 projective morphisms $Z \to ? \to S$.

Because we insist that the game always remain within the Mori category we also assume that

(2) $Z$ has $\mathbb{Q}$-factorial terminal singularities,

(3) for each ray $R \subset \overline{NE}(Z/S)$, either $K_Z \cdot R < 0$, or else the contraction $\varphi_R$ of $R$ is (projective and) small.

Finally, since we are ultimately interested in links between Mori fibre spaces, we assume that

(4) $\dim S < \dim Z$, and the Kodaira dimension $\kappa(Z/S)$ of $Z$ over $S$ is $-\infty$.

Configurations can be classified according to the type of the contractions of $P$, $Q$ into 9 types $df, dd, ds, ss, \ldots$, where we write $f$ = fibre type, $d$ = divisorial, $s$ = small. For instance, a configuration of type $ds$ is one where $\varphi_P$ is an extremal divisorial contraction in the Mori category, and $\varphi_Q$ is a projective small contraction.

A move in the 2-ray game consists of creating a configuration $s*$ starting from a given configuration $*s$. Given

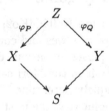

with $\varphi_Q$ small, the new configuration

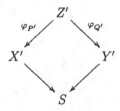

is uniquely determined by the requirements

(a) $X' = Y$ and $\varphi_{P'}\colon Z' \to X'$ is a small contraction that is the flip of $Q$. (Or the "opposite" of $Q$; here I use "flip" in the generalised sense of [Ko2], Chapter 6. It can be a flip, or equally well, a flop or inverse flip of the Mori category.)

(b) If $P' \subset \overline{\mathrm{NE}}(Z'/S)$ is the flipped ray (whose contraction is $\varphi_{P'}$), we call the other ray $Q' \subset \overline{\mathrm{NE}}(Z'/S)$; this exists simply because $\overline{\mathrm{NE}}(Z'/S) \subset \mathbb{R}^2$ is a two dimensional closed cone. Then $Z' \to Y'$ is the contraction of $Q'$.

When $K_Z \cdot Q > 0$, there are several ways in which the game may end in failure. First, there is no guarantee that the flip $Z' \to X'$ of $Q$ exists. Next, referring back to the above conditions (1–4), even if $Z'$ exists, there is no guarantee that it has terminal singularities (2), that its other ray $Q'$ can be contracted or, if $Q'$ can be contracted, that it satisfies condition (3). The game only continues if the new configuration exists and satisfies conditions (2), (3); (1) and (4) are automatic.

A game with more than one move starts with a configuration $ds$ or $fs$, and a winning game terminates with a configuration $sd$ or $sf$. There are thus four types of winning games, modelled on the four possible one-move games $dd$, $df$, $fd$, $ff$. We call these Type I, II, III and IV. A winning game is also a winning game if we play all the moves in reverse. In this sense a game of Type III is the inverse of a game of Type II.

Under the above assumptions consider for instance a winning game of Type I. By definition, the game begins with a configuration

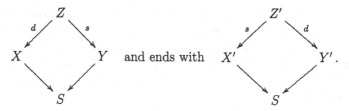

Since $Z \to X$ is a divisorial contraction in the Mori category, it follows that $X$ has $\mathbb{Q}$-factorial terminal singularities and, by assumption (4), $X \to S$ is a Mori fibre space. Similarly, $Y'$ has $\mathbb{Q}$-factorial terminal singularities and $Y' \to S$ is also a Mori fibre space. We call the resulting birational map

$$\varphi\colon X \dashrightarrow Y'$$

a *link of Type* I.

Links of Type II, III and IV are defined in a similar way, modelled on winning 2-ray games of Type II, III, IV. We can visualise the four types of links as in Figure 2.2.1:

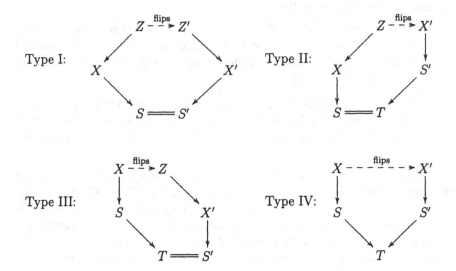

Figure 2.2.1: The links of the Sarkisov program

## 2.3   Construction of the untwisting link

Fix Mori fibre spaces $X \to S$, $X' \to S'$ and a birational map $\varphi \colon X \dashrightarrow X'$. If $\varphi$ issquare biregular there is nothing to do. In this section, assuming that $\varphi$ is not square biregular, we show how to untwist $\varphi$ by making a Mori fibre space $X_1 \to S_1$ and a link $\psi_1 \colon X \dashrightarrow X_1$ so that $\deg \varphi\psi_1^{-1} < \deg \varphi$.

We use the following terminology. Let $U \to V$ be a projective morphism. The Mori cone $\overline{\mathrm{NE}}(U/V) = \overline{\mathrm{NE}}_1(U/V) \subset N_1(U/V)$ is by definition the real closure of the cone of effective curves $C$ contained in fibres of $U \to V$. The dual cone $\overline{\mathrm{NE}}^1(U/V) \subset N^1(U/V)$ is the cone of *nef* classes. The real closure $\overline{\mathrm{NM}}^1(U/V) \subset N^1(U/V)$ of the cone $\mathrm{NM}^1(U/V)$ of effective divisors is called the quasieffective cone.

### Theorem 2.4 (Noether–Fano–Iskovskikh inequalities)

*(1) $\mu \geq \mu'$ and $\mu = \mu'$ if and only if $\varphi$ is square.*

*(2) If $K + \frac{1}{\mu}\mathcal{H}$ is canonical and quasieffective then $\mu = \mu'$ and, in particular, $\varphi$ is square.*

*(3) If $K + \frac{1}{\mu}\mathcal{H}$ is canonical and nef then $X \dashrightarrow X'$ is a Sarkisov isomorphism.* $\square$

A proof can be found in [C2], 4.2; the statement there that corresponds to Theorem 2.4, (1) is slightly weaker, but the claim here is easy enough to prove.

**Corollary 2.5** *If $\varphi$ is not a Sarkisov isomorphism then either:*

*(1) $K + \frac{1}{\mu}\mathcal{H}$ is not canonical, or*

*(2) $K + \frac{1}{\mu}\mathcal{H}$ is canonical but not nef.* $\square$

The construction of the untwisting link proceeds in 2 different ways, depending on whether we are in Case (1) or (2) of Corollary 2.5. It is traditional and convenient to honour the first case with the following definitions.

**Definition 2.6** Let $X \to S$ be a Mori fibre space and $\mathcal{H}$ a mobile linear system on $X$. Necessarily

$$\mathcal{H} \equiv -\mu K_X + A,$$

where $A$ is a pullback from $S$. Let $c = c(X, \mathcal{H})$ be the canonical threshold of the pair $X, \mathcal{H}$ and assume that

$$c < \frac{1}{\mu}.$$

(1) A *maximal extraction* is an extremal divisorial contraction

$$f \colon E \subset Y \to \Gamma \subset X$$

extracting a valuation $E$ for which $m_E(\mathcal{H}) = \frac{a_E}{c} > \mu a_E(K_X)$, or equivalently $K_Y + c\mathcal{H}_Y = f^*(K_X + c\mathcal{H})$. It is a theorem that a maximal extraction always exists; see [C2], Theorem 2.10, where this is called an *extremal blowup* of the pair $X, \mathcal{H}$.

(2) An algebraic valuation $E \subset Z \to X$ of $X$ is a *maximal singularity* of $\mathcal{H}$ (or of $X$ itself) if $E$ is the exceptional divisor of a maximal extraction. A *maximal centre* is the centre $C_E(X)$ on $X$ of a maximal singularity. In [CPR], this is called a *strong* maximal singularity.

**Theorem 2.7** *Assume that $\varphi \colon X \dashrightarrow X'$ is not a Sarkisov isomorphism; then there exist a Mori fibre space $X_1 \to S_1$ and a link*

$$\psi_1 \colon X \dashrightarrow X_1$$

*such that*

$$\deg \varphi \psi_1^{-1} < \deg \varphi.$$

**Proof**    I briefly explain how to construct the link $\psi_1$, omitting the rather delicate verification that the Sarkisov degree decreases. The construction uses the only known general method to guarantee from the start that a 2-ray game can be played to the end, namely when the 2-ray game is a minimal model program for a (log) canonical divisor $K + B$.

According to Corollary 2.5, there are 2 cases:

CASE 1: $K + \frac{1}{\mu}\mathcal{H}$ IS NOT CANONICAL    There is then a canonical threshold $c < \frac{1}{\mu}$, a maximal centre $\Gamma$ and a maximal extraction $f\colon E \subset Z \to \Gamma \subset X$ with

$$K_Z + c\mathcal{H}_Z = f^*(K_X + c\mathcal{H}_X).$$

The $K_Z + c\mathcal{H}_Z$ minimal model program over $S$ is a winning 2-ray game, leading to a link of Type I or II. It is shown in [C2], Theorem 5.4 that untwisting (strictly) decreases the Sarkisov degree.

CASE 2: $K + \frac{1}{\mu}\mathcal{H}$ IS CANONICAL BUT NOT NEF    The link is manufactured by first choosing a suitable contraction $S \to T$ as follows. Let $P \subset \overline{NE}\,X$ be the extremal ray corresponding to the Mori fibre space structure $X \to S$. By the cone theorem for $K + \frac{1}{\mu}\mathcal{H}$, for instance, there exists an extremal ray $Q \subset \overline{NE}\,X$ with $(K + \frac{1}{\mu}\mathcal{H}) \cdot Q < 0$. The contraction $X \to T$ of the 2-dimensional face $F = P + Q$ of $\overline{NE}\,X$ exists, for example by the contraction theorem for $K + \left(\frac{1}{\mu} \pm \varepsilon\right)\mathcal{H}$ and variations on it, and both the Mori fibration $X \to S$ and the contraction $X \to Y$ of $Q$ factorise the morphism $X \to T$, giving a configuration

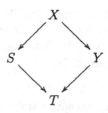

of the 2-ray game. Running a minimal model program against the divisor $K + \frac{1}{\mu}\mathcal{H}$ over $T$ wins the 2-ray game, and we get a Mori fibration $X_1 \to S_1$ and a link $\psi_1\colon X \dashrightarrow X_1$ that is of Type III if $S_1 = T$ or of Type IV if $S_1 \to T$ is an extremal contraction. When I wrote [C2] I did not realise that this kind of untwisting decreases the Sarkisov degree, but in fact it does.    □

It is not known if the Sarkisov degree is a discrete invariant in general. Nevertheless, it can be shown by a somewhat indirect and nonconstructive argument that, in a chain factoring a given birational map, the degree cannot decrease infinitely many times ([C2], pp. 246–248); thus the Sarkisov program always terminates. In all applications of the theory known to me termination is obvious, but it seems that we still lack a compelling reason why it works in complete generality.

**Problem 2.8** Find an effective proof of termination of the Sarkisov program. In other words, find an *a priori* bound on the number of links needed to factor $\varphi$, ideally depending only on the Sarkisov degree.

**Problem 2.9** Extend the general framework of the Sarkisov program to factor the *relations* in the Sarkisov category in terms of "elementary relations" or "2-links" (as yet undefined). There are a handful of ad hoc cases in the literature ([I1], [P3], [IP]), but a satisfactory general treatment is still lacking.

# 3 Singularities of linear systems

This chapter is the technical core of the whole paper. We consider a surface or 3-fold germ $P \in X$ together with a mobile linear system $\mathcal{H}$ on $X$. Assuming a condition typically something like $K_X + \frac{1}{\mu}\mathcal{H}$ not canonical, we derive estimates on the singularity of the cycle $Z = H_1 \cdot H_2$ (where $H_i \in \mathcal{H}$ are two general members) at the point $P$—typically something like $\mathrm{mult}_P Z > 4\mu^2$.

## 3.1 Linear systems on surfaces

In this section I prove the following strange-looking and deceptively simple result which is used in combination with Shokurov connectedness in the proofs of many criteria for birational rigidity.

Suppose that $P \in \Delta_1 + \Delta_2 \subset S$ is the analytic germ of a normal crossing curve on a smooth surface, that is, isomorphic to $0 \in (xy = 0) \subset \mathbb{A}^2$. Let $\mathcal{L}$ be a mobile linear system on $S$ and denote by $\mathcal{L}^2$ the local intersection multiplicity $(L_1 \cdot L_2)_P$ at $P$ of two general members $L_1, L_2 \in \mathcal{L}$.

**Theorem 3.1** *Fix rational numbers $a_1, a_2 \geq 0$ and suppose that*

$$K_S + (1 - a_1)\Delta_1 + (1 - a_2)\Delta_2 + \frac{1}{\mu}\mathcal{L}$$

*is not log canonical for some $\mu \in \mathbb{Q}$, $\mu > 0$.*

*(1) If either $a_1 \leq 1$ or $a_2 \leq 1$ then*

$$\mathcal{L}^2 > 4a_1 a_2 \mu^2.$$

*(2) If both $a_i > 1$ then*

$$\mathcal{L}^2 > 4(a_1 + a_2 - 1)\mu^2.$$

**Proof**   For ease of notation, write

$$D = (1 - a_1)\Delta_1 + (1 - a_2)\Delta_2 + \frac{1}{\mu}\mathcal{L}.$$

(Thus $D$ is a *subboundary* in the sense of Shokurov.) By assumption, there is a geometric valuation $E$ of $S$ with discrepancy

$$a_E(K_S + D) < -1.$$

STEP 1   I first show that (1) implies (2). Let $m_i = m_E(\Delta_i)$ be the multiplicity of $\Delta_i$ along $E$ and assume, say, that $m_1 \geq m_2$. Then

$$-1 > a_E(K_S + D) = a_E(K_S) - m_E(D)$$

$$= a_E(K_S) + m_1(a_1 - 1) + m_2(a_2 - 1) - \frac{m_E(\mathcal{L})}{\mu}$$

$$\geq a_E(K_S) + m_2(a_1 + a_2 - 2) - \frac{m_E(\mathcal{L})}{\mu}$$

$$= a_E\left(K_S + (2 - a_1 - a_2)\Delta_2 + \frac{1}{\mu}\mathcal{L}\right).$$

This means that the divisor

$$K_S + D_2 = K_S + (2 - a_1 - a_2)\Delta_2 + \mathcal{L}$$

is not log canonical, and (2) for $K_S + D$ is (1) for $K_S + D_2$.

STEP 2   We assume that $a_1 \leq 1$ and prove that (1) holds by descending induction on the discrepancy $a_E(K_S)$. Let $F \subset T \to P \in S$ be the blowup of the maximal ideal of $P$. We can write

$$K_S + D = K_T + (1 - a_1)\Delta_1' + (1 - a_2)\Delta_2' + \left(1 - a_1 - a_2 + \frac{m}{\mu}\right)F + \frac{1}{\mu}\mathcal{L}'$$

$$= K_T + D_T,$$

where $\Delta_i'$, $\mathcal{L}'$ denote the proper transforms, $m = m_F(\mathcal{L})$ and $D_T$ is defined by the formula. We have $a_E(K_T + D_T) < -1$ and we discuss four cases, depending on the position of the centre $C_E T$ of $E$ on $T$:

(a) $C_E T \in F \cap \Delta_1'$,

(b) $C_E T \in F \cap \Delta_2'$,

(c) $C_E T \in F \setminus \{\Delta_1' + \Delta_2'\}$,

(d) $C_E T = F$, that is, $E = F$.

By Step 1 and the inductive assumption, in the first three cases we may assume that the result holds for $K_T + D_T$, since $a_E(K_T) < a_E(K_S)$.

CASE (a)   By induction

$$\mathcal{L}^2 > 4a_1\left(a_1 + a_2 - \frac{m}{\mu}\right)\mu^2 + m^2$$
$$= 4a_1^2\mu^2 - 4a_1 m\mu + m^2 + 4a_1 a_2\mu^2$$
$$= (2a_1\mu - m)^2 + 4a_1 a_2\mu^2 \geq 4a_1 a_2\mu^2.$$

CASE (b)   If either $a_2 \leq 1$ or $a_1 + a_2 - \frac{m}{\mu} \leq 1$, same as (a), otherwise we assume that $a_2 > 1$ and $a_1 + a_2 - \frac{m}{\mu} > 1$.

$$\mathcal{L}^2 > 4\left(a_2 + a_1 + a_2 - \frac{m}{\mu} - 1\right)\mu^2 + m^2$$
$$> 4a_2\mu^2 \geq 4a_1 a_2\mu^2.$$

CASE (c)   By induction

$$\mathcal{L}^2 > 4\left(a_1 + a_2 - \frac{m}{\mu}\right)\mu^2 + m^2$$
$$\geq 4a_1\left(a_1 + a_2 - \frac{b}{\mu}\right)\mu^2 + b^2$$
$$\geq 4a_1 a_2\mu^2,$$

as in Case (a).

CASE (d)   By assumption $m > (a_1 + a_2)\mu$, so that

$$\mathcal{L}^2 \geq m^2 > (a_1 + a_2)^2\mu^2 \geq 4a_1 a_2\mu^2. \quad \square$$

## 3.2   Shokurov connectedness and its implications

I open this section by recalling the statement of Shokurov's connectedness theorem, taken from [Ko2], 17.4 (see [CPR], proof of Theorem 5.3.2 for an informal discussion) and then state some easy consequences. In closing, I tie in with the previous section and prove that a smooth point on a quartic 3-fold cannot be a maximal centre. Together with some easy arguments from [IM], §5, which I do not reproduce here, this implies that a smooth quartic 3-fold is birationally rigid.

**Theorem 3.2 (Shokurov, Kollár [Ko2], 17.4)** *Let $X$ and $Z$ be normal varieties and $h\colon X \to Z$ a proper morphism such that $h_*\mathcal{O}_X = \mathcal{O}_Z$. Assume that a $\mathbb{Q}$-divisor $D = \sum d_i D_i$ on $X$ satisfies*

*(1) if $d_i < 0$ then $\operatorname{codim}_{h(D_i)} Z \geq 2$;*

*(2) $-(K_X + D)$ is $\mathbb{Q}$-Cartier, $h$-nef and $h$-big.*

*Let $g\colon Y \to X$ be a resolution of singularities such that the support of $g^{-1}D \subset Y$ is a divisor with global normal crossings. We can write*

$$K_Y = g^*(K_X + D) + \sum a_E(K_X + D)E.$$

*Let $f = h \circ g$:*

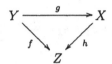

*Then*

$$A = \sum_{a_E(K_X+D)\leq -1} E$$

*is connected in a neighbourhood of any fibre of $f$.*

**Remark 3.3** The image $g(A)$ of $A$ in $X$ is sometimes called the *locus of log canonical singularities* of $K + D$, and denoted by

$$g(A) = \operatorname{LC}(X, K_X + D).$$

The theorem implies that this set is connected in a neighbourhood of any fibre of $h$.

**Corollary 3.4** *Let $P \in S_1 + S_2 \subset X \cong 0 \in (xy = 0) \subset \mathbb{A}^3$ be a smooth 3-fold germ, $\mathcal{H}$ a mobile linear system on $X$, and $0 \leq a_1, a_2 \leq 1$ rational numbers. Assume that*

$$K_X + (1 - a_1)S_1 + (1 - a_2)S_2 + \frac{1}{\mu}\mathcal{H}$$

*is not canonical at $P$ for some $\mu \in \mathbb{Q}$, $\mu > 0$.*

*(1) If $P \in S$ is a hyperplane section through $P$, $\Delta_1 + \Delta_2 = (S_1 + S_2)|_S$ and $\mathcal{L} = \mathcal{H}|_S$ then*

$$K_S + (1 - a_1)\Delta_1 + (1 - a_2)\Delta_2 + \frac{1}{\mu}\mathcal{L}$$

*is not log canonical.*

*(2) If $Z = H_1 \cap H_2$ is the intersection of two general members of $\mathcal{H}$ then*

$$\mathrm{mult}_P Z > 4a_1 a_2 \mu^2.$$

**Proof**  By assumption there is a valuation $E$ with centre $C_E X = P$ with

$$a_E\left(K_X + (1-a_1)S_1 + (1-a_2)S_2 + \frac{1}{\mu}\mathcal{H}\right) < 0.$$

Clearly $m_E(S)$ is a strictly positive integer, so

$$a_E\left(K_X + S + (1-a_1)S_1 + (1-a_2)S_2 + \frac{1}{\mu}\mathcal{H}\right)$$
$$= a_E\left(K_X + (1-a_1)S_1 + (1-a_2)S_2 + \frac{1}{\mu}\mathcal{H}\right) - m_E(S) < -1$$

and $K_X + S + (1-a_1)S_1 + (1-a_2)S_2 + \frac{1}{\mu}\mathcal{H}$ is not log canonical. Using 3.2 as in [Ko2], 17.7, we conclude that $K_S + (1-a_1)\Delta_1 + (1-a_2)\Delta_2 + \frac{1}{\mu}\mathcal{L}$ is not log canonical (this statement is usually called "inversion of adjunction").

To prove (2), let $P \in S \subset X$ be a general hyperplane section, so that $\mathcal{L} = \mathcal{H}_{|S}$ is free from base curves and $\mathrm{mult}_P Z = \mathcal{L}_S^2$, and apply 3.1.  □

I stress that it is crucial in Corollary 3.4 to assume that $a_1, a_2$ are *both* $\leq 1$; indeed the assumption $d_i \geq 0$ is absolutely crucial to the validity of 3.2.

For applications to rigidity criteria for Fano complete intersections I need the following strange-looking corollary:

**Corollary 3.5** *Let $P \in X$ be a smooth 3-fold germ, and $\mathcal{H} \subset X$ a mobile linear system on $X$. Assume that*

$$K_X + \frac{1}{\mu}\mathcal{H}$$

*is not canonical at $P$ and let $h\colon F \subset Y \to P \in X$ be the blowup of the maximal ideal $m_P$ of $P$. Then either*

*(1) $m = m_F(\mathcal{H}) > 2\mu$, or*

*(2) there is a line $\Gamma \subset F \cong \mathbb{P}^2$ such that*

$$K_Y + \left(\frac{m}{\mu} - 1\right)F + \frac{1}{\mu}\mathcal{H}$$

*is not log canonical at the generic point of $\Gamma$.*

**Proof**  If (1) fails, consider a general hyperplane section $P \in S \subset X$ and let $T \subset Y$ be its proper transform. We may write

$$h^* \left( K_X + S + \frac{1}{\mu} \mathcal{H} \right) = K_Y + T + \left( \frac{m}{\mu} - 1 \right) F + \frac{1}{\mu} \mathcal{H}_Y,$$

with

$$\frac{m}{\mu} - 1 \leq 1.$$

I assume, for simplicity, that $<$ holds (otherwise work with $(1 - \varepsilon)\mathcal{H}$ in place of $\mathcal{H}$). By 3.4, we know that

$$K_T + \left( \frac{m}{\mu} - 1 \right) F_{|T} + \frac{1}{\mu} \mathcal{H}_Y{}_{|T}$$

is not log canonical. Also, applying the connectedness theorem to the morphism

$$h \colon T \to S,$$

we conclude that the log canonical locus

$$\mathrm{LC}\left( T, K_T + \left( \frac{m}{\mu} - 1 \right) E_{|T} + \frac{1}{\mu} \mathcal{H}_Y{}_{|T} \right)$$

is a single isolated point, from which (2) follows.  □

**Theorem 3.6 (Iskovskikh, Manin [IM])** *Let $X = X_4 \subset \mathbb{P}^4$ be a quartic 3-fold and $P \in X$ a nonsingular point. Then $P$ is not a maximal centre.*

**Proof**  If the conclusion failed, we would have a mobile linear system $\mathcal{H} \subset |\mathcal{O}_X(\mu)|$ on $X$ with

$$K_X + \frac{1}{\mu} \mathcal{H} \quad \text{not canonical at } P.$$

If $P \in S \in |\mathcal{O}_X(1)|$ is a hyperplane section through $P$ and $\mathcal{L} = \mathcal{H}_{|S}$, we know from 3.4 that

$$K_S + \frac{1}{\mu} \mathcal{L}$$

is not log canonical at $P$. Now, if $S$ is chosen suitably, $\mathcal{L}$ is free from base curves and, applying Theorem 3.1, we obtain

$$4\mu^2 = L_1 \cdot L_2 \geq (L_1 \cdot L_2)_P > 4\mu^2 \quad \text{for general } L_1, L_2 \in \mathcal{L},$$

a contradiction.  □

## 3.3  Divisorial contractions

**Definition 3.7** A 3-fold *divisorial contraction* is a contraction

$$f \colon E \subset Z \to \Gamma \subset X,$$

where

(1) $E = f^{-1}\Gamma$, $f \colon Z \setminus E \to X \setminus \Gamma$ is an isomorphism;

(2) $Z$ has terminal singularities and $P \in \Gamma \subset X$ is the germ of a 3-fold terminal singularity (it *may be* that $P = \Gamma$);

(3) $E$ is an irreducible divisor and $-K_Z$ is $f$-ample.

**Problem 3.8** Classify 3-fold divisorial contractions up to local analytic isomorphism (over the base).

This is a problem of 3-fold biregular classification, analogous to the classification of terminal singularities. It seems clear that it is a fundamental problem, having many potential applications to birational geometry. I now present the known special cases of terminal quotient singularities and ordinary nodes. Both results are extremely useful in the applications to the 95 Fano hypersurfaces and to singular quartics.

The following result is due to Kawamata [Ka].

**Theorem 3.9 (Kawamata [Ka])** *Let* $P \in X \cong \frac{1}{r}(1, a, -a)$ *be a 3-fold terminal quotient singularity, with* $r \geq 2$, *and*

$$f \colon E \subset Z \to \Gamma \subset X$$

*an extremal divisorial contraction. Then* $f(E) = P$ *and* $f$ *is the weighted blowup with weights* $(1, a, r - a)$.

**Proof** I only explain the idea, which is quite simple. There is a unique valuation $F$ with centre on $X$ such that $a_F = \frac{1}{r}$. Now suppose that

$$E \subset Z \to P \in X$$

is an extraction with $E \neq F$; then $a_E \geq \frac{2}{r}$. The crucial point here is that $F$ is still there, and has a centre somewhere on $Z$. An easy calculation yields

$$a(F, K_Z) \leq 0.$$

In other words, $Z$ cannot have terminal singularities. $\quad\square$

**Theorem 3.10** *Let* $P \in X \cong (xy + zt = 0) \subset \mathbb{A}^4$, *and*

$$f \colon E \subset Z \to P \in X$$

*a divisorial contraction. Assuming that* $f(E) = P$, $f$ *is the blowup of the maximal ideal* $m_P \subset \mathcal{O}_X$.

**Proof**    Let $\mathcal{H}_Z \subset |-\mu K_Z|$ be a finite dimensional very ample linear system and $\mathcal{H}_X = f_*(\mathcal{H}_Z)$. Then

$$K_Z + \frac{1}{\mu}\mathcal{H}_Z = f^*\left(K_X + \frac{1}{\mu}\mathcal{H}_X\right),$$

so that

$$m_E(\mathcal{H}_X) = \mu a_E(K_X),$$

while

$$m_F(\mathcal{H}_X) < \mu a_F(K_X) \quad \text{for all valuations } F \neq E.$$

Letting $\varepsilon\colon F \subset Y \to P \in X$ be the blowup of the maximal ideal $m_P$, we argue that $m_F(\mathcal{H}_X) \geq \mu$, thereby proving the result. We know that

$$a_E\left(K_Y + \left(\frac{m_F}{\mu} - 1\right)F + \frac{1}{\mu}\mathcal{H}_Y\right) = 0.$$

Choose a reducible hyperplane section $P \in S_1 + S_2 \subset X$ isomorphic to $0 \in (t = 0) \subset (xy + zt = 0)$, and satisfying the following conditions:

(1) the centre $C_E Y$ of $E$ on $Y$ does not lie on the proper transform $S_1' + S_2'$ of $S_1 + S_2$ on $Y$;

(2) the curve $\Gamma = S_1' \cap S_2'$ is disjoint from the general member of $\mathcal{H}_Y$.

We have

$$\varepsilon^*\left(K_X + S_1 + S_2 + \frac{1}{\mu}\mathcal{H}_X\right) = K_Y + S_1' + S_2' + \frac{m_F}{\mu}F + \frac{1}{\mu}\mathcal{H}_Y$$

and

$$\mathrm{LC} = \mathrm{LC}\left(Y, K_Y + S_1' + S_2' + \frac{m_F}{\mu}F + \frac{1}{\mu}\mathcal{H}_Y\right) \supset S_1' + S_2' + C_E Y.$$

Now $F \subset Y$ can be contracted along the 2 rulings of $F \cong \mathbb{P}^1 \times \mathbb{P}^1$

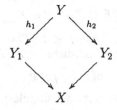

so that passing from $Y_1$ to $Y_2$ is the familiar flop. By Remark 3.3, LC is connected in a neighbourhood of every fibre of $h_1$ and every fibre of $h_2$. The only possibility is that $\mathrm{LC} = F$, or, in other words, that $C_E Y = F$ and $m_F \geq \mu$. $\square$

For smooth points all we have is the following

**Conjecture 3.11** *Let* $f\colon E \subset Z \to 0 \in \mathbb{A}^3$ *be a divisorial contraction. Then* $f$ *is a weighted blowup with weights* $1, m_1, m_2$ *for some* $(m_1, m_2) = 1$.

I am convinced that the method of proof of Theorem 3.10 can eventually be used to tackle this conjecture.

## 3.4   Linear systems on 3-folds

Let $x \in X$ be a 3-fold germ, $\mathcal{H}$ a mobile linear system on $X$ and $Z = H_1 \cdot H_2$ the cycle theoretic intersection of 2 general members $H_1, H_2 \in \mathcal{H}$. A special case of Corollary 3.4 states that

$$K + \frac{1}{\mu}\mathcal{H} \quad \text{not canonical} \quad \Longrightarrow \quad \text{mult}_x Z > 4\mu^2.$$

In Section 3.2 I reduce this to the surface result Theorem 3.1 by means of Shokurov connectedness (Theorem 3.2). Recall that the surface case 3.1 has a very elementary proof by induction on the discrepancy (with respect to the canonical class) of a valuation $E$ with $a_E(K_S + D) < -1$ (in the notation introduced at the beginning of the proof of 3.1). The purpose of this section is to give a proof of the 3-fold case in the same spirit as in the surface case, that is, an elementary proof by induction on the discrepancy of a valuation $E$ with $a_E(K_X + \frac{1}{\mu}\mathcal{H}) < 0$. There are several good reasons for wanting to do this:

(1) The proof by reduction to surfaces uses Shokurov's connectedness principle 3.2 which in turn depends on the Kodaira vanishing theorem. The method of Iskovskikh and Manin is more "elementary" and works in any characteristic.

(2) In 3.4, (2), I insisted that $a_1$, $a_2$ are both $\leq 1$. This assumption is indeed necessary to apply the connectedness theorem. However, the surface statement Theorem 3.1 does not need this. The discussion here suggests that the following might be true. Let $x \in S \subset X$ be a surface through $x$ (smooth, say). Then for $\lambda \geq 0$, if $K_X - \lambda S + \frac{1}{\mu}\mathcal{H}$ is not canonical then $\text{mult}_x Z > 4(1 + \lambda)\mu^2$. This kind of statement may look rather artificial but it is in fact quite natural and it would be very useful in the study of rigidity of Del Pezzo fibrations. Unfortunately I don't know if it is true; I prove a technical statement which is similar and sufficient to establish the application to the rigidity theorem for Del Pezzo fibrations in Section 5.1.

**Theorem 3.12** *Let* $x \in \sum S_i \subset X$ *(for $i \geq 1$) be the germ of a smooth normal crossing surface* $\sum S_i$ *in a 3-fold* $X$.

Let $\mathcal{H}$ be a mobile linear system on $X$ and $Z = H_1 \cdot H_2$ the cycle theoretic intersection of two general members $H_1, H_2 \in \mathcal{H}$. Write

$$Z = Z_{\mathrm{pr}} + \sum Z_{S_i},$$

where the support of $Z_{S_i}$ is contained in $S_i$ and $Z_{\mathrm{pr}}$ intersects $\sum S_i$ properly. Note that $Z$ may have components contained in several of the $S_i$, and as a consequence there may be a choice in the decomposition $Z = Z_{\mathrm{pr}} + \sum Z_{S_i}$. Let $\lambda_i \geq 0$ be rational numbers and assume that

$$K_X - \sum \lambda_i S_i + \frac{1}{\mu}\mathcal{H} \quad \text{is not canonical.}$$

Then there are positive rational numbers $0 \leq t_i \leq 1$ (with $0 < t_i$ if $Z_{S_i}$ is nonempty) such that

$$\mathrm{mult}_x Z_{\mathrm{tr}} + \sum t_i \, \mathrm{mult}_x Z_{S_i} > 4\Big(1 + \sum \lambda_i t_i\Big)\mu^2.$$

**Remark 3.13** In fact the argument proves that, more generally, for any number of surfaces $\sum S_i \subset X$, possibly singular, $K_X - \sum \lambda_i S_i + \frac{1}{\mu}\mathcal{H}$ not canonical implies that there are rational numbers $0 \leq t_i \leq 1$ (with $0 < t_i$ if $Z_{S_i}$ is nonempty) such that

$$\mathrm{mult}_x Z_{\mathrm{tr}} + \sum t_i \, \mathrm{mult}_x Z_{S_i} > 4\Big(1 + \sum \lambda_i t_i \nu_i\Big)\mu^2,$$

where $\nu_i = \mathrm{mult}_x S_i$ is the multiplicity of $S_i$ at $x$.

**Proof** The proof is very similar to the proof of Theorem 3.1 and I sketch it briefly, leaving some of the details and calculations to the reader (the main difficulty here is to make the statement, not the proof).

    Let

$$\varepsilon \colon S_0' \subset X' \to x \in X$$

be the blowup of the maximal ideal of $x \in X$. In other words, I depart from usual practice, denoting by $S_0'$ the exceptional divisor of the blowup. Write $S_i' = \varepsilon_*^{-1} S_i$ (for $i \geq 1$) and $\mathcal{H}' = \varepsilon_*^{-1}\mathcal{H}$. If $m = m_{S_0'}\mathcal{H}$, we have that

$$\varepsilon^*\Big(K_X - \sum_{i \geq 1} \lambda_i S_i + \frac{1}{\mu}\mathcal{H}\Big) = K_{X'} - \Big(-\frac{m}{\mu} + 2 + \sum_{i \geq 1}\lambda_i\Big)S_0' - \sum_{i \geq 1}\lambda_i S_i' + \frac{1}{\mu}\mathcal{H}'$$

is not canonical. By induction, we may assume that the statement holds on $X'$. The plan is to massage the ensuing inequality until we get the result for $X$.

We consider 3 cases, depending on the nature of the centre $C_E X' \subset X'$ of a valuation $E$ with negative discrepancy $a_E(K_X - \sum \lambda_i S_i + \frac{1}{\mu}\mathcal{H}) < 0$:

CASE 1: $E = S_0'$ This is obvious.

CASE 2: $C_E X' \subset X'$ IS A CURVE In this case, the statement on $X$ follows easily by applying the surface statement 3.1 to $C_E X' \subset X'$. Details are left to the reader.

CASE 3: $C_E X' = x' \in X'$ IS A CLOSED POINT I treat this case in some detail. Let $Z' = H_1' \cdot H_2'$ be the intersection of 2 general members of $\mathcal{H}'$; note that this is neither $\varepsilon_*^{-1} Z$ nor $\varepsilon^* Z$. As before for $Z$, write

$$Z' = Z'_{\mathrm{pr}} + \sum_{i \geq 1} Z'_{S_i'}.$$

I need the following, which are easy to see:

(1) $\mathrm{mult}_x Z = \deg Z'_{S_0'} + m^2 \geq \mathrm{mult}_{x'} Z'_{S_0'} + m^2$,

(2) $\mathrm{mult}_x Z_{S_i} \geq \mathrm{mult}_{x'} Z'_{S_i'}$ if $i \geq 1$,

(3) $\mathrm{mult}_x Z_{\mathrm{pr}} \geq \mathrm{mult}_{x'} Z'_{\mathrm{pr}}$.

There are rational numbers $0 \leq t_i' \leq 1$ such that

$$\mathrm{mult}_{x'} Z'_{\mathrm{pr}} + \sum_{i \geq 0} t_i' \mathrm{mult}_{x'} Z'_{S_i'} \geq 4 \left( 1 + \sum_{i \geq 0} \lambda_i' t_i' \right) \mu^2,$$

where, of course, I have set $\lambda_i' = \lambda_i$ if $i \geq 1$ and

$$\lambda_0' = -\frac{m}{\mu} + 2 + \sum_{i \geq 1} \lambda_i \geq 0$$

(if it were $< 0$, we would be in Case 1). The statement follows by setting

$$t_i = \frac{t_i' + t_0'}{1 + t_0'} \quad \text{for } i > 0.$$

Indeed, first of all, by (1–3) above and the inductive hypothesis,

$$\text{mult}_x \, Z_{\text{pr}} + \sum_{i \geq 1} t_i \, \text{mult}_x \, Z_{S_i}$$

$$= \frac{t_0'}{1 + t_0'} \left( \text{mult}_x \, Z_{\text{pr}} + \sum_{i \geq 1} \text{mult}_x \, Z_{S_i} \right)$$

$$+ \frac{1}{1 + t_0'} \left( \text{mult}_x \, Z_{\text{pr}} + \sum_{i \geq 1} t_i' \, \text{mult}_x \, Z_{S_i} \right)$$

$$\geq \frac{t_0'}{1 + t_0'} (\text{mult}_{x'} \, Z_{S_0'} + m^2) + \frac{1}{1 + t_0'} \left( \text{mult}_{x'} \, Z_{\text{pr}}' + \sum_{i \geq 1} t_i' \, \text{mult}_{x'} \, Z_{S_i'} \right)$$

$$= \frac{1}{1 + t_0'} \left( \text{mult}_{x'} \, Z_{\text{pr}}' + \sum_{i \geq 0} t_i' \, \text{mult}_{x'} \, Z_{S_i'} \right) + \frac{t_0'}{1 + t_0'} m^2$$

$$\geq \frac{4}{1 + t_0'} \left( 1 + \sum_{i \geq 0} \lambda_i' t_i' \right) \mu^2 + \frac{t_0'}{1 + t_0'} m^2.$$

An easy estimate allows us to complete the proof (recall the definition of $\lambda_i'$):

$$\frac{4}{1 + t_0'} \left( 1 + \sum_{i \geq 0} \lambda_i' t_i' \right) \mu^2 + \frac{t_0'}{1 + t_0'} m^2$$

$$= \frac{4}{1 + t_0'} \left( 1 + \sum_{i \geq 1} (t_0' + t_i') \lambda_i \right) \mu^2 + \frac{t_0'}{1 + t_0'} \left( 4 \left( 2 - \frac{m}{\mu} \right) \mu^2 + m^2 \right)$$

$$\geq 4 \mu^2 \left( 1 + \sum_{i \geq 1} t_i \lambda_i \right),$$

where I have used $4 \left( 2 - \frac{m}{\mu} \right) \mu^2 + m^2 \geq 4 \mu^2$.  □

# 4    Conic bundles

This chapter is devoted to the study of conic bundles. In the first section I give several proofs and generalisations of the known rigidity theorem, first shown by Sarkisov [Sa1], [Sa2]. Then I state some natural conjectures and discuss possible approaches to them.

## 4.1    Rigid Conic bundles

**Definition 4.1** A *conic bundle* is a 3-fold Mori fibre space $X \to S$ with (generic) fibre of dimension 1. In particular, it is always automatically assumed that rank NS $X$ − rank NS $S = 1$. The generic fibre $X_\eta$ is a conic over the rational function field $K(S)$ of $S$.

A conic bundle $X \to S$ is *standard* if the total space $X$ is smooth. It is easy to see that then $S$ must itself be smooth and, given any $P \in S$, there is an analytic neighbourhood $P \in U$ and analytic coordinates $s_1, s_2$ on $U$ such that $X_{|_U} \subset \mathbb{P}^2 \times U$ is isomorphic to one of the following models:

(a) $x_0^2 + x_1^2 + x_2^2 = 0$, or

(b) $x_0^2 + x_1^2 + s_2 x_2^2 = 0$, or

(c) $x_0^2 + s_1 x_1^2 + s_2 x_2^2 = 0$.

It is easy to see that any conic bundle can be put in standard form; there is therefore no real loss of generality in restricting attention to standard conic bundles.

**Theorem 4.2 (Sarkisov [Sa1])** *Let $X \to S$ be a standard conic bundle. Denoting by $\Delta \subset S$ the discriminant of the conic bundle, we assume that $4K_S + \Delta$ is quasieffective. If $X' \to S'$ is another Mori fibre space, every birational map $\varphi: X \dashrightarrow X'$ is square.*

To understand the result, it is useful to consider the following invariant.

**Definition 4.3** Let $X \to S$ be a conic bundle. The *effective threshold* $\tau = \tau(S, \Delta)$ is the rational number

$$\tau = \sup\{s \mid sK_S + \Delta \geq 0\}.$$

**Proposition 4.4** *Let $X \to S$ and $X' \to S'$ be conic bundles. Assume that $X \to S$ is square birational to $X' \to S'$. Assume that*

*(1) $X \to S$ is standard, and*

*(2) $\tau \geq 1$.*

*Then $\tau' \geq \tau$. In particular, if in addition $X' \to S'$ is also standard then $\tau = \tau'$.*

**Proof** The proof is a formal consequence of the following two observations:

(a) Given a standard conic bundle $X \to S$ and the blowup $C \subset S' \to P \in S$ of the maximal ideal of a point $P \in S$, there is a (nonunique) standard conic bundle $X' \to S'$, square birational to $X \to S$, with $X'_{|_{S' \setminus C}} = X_{|_{S \setminus P}}$.

(b) For a conic bundle $X \to S$, the discriminant only depends on the generic fibre $X_\eta$. $\square$

**Example 4.5** It is important to understand that $\tau$ is not a square birational invariant for general (possibly nonstandard) conic bundles. Here is an example. Let $Z \subset \mathbb{P}^6$ be the cone over the Veronese surface and let $p\colon Z \dashrightarrow \mathbb{P}^2$ be the projection from a general 3-plane $H \subset \mathbb{P}^6$. The blowup $X' \to Z$ of the 4 points $H \cap X$ resolves the singularities of $p$, giving rise to a conic bundle $X' \to \mathbb{P}^2$

Note that $X'$ has an index 2 quotient singular point $Q \in X'$ and the discriminant $\Delta' \subset \mathbb{P}^2$ consists of 3 lines meeting in a common point $P \in \mathbb{P}^2$; hence $\tau' = 1$. The fibre over $P$ is the sum $\Gamma_1 + \Gamma_2 + \Gamma_3 + \Gamma_4$ of 4 curves $\Gamma_i \cong \mathbb{P}^1$ meeting at $Q$ and $-K_{X'} \cdot \Gamma_i = \frac{1}{2}$. If $C \subset \mathbb{F}_1 \to P \in \mathbb{P}^2$ is the blowup of $P$, $W \to X'$ the blowup of $Q$, and $W \dashrightarrow X$ the flop of the proper transforms of the curves $\Gamma_i$, we have a link of Type II

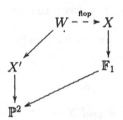

to a standard conic bundle $X \to \mathbb{F}_1$ whose discriminant $\Delta \subset \mathbb{F}_1$ consists of 3 disjoint fibres; hence $\tau = 0$.

It is easy to imagine a link $X \dashrightarrow X'$ which looks like that just constructed in a neighbourhood of the contraction $C \subset S \to P \in S'$ of a $-1$-curve. For such a link we would have $3K_S + \Delta = 0$. In other words,

$$\alpha K_S + \Delta = \alpha K_{S'} + \Delta' + (\alpha - 3)C \quad \text{for all } \alpha,$$

so that $\tau \geq 3$ implies $\tau' = \tau$ but, if $\tau < 3$ (or $\tau' < 3$), it might happen that $\tau < \tau' < 3$. The reader is invited to construct arbitrarily complicated examples using this idea.

I leave the following well known lemma as an exercise.

**Lemma 4.6** *Let* $\pi\colon X \to S$ *be a conic bundle. Then* $-\pi_*(K_X^2) = 4K_S + \Delta$. $\square$

**Proof of Theorem 4.2**  I first prove the result in the case of surfaces. In other words, I assume that $X$ is a smooth surface, defined over an algebraically nonclosed field $k$, and $\pi\colon X \to S$ a conic bundle structure. I think that this is the natural context for the theorem: the proof generalises easily to 3-folds (and in fact also to higher dimensions, see below). In closing, I give a second proof for the surface case which, while technically more sophisticated than the first, has the advantage that it generalises well to the case of Del Pezzo fibrations (Chapter 5).

I argue by contradiction in each case, assuming a Mori fibre space $X' \to S'$ and a nonsquare map

$$
\begin{array}{ccc}
X & \overset{\varphi}{-\!\!-\!\!\rightarrow} & X' \\
\pi\downarrow & & \downarrow \\
S & & S'.
\end{array}
$$

Choose a very ample complete linear system

$$\mathcal{H}' = |-\mu' K' + A'|$$

on $X'$, where $A'$ is the pullback of an ample divisor on the base $S'$. The proper transform $\mathcal{H}$ on $X$ is a mobile linear system and, because $X \to S$ is a Mori fibre space, there is a rational number $\mu$ (with $2\mu \in \mathbb{Z}$) for which

$$K + \frac{1}{\mu}\mathcal{H} = A$$

is a pullback from the base $S$. The assumption that $\varphi$ is not square, together with the Noether–Fano–Iskovskikh inequalities (Theorem 2.4, (1)), implies that $\mu > \mu'$. Basically, all the proofs rest on the following idea. Assume that $K + \frac{1}{\mu}\mathcal{H}$ is canonical. Then, by Theorem 2.4, (2), $A$ is not quasieffective. On the other hand a small calculation using Lemma 4.6 gives that

$$A = \pi_* \frac{1}{\mu^2} Z + (4K_S + \Delta) \quad \text{is quasieffective,}$$

where $Z = H_1 \cdot H_2$ is the intersection of 2 general members of $\mathcal{H}$, a contradiction. Now $K + \frac{1}{\mu}\mathcal{H}$ need not be canonical in general; there are two ways around this. One way (the first proof) is to run the Sarkisov program until it becomes canonical; the other (the second proof) is to study $A$ more closely and show that it can only be large at the expense of $\mathcal{H}$ becoming too singular.

(1) SURFACES (FIRST PROOF)  By the Noether–Fano–Iskovskikh inequalities (Theorem 2.4, (3)), running the Sarkisov program gives a chain of square links

$$X/S \dashrightarrow X_1/S \dashrightarrow \cdots \dashrightarrow X_n/S$$

until

$$K_n + \frac{1}{\mu}\mathcal{H}_n = A_n \quad \text{is not nef;}$$

this amounts to saying that

$$\deg A_n < 0.$$

We now choose general members $H_{1,n}, H_{2,n} \in \mathcal{H}$ and calculate the *effective cycle* $Z_n = H_{1,n} \cdot H_{2,n}$ by

$$\frac{1}{\mu^2}Z_n = (-K_n + A_n)^2 = K_n^2 - 2K_n \cdot A_n.$$

Taking the direct image under $\pi_n \colon X_n \to S$ gives

$$A_n = \pi_*\frac{1}{\mu^2}Z_n + (4K_S + \Delta_n).$$

The assumption and the birational invariance of $\tau$ imply that $4K_{S_n} + \Delta_n$ is quasieffective, and hence also $A_n$, a contradiction.

(2) 3-FOLDS  The proof is essentially the same. By the Noether–Fano–Iskovskikh inequalities (Theorem 2.4, (2)), running the Sarkisov program gives a chain of square links

$$X/S \dashrightarrow X_1/S_1 \dashrightarrow \cdots \dashrightarrow X_n/S_n$$

until $K_n + \frac{1}{\mu}\mathcal{H}_n = A_n$ is *not quasieffective*. As before, writing

$$A_n = \pi_*\frac{1}{\mu^2}Z_n + (4K_S + \Delta_n)$$

shows that $A_n$ is quasieffective, a contradiction.

(3) SURFACES (SECOND PROOF)  In this proof I obtain a contradiction from the cycle $Z = H_1 \cdot H_2$ by arguing directly on $X$, without first running the Sarkisov program. In doing so, I use the assumption in the form $K^2 \le 0$. Here is the crucial point.

I claim that there are maximal singularities $E_i$ with centres $x_i \in F_i \subset X$ lying in fibres $F_i$ over distinct points $s_i \in S$, and positive rational numbers $\lambda_i > 0$, such that

(1) $a_{E_i}\left(K - \lambda_i F_i + \frac{1}{\mu}\mathcal{H}\right) = a_{E_i}(K) + \lambda_i m_{E_i}(F_i) - \frac{1}{\mu}m_{E_i}(\mathcal{H}) = 0$ and

(2) $\sum \lambda_i > \deg A.$

This is a small variation on the Noether–Fano–Iskovskikh inequalities. Indeed, choose a common resolution

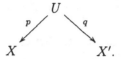

Let $F_i \subset X$ be all the fibres containing the centre of some maximal singularity of $\mathcal{H}$ and $E_{ij} \subset U$ the $p$-exceptional divisors with centre $C_{E_{ij}} X \in F_i$. Above a small neighbourhood of $F_i \subset X$ we may write

$$K_U = K + \sum a_{ij} E_{ij},$$
$$\mathcal{H}_U = \mathcal{H} - \sum m_{ij} E_{ij},$$
$$F_i' = F_i - \sum c_{ij} E_{ij},$$

where $\mathcal{H}_U$ and $F_i'$ are the proper transforms. By choice, for each $i$ there is some $j = j(i)$ for which

$$m_{ij(i)} > \mu a_{ij(i)}.$$

Finally, suppose that

$$\lambda_i = \max_j \left\{ \frac{m_{ij} - \mu a_{ij}}{\mu c_{ij}} \right\} > 0.$$

Then

$$K_U - \sum \lambda_i F_i' + \frac{1}{\mu} \mathcal{H}_U = K - \sum \lambda_i F_i + \frac{1}{\mu} \mathcal{H} + \sum \alpha_{ij} E_{ij},$$

with

$$\alpha_{ij} = a_{ij} + \lambda_i c_{ij} - \frac{1}{\mu} m_{ij} \geq 0,$$

and for each $i$ there is at least one index $j(i)$ for which $\alpha_{ij(i)} = 0$. To show that $\sum \lambda_i > \deg A$, I prove the equivalent statement:

$$D = K - \sum \lambda_i F_i + \frac{1}{\mu} \mathcal{H} \quad \text{is not quasieffective.}$$

From the last 4 displayed equations, if $D$ were quasieffective then

$$K_U + \frac{1}{\mu} \mathcal{H}_U$$

would also be (recall that $K_X + \frac{1}{\mu}\mathcal{H}$ is canonical away from $\bigcup F_i$), and then also

$$q_*\left(K_U + \frac{1}{\mu}\mathcal{H}_U\right) = K' + \frac{1}{\mu}\mathcal{H}' = K' + \frac{1}{\mu'}\mathcal{H}' - \left(\frac{1}{\mu'} - \frac{1}{\mu}\right)\mathcal{H}'$$

would be quasieffective; this is a contradiction because by Theorem 2.4, (1),

$$\frac{1}{\mu'} - \frac{1}{\mu} > 0.$$

This proves the claim.

I now prove that the claim implies the result. Indeed,

$$K + \sum(1 - \lambda_i)F_i + \frac{1}{\mu}\mathcal{H}$$

is not log terminal at $x_i = C_{E_i}X$ and, by Theorem 3.1,

$$K^2 + 4\deg A = (-K + A)^2 = \frac{1}{\mu^2}H_1 \cdot H_2 = \frac{1}{\mu^2}\deg Z \geq 4\sum\lambda_i > 4\deg A,$$

a contradiction, given that $K^2 \leq 0$ by assumption. $\square$

It seems to me that the rationally connected fibrations of Kollár, Miyaoka and Mori [KMM] provide a reasonable framework to study birational rigidity of higher dimensional varieties.

**Definition 4.7**   (1) A variety $V$ is *rationally connected* (*RC* for short) if any 2 general points on $V$ can be joined by a rational curve.

(2) A *rationally connected fibration* (or *RC fibration*), written $X \supset U \xrightarrow{\pi} Z$, is a variety $X$, together with a smooth open subset $X \supset U$ and a proper morphism $\pi\colon U \to Z$, such that all fibres of $\pi$ are RC varieties.

(3) A RC fibration is *extremal* (or an *ERC fibration*) if rank $N^1(U/Z) = 1$.

**Exercise 4.8** Use the above ideas to prove the following statement. Let $X \supset U \xrightarrow{\pi} Z$ be an ERC fibration (with everything defined over an arbitrary base field). Assume that the following 2 conditions hold:

(1) $\dim X - \dim Z = 1$;

(2) if $\Delta \subset Z$ is the discriminant, there is a compactification $(\overline{Z}, \overline{\Delta}) \supset (Z, \Delta)$ such that $\mathrm{codim}_{\overline{Z}\backslash Z}\overline{Z} \geq 2$ and

$$K_{\overline{Z}} + \frac{1}{4}\overline{\Delta} \geq 0$$

is canonical and effective.

Let $X' \supset U' \xrightarrow{\pi'} Z'$ be another ERC fibration and $\varphi \colon X \dashrightarrow X'$ a birational map. Then $\varphi$ is square, that is, it is an isomorphism on generic fibres.

**Remark 4.9** If $W$ is a variety, let $\mathrm{MV}^1 W \subset N^1 W$ be the real cone generated by the classes of mobile linear systems. If $X \to S$ is a Mori fibre space, it seems natural to consider the cone

$$\mathrm{KE}^1 S = \{A \in N^1 S \mid -K_X + A \in \mathrm{MV}^1 X\} \subset N^1 S.$$

It seems reasonable to expect that $\mathrm{KE}^1 S \subset \mathrm{NE}^1 S$ implies that $X \to S$ is birationally rigid. It is of course not clear what the assumption means, nor that it is invariant under square birational operations. The argument just given for conic bundles says that $4K_S + \Delta \geq 0$ implies $\mathrm{KE}^1 S \subset \mathrm{NE}^1 S$, while Chapter 5 says that $K_X^2 \notin \mathrm{NE}^2$ implies $\mathrm{KE}^1 S \subset \mathrm{NE}^1 S$.

Does the cone $\mathrm{KE}^1 S$, or some other cone related to it, satisfy a general structure theorem?

## 4.2 Open questions

The following conjectures are due to Iskovskikh ([I2], [I3]).

**Conjecture 4.10 (Iskovskikh [I2])** *Let $X \to S$ be a standard conic bundle. If $X$ is rational, $\tau \leq 2$.*

The near-converse is easy: if $\tau \leq 2$, $X$ is either rational or birational to a cubic 3-fold (and there is an excellent biregular criterion to say which is which).

**Conjecture 4.11 (Iskovskikh [I2])** *If $X$ is rational, there is a birational map $X \dashrightarrow \mathbb{P}^3$ transforming fibres into conics.*

It is easy to see that the 2 conjectures are equivalent. This is quite remarkable. Conjecture 4.10 is a statement about conic bundles: if the discriminant is large, the conic bundle is not rational. Conjecture 4.11 on the other hand is a statement on the size of the Cremona group of $\mathbb{P}^3$: any net of rational curves in $\mathbb{P}^3$ is birational to a net of conics. It is natural to ask whether the conjectures have a counterpart for Del Pezzo fibrations. If so, this casts a new light on the meaning of rigidity: $\mathbb{P}^3$ almost looks like a rigid variety because it has a *large* birational group.

In the remainder of this section, I outline a possible approach to Conjecture 4.10. As an intermediate step, I would like to pose Problem 4.13 below as a challenge. First, note the following result:

**Theorem 4.12** *Let $X \to S$ be a surface conic bundle defined over a field $k$. If $3K_S + \Delta \geq 0$ then $X \to S$ is birationally rigid.*

**Proof** The idea is as follows. Assume that there is a link $X/S \dashrightarrow X'/S'$ of Type III or IV. This means that $X$ is a Del Pezzo surface of degree

$$K^2 = 8 - \deg \Delta \leq 2$$

having 2 extremal rays $R_1$ and $R_2$; suppose that $R_1$ corresponds to the Mori fibre structure $X/S$. Let $\tau \colon X \to X$ be the Geiser or Bertini involution of $X$. I claim that $\tau R_1 = R_2$. This proves the statement since it follows that $\tau$ is a square birational map from $X/S$ to $X'/S'$. If $d = 2$ let $\pi \colon X \to \mathbb{P}^2$ be the 2-to-1 cover given by the complete linear system $|-K_X|$, and if $d = 1$ let $\pi \colon X \to Q$ be the 2-to-1 cover of the singular quadric cone $Q \subset \mathbb{P}^3$ given by the complete linear system $|-2K_X|$. In either case $\tau$ is the involution exchanging the sheets of $\pi$, and in either case $\overline{NE}^\tau(X) = \overline{NE}(X^\tau) = \mathbb{Z}$, which implies $\tau R_1 = R_2$. $\square$

**Problem 4.13** Do something similar for 3-folds. I am almost tempted to conjecture that if $X \to S$ is a standard (3-fold) conic bundle and if $\tau \geq 3$ then $X \to S$ is birationally rigid.

In closing, I wish to outline an approach to improving the notion of standard models of conic bundles. As we know, any conic bundle $X \to S$ in the Mori category is birational to a standard conic bundle, but this involves blowing up the surface $S$. It would be desirable, even at the expense of introducing some singularities, to construct distinguished biregular models of conic bundles over $\mathbb{P}^2$ or minimally ruled surfaces. In addition to being an interesting question in its own right, this might also be relevant for Conjectures 4.10 and 4.11.

Let $X \to S$ be a standard conic bundle with $1 \leq \tau < 4$ (if $\tau < 1$ it is easy to show that $X$ is rational). Now $K_S + \Delta$ is quasieffective and we assume, as we may (see for example [I3], Lemma 4), that $K_S + \Delta$ is nef.

We first run an (ordinary) minimal model program for the base surface $S$ as follows. The nef threshold is

$$t = t(S, \Delta) = \max\{s \mid sK + \Delta \text{ is nef}\}.$$

Clearly $t \leq \tau$. We define the chain

$$(S, t) = (S_0, t_0) \to \cdots (S_i, t_i) \to (S_{i+1}, t_{i+1}) \cdots \to (S_n, t_n) = (S', \tau)$$

inductively by setting $t_i = t(S_i, \Delta_i)$, letting $\sigma_i \colon S_i \to S_{i+1}$ be the contraction of an extremal rational curve $C_i$ with $(t_i K_i + \Delta_i) \cdot C_i = 0$ (clearly $K_i \cdot C_i < 0$ and $C_i$ is a $-1$-curve) and setting $\Delta_{i+1} = \sigma_i(\Delta_i)$. The program ends when the nef threshold finally catches up with the effective threshold $t_n = \tau$, and we meet a contraction $S_n = S' \to T$ of fibre type. Since $S' \to T$ is $\mathbb{P}^2$ or a minimally ruled surface $\mathbb{F}_m$, we get the following result.

**Corollary 4.14** *If* $\tau < 4$, *then:*

$$\tau \in \left\{ \frac{1}{2}, \frac{2}{2}, \ldots, \frac{7}{2}, \frac{1}{3}, \frac{2}{3}, \ldots, \frac{11}{3} \right\}.$$

The typical cases are conic bundles over $\mathbb{P}^2$ with discriminant of degree $< 12$, or over a ruled surface of degree $< 8$ over the base.

As a second step, I propose the following analog of the results of [C1] and [Ko3].

**Conjecture 4.15**   *(1) If* $\sigma_i \colon S_i \to S_{i+1}$ *is a contraction with* $t_i \leq 2$, *there is a conic bundle* $X_{i+1} \to S_{i+1}$ *square birational to* $X_i \to S_i$. *Moreover,* $X_{i+1}$ *has terminal singularities of index 1.*

*(2) If* $\tau \leq 3$, *the above procedure eventually makes a 3-fold* $X''$, *with terminal singularities of index 1, and a conic bundle* $X'' \to S''$ *over a smooth surface* $S''$, *where* $3K_{S''} + \Delta''$ *is nef. I hope that one can do this with* $S''$ *a Del Pezzo surface or a conic bundle* $S'' \to T''$.

*(3) If* $3 < \tau < 4$, *something nice can be done, e.g., a distinguished model with terminal singularities of index one of* $\{1, 2, 3, 4, 6\}$.

I hope that the singularities on the models (2) can be classified and will turn out to be mild enough so that one can attach an intermediate Jacobian and use it to show that $X$ is not rational, as in the classical cases Beauville ([Be1], [Be2]) and Shokurov [Sh].

# 5   Del Pezzo fibrations

In this chapter, devoted to Del Pezzo fibrations, I prove Pukhlikov's rigidity criterion for fibrations of degree 1 and 2, explain Kollár's method of constructing semistable models of fibrations of cubic surfaces—this is a kind of analog of the standard conic bundles of Chapter 4—and, in the final section, raise some open questions.

## 5.1   Rigid Del Pezzo fibrations

**Theorem 5.1 (Pukhlikov [P5])** *Let* $X \to S$ *be a 3-fold Del Pezzo fibration of degree* $d \leq 2$ *and assume that*

*(1)* $X$ *is smooth, and*

*(2) the 1-cycle* $K_X^2 \in N_1 X$ *does not lie in the interior* $\text{Int } \overline{NE} \, X$ *of the Mori cone.*

*Then* $X$ *is birationally rigid over* $S$.

**Proof**  The proof is essentially the same as the second proof of the rigidity criterion for conic bundle surfaces, given in Chapter 4, with Theorem 3.12 replacing Theorem 3.1 in the endgame. From now on until the final calculation in the last Step 4, I only assume $d \leq 3$.

The proof is by contradiction, assuming a Mori fibre space $X' \to S'$ and a map $\varphi \colon X \dashrightarrow X'$

$$
\begin{array}{ccc}
X & \dashrightarrow & X' \\
{\scriptstyle \pi}\downarrow & & \downarrow \\
S & & S'
\end{array}
$$

which is not birational on generic fibres. Choose a very ample complete linear system

$$\mathcal{H}' = |-\mu'K' + A'|$$

on $X'$, where $A'$ is the pullback of a divisor ample on the base $S'$. The proper transform $\mathcal{H}$ on $X$ is a mobile linear system and, because $X \to S$ is a Mori fibre space, there is a rational number $\mu$ such that

$$K + \frac{1}{\mu}\mathcal{H} = A$$

is a pullback from the base $S$. The assumption $\varphi$ not square, together with the Noether–Fano–Iskovskikh inequalities (Theorem 2.4, (1)), implies $\mu > \mu'$.

STEP 1   Using well known properties of the generic fibre (see for instance [C2], Appendix), perhaps after composing $\varphi$ with a birational selfmap

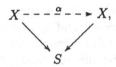

$$
X \dashrightarrow^{\alpha} X,
$$
$$
\searrow \quad \swarrow
$$
$$
S
$$

I may assume that no maximal centre is a curve dominating $S$.

I now claim that no (vertical) curve in $X$ can be a maximal centre. Indeed, if for example $E$ is a maximal singularity having centre $C_E X = \Gamma \subset F_t$, a line contained in the fibre $F_t$ over $t \in S$, let $C \subset F_t$ be a general conic; then

$$2\mu = \mathcal{H} \cdot C \geq 2\operatorname{mult}_C \mathcal{H} > 2\mu,$$

which is a contradiction. The case $\deg \Gamma \geq 2$ is easier and left to the reader.

STEP 2   I claim that there are maximal singularities $E_i$, having centres $x_i \in F_i \subset X$ lying in fibres $F_i$ over distinct points $s_i \in S$, and positive rational numbers $\lambda_i > 0$ such that

(1) $a_{E_i}\left(K - \lambda_i F_i + \frac{1}{\mu}\mathcal{H}\right) = a_{E_i}(K) + \lambda_i m_{E_i}(F_i) - \frac{1}{\mu}m_{E_i}(\mathcal{H}) = 0$, and

(2) $\sum \lambda_i > \deg A$.

The proof is identical to that of the corresponding claim in the proof of Theorem 4.2 (surfaces, second proof) and is therefore omitted.

STEP 3   Let $H_1$, $H_2$ be general members of $\mathcal{H}$ and $Z = H_1 \cdot H_2$. By Step 1, Theorem 3.12 and Remark 3.13, there are numbers $0 < t_i \leq 1$ such that

$$\text{mult}_{x_i} Z_{\text{hor}} + t_i \, \text{mult}_{x_i} Z_{F_i} \geq 4(1 + \lambda_i t_i)\mu^2,$$

where I hope that the notation is self-explanatory: $Z_{\text{hor}}$ is the "horizontal" part of $Z$ and the support of $Z_{F_i}$ is contained in the $i$th fibre $F_i$. Write $Z_{\text{vert}}$ for the "vertical" component of $Z$, so that

$$Z = Z_{\text{vert}} + Z_{\text{hor}}.$$

If $F$ is a fibre, $Z_{\text{hor}} \cdot F = d\mu^2$, hence

$$\text{mult}_{x_i} Z_{\text{hor}} \leq d\mu^2;$$

since $d \leq 2 \leq 4$, this certainly implies that

$$\text{mult}_{x_i} Z_{F_i} \geq 4\lambda_i \mu^2.$$

STEP 4   Our assumptions entail an *a priori* bound on the size of the vertical component $Z_{\text{vert}}$. Indeed if

$$Z_{\text{vert}} = \gamma[\text{line}]$$

then

$$\begin{aligned} Z = Z_{\text{hor}} + Z_{\text{vert}} &= \mu^2(-K + A)^2 \\ &= \mu^2 K^2 - 2\mu^2 K \cdot A = \mu^2 K^2 + 2d\mu^2 \deg A[\text{line}] \end{aligned}$$

and

$$\mu^2 K^2 = Z_{\text{hor}} + (\gamma - 2d\mu^2 \deg A)[\text{line}].$$

By assumption $K^2$ is not in the interior of $\overline{\text{NE}}$, hence

$$\gamma \leq 2d\mu^2 \deg A.$$

Combining with the previous step, and *using $d \leq 2$*, we have a contradiction:

$$\gamma \geq \sum \text{mult}_{x_i} Z_{F_i} \geq 4 \sum \lambda_i \mu^2 > 4\mu^2 \deg A. \quad \square$$

**Remark 5.2** Pukhlikov [P6] proves that the same statement also holds for fibrations of cubic surfaces—with minor restrictions on the nature of the singular fibres.

## 5.2    Semistable models

In this section, I copy from Kollár [Ko3] the theory of semistable models for hypersurfaces, which improves upon [C1]. Let $\mathcal{O}$ be a DVR with parameter $p \in \mathcal{O}$ and field of fractions $K$.

**Definition 5.3** A *weight system* $(\mathbf{x}, \mathbf{w})$ on $\mathcal{O}[y_0, \ldots, y_n]$ is a choice of coordinates

$$(x_0, \ldots, x_n)^t = M(y_0, \ldots, y_n)^t, \quad \text{where } M \in \mathrm{SL}(n+1, \mathcal{O})$$

and weights $x_i \mapsto w_i = w(x_i) \in \mathbb{R}$.

Let $f_K \in K[y_0, \ldots, y_n]$ be a polynomial. One can always find an integer $s$ such that $f = p^{-s} f_K \in \mathcal{O}[y_0, \ldots, y_n]$ ($f$ is called an $\mathcal{O}$-model of $f_K$). The largest such $s$ is the *multiplicity* of $f_K$ at $p$; it is denoted $\mathrm{mult}_p f_K$.

**Definition 5.4** Let $f \in \mathcal{O}[y_0, \ldots, y_n]$ be a homogeneous polynomial and $X \subset \mathbb{P}^n_{\mathcal{O}}$ the hypersurface defined by the equation $f = 0$.

(1) A weight system $(\mathbf{x}, \mathbf{w})$ over $\mathcal{O}$ is called

$$\begin{cases} \textit{properly stable} & \text{if } \mathrm{mult}_p f(p^{\mathbf{w}}\mathbf{x}) < \dfrac{\deg f}{n+1} \sum_i w_i, \\[2ex] \textit{semistable} & \text{if } \mathrm{mult}_p f(p^{\mathbf{w}}\mathbf{x}) \leq \dfrac{\deg f}{n+1} \sum_i w_i, \\[2ex] \textit{unstable} & \text{if } \mathrm{mult}_p f(p^{\mathbf{w}}\mathbf{x}) > \dfrac{\deg f}{n+1} \sum_i w_i. \end{cases}$$

(2) $f$ (or $X$) is called *properly stable* (respectively *semistable*) if every weight system is properly stable (or semistable).

(3) $f$ (or $X$) is called *unstable* if there is an unstable weight system.

**A procedure to find semistable models.**    We start with a homogeneous polynomial $f_K \in K[y_0, \ldots, y_n]$.

(1) Find any $\mathcal{O}$-model $f_1$ of $f_K$.

(2) Assume that we already have $f_j$. If $f_j$ is semistable we are done.

(3) Otherwise there is a weight system $(\mathbf{x}, \mathbf{w})$ which is unstable on $f_j$. Set

$$f_{j+1} = p^{-s} f_j(p^{w_0} x_0, \ldots, p^{w_n} x_n), \quad \text{where } s = \mathrm{mult}_p f_j(p^{\mathbf{w}}\mathbf{x}),$$

and go back to (2).

The next statement is a special case of the main result of [Ko3].

**Theorem 5.5 (Kollár [Ko3])** *Let* $f_K \in K[x_0, \ldots, x_n]$ *be a homogeneous polynomial. If the hypersurface* $X_K = (f_K = 0) \subset \mathbb{P}^n_K$ *is nonsingular,* $f_K$ *has a semistable model over* $\mathcal{O}$ *if and only if* $f_K$ *is semistable over* $\overline{K}$. *Moreover,* $f_K$ *has only finitely many semistable models over* $\mathcal{O}$ *up to the action of* $\mathrm{SL}(n + 1, \mathcal{O})$.

**Proof** The idea is that the procedure stops, because some invariant goes down every time the procedure is applied. $\square$

**Theorem 5.6 (Kollár [Ko3])** *In order to check semistability of a family of cubic surfaces it is sufficient to use weight systems with the following 5 weight sequences*

$$(0,0,0,1), \quad (0,0,1,1), \quad (0,1,1,1), \quad (0,1,2,2), \quad (0,2,2,3). \quad \square$$

This implies that for cubic forms in 4 variables over $\mathcal{O}$ the above procedure becomes an effective algorithm.

## 5.3   Open questions

**Example 5.7** Let $\mathcal{O} = \mathbb{C}[t]$ and consider the cubic form

$$f_1 = x_0^3 + x_1^3 + x_2^2 x_3 + t^6 x_3^3.$$

It is easy to see that it is semistable with respect to every weight system where all the weights are 0 or 1. $f_1$ is unstable with weights $(0, 0, 1, -2)$ and we obtain the properly stable cubic form $f_2 = x_0^3 + x_1^3 + x_2^2 x_3 + x_3^3$, which is the unique semistable model.

The above is also an example of a square link of Del Pezzo fibrations of degree 3 that is centred at an Eckardt point of a fibre.

**Remark 5.8** I have not as yet been able to carry the necessary calculations to conclusion, but I have the feeling that square links between Del Pezzo fibrations are relatively common.

For example, let $X \to \Delta$ be a smooth Del Pezzo fibration of degree 2 over a small disc $\Delta$ with central fibre $X_0$. The anticanonical linear system defines a finite 2-to-1 morphism $\pi \colon X_0 \to \mathbb{P}^2$ ramified along a smooth quartic curve $B \subset \mathbb{P}^2$. Let $Q \in B$ be a point lying on 1 of the 28 bitangents of $B$ and $P \in X$ the inverse image. In suitable coordinates around $P \in X$ the fibration is defined by the function

$$f = x + y^2 + z^3.$$

Let $Y \to X$ be the weighted blowup with weights $3, 2, 1$. It seems quite possible to me that the 2-ray game starting with $Y \to X \to \Delta$ can be played to the very end, thus making a link.

One should also be able to do this for Del Pezzo fibrations of degree 3 and points $P \in X_0$ where the curve $T_P X_0 \cap X_0$ is a cuspidal plane cubic.

These examples reflect the indirect nature of the proof of Theorem 5.1. I would like to pose the following problems.

**Problem 5.9**    (1) Set up a notion of semistable models for Del Pezzo fibrations of degree 2 and 1, and prove that they exist. The results of [C1] are encouraging.

(2) Prove 1.6—including the $d = 3$ case—under the more general assumption that $X \to \mathbb{P}^1$ is semistable at every point $t \in \mathbb{P}^1$.

(3) Does there exist an analog for Del Pezzo fibrations of the invariant $\tau$ of conic bundles? that is, a rational number (perhaps obtained as a threshold) that is minimal on a semistable model, detects rigidity, and perhaps explains the role of the $K^2$ assumption in Theorem 5.1? For instance, assume that the $K^2$ condition holds for a semistable $X/S$; does it then hold for every $X'/S'$ square to $X/S$?

# 6    Fano 3-folds

I review the main known rigidity theorems for Fano 3-folds: hypersurfaces and complete intersections. In the last section, I indicate possible future directions. For more information and further discussion, see [CPR].

## 6.1    Rigid Fano 3-fold hypersurfaces

Anthony Iano-Fletcher wrote a list of 95 families of 3-fold Fano weighted hypersurfaces $X = X_d \subset \mathbb{P}^4_w$ with $-K_X = \mathcal{O}(1)$.

**Theorem 6.1 (Corti, Pukhlikov, Reid [CPR])** *Assume that $X$ is a general member of any of the 95 families. Then*

*(1) $X$ is birationally rigid,*

*(2) $\operatorname{Bir} X / \operatorname{Aut} X$ is generated by a finite number of explicit rational involutions.*

Rather than give an overview of the whole proof, which is too long to summarise here, I treat a special case in detail.

Let $X = X_5 \subset \mathbb{P}(1,1,1,1,2)$, $Q = (0,0,0,0,1) \in X$. We can write the equation of $X$ as

$$F(y, x_i) = y^2 x_0 + y f_3(x_i) + f_5(x_i) = 0.$$

Consider the projection

$$\mathbb{P}(1,1,1,1,2) \dashrightarrow \mathbb{P}^3$$

on the first 4 coordinates: this is well defined on the unique divisorial contraction $\pi = \pi_Q \colon E \subset Y \to Q \in X$ at $Q$, $-K_Y = -K_X - \frac{1}{2}E$ is nef and the anticanonical model of $Y$ is the variety

$$\overline{Y} = \mathrm{Proj} \bigoplus_{n \geq 0} H^0(-nK_Y) = \overline{Y}_6 \subset \mathbb{P}(1,1,1,1,3).$$

All of this can be seen explicitly by completing the square in the defining equation of $X$

$$x_0 F(y, x_i) = \left(x_0 y + \frac{1}{2}f_3\right)^2 + x_0 f_5 - \frac{1}{4}f_3^2.$$

The equation of $\overline{Y}_6 \subset \mathbb{P}(1,1,1,1,3)$ is then

$$G(z, x_i) = z^2 + x_0 f_5 - \frac{1}{4}f_3^2 = 0,$$

and the birational map $X \dashrightarrow \overline{Y}$ is given by

$$z = x_0 y + \frac{1}{2}f_3.$$

The exceptional set of the anticanonical map $\varphi \colon Y \to \overline{Y}$ consists of the proper inverse images of the 15 lines $\Gamma_i \subset X$ for $i = 1, 2, \ldots, 15$, given by

$$x_0 = f_3(x_i) = f_5(x_i) = 0.$$

Each of these lines on $X$ has anticanonical degree

$$-K_X \cdot \Gamma = \frac{1}{2}.$$

Being a double cover of $\mathbb{P}^3$, $\overline{Y}$ has a rational biregular involution $\sigma \colon \overline{Y} \to \overline{Y}$. The corresponding birational involution $\tau$ of $X$ is a link; indeed it can be written as

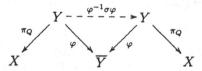

where $\varphi^{-1}\sigma\varphi$ is the flop of the 15 proper transforms $\Gamma_i' \subset Y$.

**Theorem 6.2** *If* $X = X_5 \subset \mathbb{P}(1, 1, 1, 1, 2)$ *is quasismooth, then*

*(1)* $X$ *is birationally rigid,*

*(2)* $\operatorname{Bir} X / \operatorname{Aut} X$ *is generated by the rational involution* $\tau$ *just described.*

**Proof**  Let $X' \to S'$ be a Mori fibre space, and $\varphi \colon X \dashrightarrow X'$ a birational map. Choose a very ample complete linear system

$$\mathcal{H}' = |-\mu' K' + A'|$$

on $X'$, where $A'$ is the pullback of a divisor ample on the base $S'$, and let $\mathcal{H} \subset |-\mu K|$ be the proper transform on $X$. In the first part of the proof (Steps 1 and 2 below) I classify the possible maximal centres of $\mathcal{H}$; in the second part (Step 3) I finish by invoking the main statement of the Sarkisov program.

STEP 1  No curve on $X$ is a maximal centre of $\mathcal{H}$. If a curve $C$ is a maximal centre, there is a maximal extraction $E \subset Y \to C \subset X$. By Exercise 2.2, (1), $c = \operatorname{mult}_C \mathcal{H} > \mu$ and, by Kawamata's Theorem 3.9, $C$ is contained in the smooth locus of $X$. In particular then $\deg C$ is a positive integer. Let $Z$ be the intersection $H_1 \cdot H_2$ of two general members $H_1$, $H_2$ of $\mathcal{H}$. Intersecting with a general surface $S \in |-K|$ gives

$$\frac{5}{2}\mu^2 = Z \cdot S \geq c^2 \deg C > \mu^2 \deg C,$$

so that $\deg C = 1$ or $2$.

If $\deg C = 1$, we can choose coordinates on $\mathbb{P}$ so that $C$ is given by $y = x_0 = x_1 = 0$. The blowup $f \colon E \subset Y \to C \subset X$ of the ideal sheaf of $C$ is the only divisorial contraction centred at $C$, and $C$ is the base locus of $|I_C(-2K)|$; therefore the class

$$M = f_*^{-1}|I_C(-2K)| = -2K - E$$

is nef on $Y$. An easy calculation shows

$$\begin{aligned}
M \cdot Z &= (-2K - E)(-\mu K - cE)^2 \\
&= 2\mu^2(-K)^3 - (4\mu c + \mu^2)(-K)^2 \cdot E + (2c^2 + 2\mu c)(-K) \cdot E^2 - c^2 E^3 \\
&= 5\mu^2 - 2c^2 - 2\mu c - c^2 < 0,
\end{aligned}$$

a contradiction. In the calculation I used that $(-K)^3 = \frac{5}{2}$, $(-K)^2 \cdot E = 0$ (the projection formula), $(-K) \cdot E^2 = -\deg C = -1$ (also by the projection formula) and $E^3 = -\deg N_C X = -\deg C + 2 - 2p_a(C) = 1$.

If $\deg C = 2$ then $C$ is given in suitable coordinates by

$$y = x_3 = x_0 x_1 + x_2^2 = 0 \quad \text{or} \quad y^2 + f_4(x_0, x_1) = x_2 = x_3 = 0.$$

In the first case, we work as before with the class $M = -2K - E$, in the second with $M = -4K - E$. In either case it is easy to check that $M \cdot Z < 0$ (I leave the numerics as an exercise to the reader) and, as above, $C$ cannot be a maximal centre.

STEP 2  No smooth point $P \in X$ is a maximal centre of $\mathcal{H}$.

Assume first that $P$ does not lie on any of the 15 lines $\Gamma_i \subset X$ described above. Then the linear system

$$|I_P(-K)|$$

isolates $P$. Intersecting with a general member $S \in |I_P(-K)|$, using Corollary 3.4, (2), we get

$$\frac{5}{2}\mu^2 = Z \cdot S \geq (Z \cdot S)_P > 4\mu^2$$

(where, as usual, $Z = H_1 \cdot H_2$ is the intersection of 2 general members of $\mathcal{H}$), which is a contradiction. Assume now that $P \in \Gamma$, one of the 15 lines. Even though the base locus of $|I_P(-K)|$ is $\Gamma$, we still choose a general surface

$$S \in |I_P(-K)|$$

and use it as a "test surface". Restricting $\mathcal{H}$ to $S$ we obtain

$$\mathcal{H}_{|S} = c\Gamma + \mathcal{L},$$

where $\mathcal{L}$ is a mobile linear system on $S$ and $c = \text{mult}_\Gamma \mathcal{H}$.

The argument now proceeds in 3 straightforward steps as usual: first, by 3.4, (1) and 3.1, two general members $L_1, L_2$ have a large intersection at $P$; second, we estimate $\mathcal{L}^2 = L_1 \cdot L_2$ from above; third, we derive a contradiction from the 2 previous steps.

(a) We are assuming that $P$ is a maximal centre; therefore, by 3.4, (1), $K_S + \frac{c}{\mu}\Gamma + \frac{1}{\mu}\mathcal{L}$ is not log canonical and by 3.1

$$(L_1 \cdot L_2)_P > 4\left(1 - \frac{c}{\mu}\right)\mu^2.$$

(b) $S$ has an ordinary double point at $Q = (0,0,0,0,1)$ and

$$(K_S \cdot \Gamma)_S = (K + S \cdot \Gamma)_X = 0,$$

so that

$$K_S + \Gamma_{|_\Gamma} = K_\Gamma + \text{Diff} = K_\Gamma + \frac{1}{2}Q$$

and

$$(\Gamma \cdot \Gamma)_S = -2 + \frac{1}{2} = -\frac{3}{2}.$$

From this we can calculate

$$L_1 \cdot L_2 = (-\mu K_{|_S} - c\Gamma)_S^2 = \frac{5}{2}\mu^2 - \mu c - \frac{3}{2}c^2.$$

(c) From the 2 previous steps, since $\mathcal{L}$ is free from base curves, we conclude that

$$4\left(1 - \frac{c}{\mu}\right)\mu^2 < \frac{5}{2}\mu^2 - \mu c - \frac{3}{2}c^2,$$

a contradiction.

STEP 3: CONCLUSION    According to the Sarkisov program there is a Mori fibre space $X_1 \to S_1$ and a link $\psi_1 \colon X \dashrightarrow X_1$ such that $\deg \varphi \psi_1^{-1} < \deg \varphi$. Now $X$ is a Fano 3-fold, so we must be in Case (1) of the Noether–Fano–Iskovskikh inequalities 2.4. Then the link starts with a maximal extraction and, by what I have proved and Theorem 3.9, the extraction in question is the Kawamata blowup $E \subset Y \to Q \in X$. The link, if it exists, is the unique answer to the 2-ray game starting with $Y \to X \to$ pt. But we know that the link $\tau$ is an answer to this 2-ray game. Therefore the link in question is $\tau$.   $\square$

## 6.2   Rigid Fano 3-fold complete intersections

Let $X = X_{2,3} \subset \mathbb{P}^5$ be a smooth complete intersection of a quadric and a cubic in $\mathbb{P}^5$. Every line $\Gamma \subset X$ is the centre of a link

$$\tau_\Gamma \colon X \dashrightarrow X$$

that can be understood as follows. Let $\pi_\Gamma \colon Y \to X$ be the blowup of $\Gamma$. The proper inverse image of the linear system $\delta(\Gamma)$ of hyperplane sections through $\Gamma$ is free on $Y$ and defines a generically 2-to-1 map $\varphi_{\delta(\Gamma)} \colon Y \to \mathbb{P}^3$

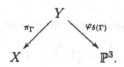

If $\sigma: Y \to Y$ is the biregular involution of $Y$ exchanging the two sheets of $\varphi_{\delta(\Gamma)}$ then we define

$$\tau_\Gamma = \pi_\Gamma \sigma \pi_\Gamma^{-1}.$$

We leave it to the reader to construct a link $\tau_\Gamma$ starting with *any conic* $\Gamma \subset X$.

In this section, I prove the following result.

**Theorem 6.3 (Iskovskikh, Pukhlikov [I1], [IP])** *If $X = X_{2,3} \subset \mathbb{P}^5$ is sufficiently general (the precise condition is stated in the beginning of the proof), then*

*(1) $X$ is birationally rigid, and*

*(2) Bir $X/$ Aut $X$ is generated by the rational involutions $\tau_\Gamma$ centred on lines and conics.*

**Proof** The logic of the proof is identical to 6.2. The same method shows that if a curve $C \subset X$ is a maximal centre then it must be a line or conic. In 6.4 below, which is the hardest part of the proof, I prove that a closed point $P \in X$ cannot be a maximal centre. The statement then follows from the Sarkisov program, in the same way as the conclusion of the proof of Theorem 6.2. $\square$

**Theorem 6.4 (Pukhlikov [IP])** *If $X = X_{2,3} \subset \mathbb{P}^5$ is sufficiently general, then no (smooth) point $P \in X$ can be a maximal centre.*

**Proof** The proof is very similar to the proof of Theorem 6.2, Step 2, and is conceptually an easy application of the technique of Chapter 3: reduction to surfaces via inversion of adjunction, and a calculation using Theorem 3.1. The new element is the use of the "bad line" of 3.5 to improve some inequalities (see Step 4 of the proof).

STEP 1. MAIN DIVISION INTO CASES In this proof I take $X$ general to mean the following. Let $P \in X$ be a point, $T_P = T_P X$ the tangent plane to $X$ at $P$, $C = T_P \cap X$ a curve of degree 6 in $T_P \cong \mathbb{P}^3$ with multiplicity 4 at $P$. Then we have one of the following 4 cases:

(a) $P \in C = \Delta \subset T_P \cong \mathbb{P}^3$ is an irreducible (rational) sextic with a 4-tuple point at $P$.

(b) $P \in C = \Delta + \Gamma \subset T_P \cong \mathbb{P}^3$ is the sum of an irreducible (rational) quintic with a triple point at $P$ and a line $P \in \Gamma$.

(c) $P \in C = \Delta + \Gamma_1 + \Gamma_2 \subset T_P \cong \mathbb{P}^3$ is the sum of an irreducible (rational) quartic with a double point at $P$ and two lines through $P$.

(d) $P \in C = \Delta + \Gamma_1 + \Gamma_2 + \Gamma_3 \subset T_P \cong \mathbb{P}^3$ is the sum of a twisted cubic and three lines through $P$.

In each case we may also assume that the singularity $P \in C \subset T_P$ is analytically equivalent to a cone over 4 general points in $\mathbb{P}^2$.

STEP 2. BASIC SET UP   For the rest of the proof, we let

$$F \subset Y \to P \in X$$

be the blowup of the maximal ideal; choose a general hyperplane section $S$ containing $C = T_P \cap X$,

$$P \in C \subset S \in |\mathcal{O}_X(1)|,$$

and let $S' \subset Y$ be the proper transform. Also denote by $\Delta'$, $\Gamma_i' \subset S'$ the proper transforms of $\Delta$, $\Gamma_i$ and by $\Phi = F|_{S'}$ the restriction of $F$ to $S'$. Then $\Delta'$, $\Gamma_i'$ and $\Phi$ all have self-intersection $-2$ on $S'$ and, in addition, it is easy to compile a multiplication table in each of the 4 cases:

(a) $\Delta' \cdot \Phi = 4$.

(b) $\Delta' \cdot \Gamma' = 1$, $\Gamma' \cdot \Phi = 1$ and $\Delta' \cdot \Phi = 3$.

(c) $\Gamma_1' \cdot \Gamma_2' = 0$, $\Delta' \cdot \Gamma_i' = 1$, $\Gamma_i' \cdot \Phi = 1$ and $\Delta' \cdot \Phi = 2$.

(d) $\Gamma_i' \cdot \Gamma_j' = 0$ (for $i \neq j$), $\Delta' \cdot \Gamma_i' = 1$, $\Gamma_i' \cdot \Phi = 1$ and $\Delta' \cdot \Phi = 1$.

Assuming that $P \in X$ is a maximal centre of a linear system $\mathcal{H} \subset |\mathcal{O}_X(\mu)|$ let $\mathcal{H}' \subset Y$ be the birational transform and

$$\mathcal{H}'|_{S'} = \mathcal{L}' + d\Delta' + \sum c_i \Gamma_i'$$

the restriction of $\mathcal{H}'$ to $S'$, where $\mathcal{L}'$ is free from base curves. We reach a contradiction in three steps as usual. First, we show that, because $\mathcal{H}$ has a high singularity at $P$, $P$ contributes a large amount to the intersection $\mathcal{L}'^2 = L_1' \cdot L_2'$ of two general members $L_1'$, $L_2'$ of $\mathcal{L}'$, which must therefore be large. Second, we use the multiplication table to estimate $\mathcal{L}'^2$ from above. Third, a contradiction is drawn from the previous steps by comparison.

In each case, write

$$\frac{1}{\mu}\mathcal{L}' \sim \varphi\Phi + \delta\Delta + \sum \gamma_i \Gamma_i,$$

with $\varphi = 2 - \frac{m_E(\mathcal{H})}{\mu} \leq 1$, $\delta = 1 - \frac{d}{\mu} \leq 1$ and $\gamma_i = 1 - \frac{c_i}{\mu} \leq 1$. Note that necessarily $\delta \geq 0$ (as observed above, $\Delta$ cannot be a maximal centre) and $\gamma_i < 0$ can only happen if $\Gamma_i$ is a maximal centre. After untwisting by finitely many involutions $\tau_\Gamma$ centred on lines and conics I may and will assume that no curve on $X$ is a maximal centre; hence $\gamma_i \geq 0$. (This manoeuvre is not logically strictly necessary but it cuts down the complexity of the calculations by a factor of 2.)

From this and the multiplication table we can calculate $\frac{1}{\mu^2}\mathcal{L}'^2$, in the four main cases, as follows:

(a) $\frac{1}{\mu^2}\mathcal{L}'^2 = -2\varphi^2 - 2\delta^2 + 8\varphi\delta$;

(b) $\frac{1}{\mu^2}\mathcal{L}'^2 = -2\varphi^2 - 2\delta^2 - 2\gamma^2 + 6\varphi\delta + 2\varphi\gamma + 2\delta\gamma$;

(c) $\frac{1}{\mu^2}\mathcal{L}'^2 = -2\varphi^2 - 2\delta^2 - 2\gamma_1^2 - 2\gamma_2^2 + 4\varphi\delta + 2\varphi\gamma_1 + 2\varphi\gamma_2 + 2\delta\gamma_1 + 2\delta\gamma_2$;

(d) $\frac{1}{\mu^2}\mathcal{L}'^2 = -2\varphi^2 - 2\delta^2 - 2\gamma_1^2 - 2\gamma_2^2 - 2\gamma_3^2$
$\qquad\qquad +2\varphi\delta + 2\varphi\gamma_1 + 2\varphi\gamma_2 + 2\varphi\gamma_3 + 2\delta\gamma_1 + 2\delta\gamma_2 + 2\delta\gamma_3$.

STEP 3. $F$ IS A MAXIMAL SINGULARITY   I do this case first as a warm-up. The assumption means that $\varphi < 0$. I draw a contradiction from the assumption that $\frac{1}{\mu^2}\mathcal{L}'^2 \geq 0$, using the expressions derived in the previous step. I do this only in Case (d) since the arithmetic in the other cases is a specialisation of this case (and, in any event, it is easier). We have

$$-\frac{1}{\mu^2}\mathcal{L}'^2 = 2\varphi^2 + 2\delta^2 + 2\gamma_1^2 + 2\gamma_2^2 + 2\gamma_3^2 -$$
$$- 2\varphi\delta - 2\varphi\gamma_1 - 2\varphi\gamma_2 - 2\varphi\gamma_3 - 2\delta\gamma_1 - 2\delta\gamma_2 - 2\delta\gamma_3$$
$$= (\varphi + \delta - \gamma_1 - \gamma_2)^2 + (\varphi + \delta - \gamma_3)^2 + (\gamma_1 - \gamma_2)^2 + \gamma_3^2 - 6\varphi\delta > 0,$$

a contradiction.

STEP 4. $F$ IS NOT A MAXIMAL SINGULARITY: $\mathcal{L}'^2$ MUST BE LARGE   In the 4 cases we get

$\qquad$ (a) $\qquad \frac{1}{\mu^2}\mathcal{L}'^2 > 8\varphi\delta$.

$\qquad$ (b) $\qquad \frac{1}{\mu^2}\mathcal{L}'^2 > \begin{cases} \text{either} & 8\varphi\delta \\ \text{or} & 4\varphi\delta + 4\varphi\gamma. \end{cases}$

$\qquad$ (c) $\qquad \frac{1}{\mu^2}\mathcal{L}'^2 > \begin{cases} \text{either} & 8\varphi\delta \\ \text{or} & 4\varphi\delta + 4\varphi\gamma_1 \\ \text{or} & 4\varphi\gamma_1 + 4\varphi\gamma_2. \end{cases}$

$\qquad$ (d) $\qquad$ Same as (c).

I prove this assuming that we are in Case (d) (the other cases are similar and left to the reader). Consider the divisor

$$K_{S'} + D = \left(K_Y + S' + (1 - \varphi)F + \frac{1}{\mu}\mathcal{H}'\right)_{|S'}$$

$$= K_{S'} + (1 - \varphi)\Phi + (1 - \delta)\Delta + \sum(1 - \gamma_i)\Gamma_i' + \frac{1}{\mu}\mathcal{L}'.$$

The crucial point here is that $P \in S$ is a double point; hence $S' \cap F \subset F \cong \mathbb{P}^2$ is a conic. Therefore, by Corollary 3.5, there are *two* points $Q_1, Q_2 \in \Phi \subset S'$ where $K_{S'} + D$ is not log canonical. The statement follows from 3.1 and a straightforward division into cases. In more detail, up to renumbering, one of the following 3 cases occurs:

**(d.1)** Neither of $Q_1, Q_2$ lies in any of the $\Gamma_i'$. By 3.1, the contribution $\mathcal{L}_{Q_i}'^2$ is either $> 4\mu^2\varphi$ or $> 4\mu^2\varphi\delta$, and hence $\geq 4\mu^2\varphi\delta$ anyway.

**(d.2)** $Q_1 \in \Gamma_1'$, while $Q_2$ is not on any $\Gamma_i'$. Then, by 3.1, $\mathcal{L}_{Q_1}'^2 > 4\mu^2\varphi\gamma_1$ and $\mathcal{L}_{Q_2}'^2 > 4\mu^2\varphi\delta$.

**(d.3)** $Q_1 \in \Gamma_1'$, $Q_2 \in \Gamma_2'$. Then, by 3.1, $\mathcal{L}_{Q_1}'^2 > 4\mu^2\varphi\gamma_1$, and $\mathcal{L}_{Q_2}'^2 > 4\mu^2\varphi\gamma_2$.

(5) $F$ IS NOT A MAXIMAL SINGULARITY: CONCLUSION   I now conclude the proof in Case (d) only, since the arithmetic in the other cases is a specialisation of this case (and, in any event, it is easier). Using the calculation for $\frac{1}{\mu^2}\mathcal{L}'^2$ in Step 2 above we get, in cases (d.1), (d.2)

$$0 > -\frac{1}{\mu^2}\mathcal{L}'^2 + 4\varphi\delta$$

$$= 2\varphi^2 + 2\delta^2 + 2\gamma_1^2 + 2\gamma_2^2 + 2\gamma_3^2 +$$

$$+ 2\varphi\delta - 2\varphi\gamma_1 - 2\varphi\gamma_2 - 2\varphi\gamma_3 - 2\delta\gamma_1 - 2\delta\gamma_2 - 2\delta\gamma_3$$

$$= (\varphi + \delta - \gamma_2 - \gamma_3)^2 + (\varphi - \gamma_1)^2 + (\delta - \gamma_1)^2 + (\gamma_2 - \gamma_3)^2,$$

a contradiction. In Case (d.3) we get

$$0 > -\frac{1}{\mu^2}\mathcal{L}'^2 + 4\varphi\gamma_1 + 4\varphi\gamma_2$$

$$= 2\varphi^2 + 2\delta^2 + 2\gamma_1^2 + 2\gamma_2^2 + 2\gamma_3^2 +$$

$$- 2\varphi\delta + 2\varphi\gamma_1 + 2\varphi\gamma_2 - 2\varphi\gamma_3 - 2\delta\gamma_1 - 2\delta\gamma_2 - 2\delta\gamma_3$$

$$= (\varphi + \gamma_1 + \gamma_2 - \delta)^2 + (\gamma_1 - \gamma_2)^2 + (\varphi - \gamma_3)^2 + (\delta - \gamma_3)^2,$$

a contradiction.   $\square$

## 6.3   Failure of rigidity

Let $X = X_4 \subset \mathbb{P}^4$ be a quartic 3-fold with a single ordinary double point $P \in X$. Projection from the node generates a link (which is a birational involution) $\tau$ of $X$. It is easy to see that there are 24 lines $P \in \Gamma_i \subset X$ each of which generates a link (which also turns out to be a birational involution) $\sigma_i \colon X \dashrightarrow X$ from the 2-ray game beginning with the extraction $E_i \subset Y \to \Gamma_i \subset X$.

**Theorem 6.5 (Pukhlikov [P3])** *Let $X = X_4 \subset \mathbb{P}^4$ be a quartic 3-fold with a single ordinary double point $P \in X$.*

*(1) $X$ is birationally rigid;*

*(2) $\tau$ and the $\sigma_i$ generate $\operatorname{Bir} X / \operatorname{Aut} X$.*

**Proof** Along the same lines as 6.2 and 6.3. Using the same method as Theorem 6.2, Step 1, it is easy to see that, if a curve $C$ is a maximal centre, then it must be a line through $P$. By 3.6 no smooth point $Q \in X$ can be a maximal centre. We then conclude verbatim as in the proof of 6.2 (conclusion) bearing in mind that, by 3.10, the only divisorial contraction $E \subset Y \to P \in X$, landing the exceptional divisor into $P$, is the blowup of the maximal ideal. $\square$

This is encouraging, but there is a point where things go wrong.

**Example 6.6** Let $X = X_4 \subset \mathbb{P}^4$ be a quartic 3-fold with a singular point $P \in X$ of the form

$$x^2 + y^2 + z^4 + t^4 = 0.$$

$\lambda x + \mu y = 0$ defines a pencil of rational surfaces on $X$. The 2-ray game beginning with the weighted blowup

$$\varepsilon \colon E \subset W \to P \in X$$

with weights 2,2,1,1 links $X$ to a Del Pezzo fibration of degree 2 with two unremovable index 2 points.

When the singularity degenerates to

$$x^2 + y^3 + z^4 + t^4 = 0,$$

the 2 singular points coalesce into an exotic index 2 terminal singularity and the 2 fibres they live in to an exotic fibre; cf. [C1].

**Example 6.7** Consider a quasismooth 3-fold $X_7 \subset \mathbb{P}(1,1,1,2,3)$ with coordinates $x_i, y, z$. In [CPR] we construct a link $\tau\colon X \dashrightarrow X$ centred at the point $P_3 = (0,0,0,1,0)$ and show that, as the coefficient of the monomial $y^2z$ degenerates to zero, the birational involution $\tau$ disappears (see [CPR], 7.4.2 for a further discussion of this phenomenon).

I find the above examples relatively encouraging, since they seem to suggest that birational rigidity is a fairly stable property. This has led me to Conjecture 1.4, that the property is open in moduli.

I would like to ask the following questions, all of which are accessible in principle using the methods of this paper, aiming to explore ways in which rigidity can break down.

(1) Determine which singular quartics (with $\mathbb{Q}$-factorial terminal singularities) are birationally rigid, and describe alternative Mori fibre space models for the ones which are not (joint work in progress with Massimiliano Mella). Extend this to all 95 Fano hypersurfaces.

(2) Extend [CPR] to all Fano 3-folds in codimension 2 complete intersections [IF], codimension 3 Pfaffians and codimension 4 (Altınok [A]).

Basically, I want to see what happens when the singularities become worse, or the structure of the anticanonical ring more complicated. The reader of [CPR] may feel overwhelmed by the large volume of calculations needed (at least at present) to pursue this type of question and wonder what the point of it all is. I believe that it is important to understand how birational rigidity fails: for example, when can it happen that a Fano 3-fold is birational to a strict Mori fibration, and what is the meaning of the condition on $K^2$ in Theorems 4.2 and 5.1? It is certainly possible that this corner of nature is just chaotic. My hope is that once sufficiently many examples are understood it will be possible to see some structure.

# References

[A]     S. Altınok, *Graded rings corresponding to polarised K3 surfaces and $\mathbb{Q}$-Fano 3-folds*, Univ. of Warwick Ph.D. thesis, Sep. 1998, 93 + vii pp.

[Be1]   A. Beauville, *Prym varieties and the Schottky problem*, Inv. Math. **41** (1977), 149–196

[Be2]   A. Beauville, *Variétés de Prym et Jacobiennes intermédiaires*, Ann. scient. Éc. Norm. Sup. **10** (1977), 309–391

[BM]   A. Bruno & K. Matsuki, *Log Sarkisov program*, Internat. J. Math. **8**, 1997, 451–494

[C1]   A. Corti, *Del Pezzo surfaces over Dedekind schemes*, Annals of Math. **144** (1996), 641–683

[C2]   A. Corti, *Factoring birational maps of 3-folds after Sarkisov*, Jour. Alg. Geom. **4**, 1995, 223–254

[CPR]  A. Corti, A. Pukhlikov & M. Reid, *Fano 3-fold hypersurfaces*, this volume, 175–258

[IF]   A. R. Iano-Fletcher, *Working with weighted complete intersections*, this volume, 73–173

[I1]   V. A. Iskovskikh *Birational automorphisms of three-dimensional algebraic varieties*, J. Soviet Math. **13**, 1980, 815–868

[I2]   V. A. Iskovskikh, *On the rationality problem for conic bundles*, Duke Math. J. **54** (1987), 271–294

[I3]   V. A. Iskovskikh, *Towards the problem of rationality of conic bundles*, Algebraic Geometry, Chicago 1989 (Proc. of the US–USSR Symp.), Springer LNM **1479**, 50–56

[IM]   V. A. Iskovskikh & Yu. I. Manin, *Three-dimensional quartics and counterexamples to the Lüroth problem*, Math. USSR Sb. **15**, 1971, 141–166

[IP]   V. A. Iskovskikh & A. V. Pukhlikov, *Birational automorphisms of multidimensional algebraic manifolds*, J. Math. Sci. **82**, 1996, 3528–3613

[KMM]  J. Kollár, Y. Miyaoka & S. Mori, *Rationally connected varieties*, J. Alg. Geom. **1**, 1992, 429–448

[Ka]   Y. Kawamata, *Divisorial contractions to 3-dimensional terminal quotient singularities*, Higher dimensional complex varieties (Trento 1994), 241–245, de Gruyter, Berlin, 1996

[Ko1]  J. Kollár, *Flips, flops, minimal models, etc.*, Surv. in Diff. Geom. **1** (1991), 113–199

[Ko2]  J. Kollár, et al., *Flips and abundance for algebraic threefolds*, Astérisque **211** (1993)

[Ko3]   J. Kollár, *Integral polynomials, equivalent to a given polynomial*, Electron. Res. Announc. Amer. Math. Soc. **3** (1997), 17–27 (electronic)

[P1]    A. V. Pukhlikov, *Birational isomorphisms of four-dimensional quintics*, Inv. Math. **87** (1987), 303–329

[P2]    A. V. Pukhlikov, *Birational automorphisms of a double space and double quadric*, Math. USSR Izv. **32**, 1989, 233–243

[P3]    A. V. Pukhlikov, *Birational automorphisms of a three-dimensional quartic with a quadratic singularity*, Math. USSR Sbornik **66** (1990), 265–284

[P4]    A. V. Pukhlikov, *Essentials of the method of maximal singularities*, this volume, 73–100

[P5]    A. V. Pukhlikov, *Birational automorphisms of three-dimensional Del Pezzo fibrations*, Warwick preprint 30/1996

[P6]    A. V. Pukhlikov, *Birational automorphisms of algebraic varieties with a pencil of cubic surfaces*, MPI preprint 1996

[Sa1]   V. G. Sarkisov, *Birational automorphisms of conic bundles*, Math. USSR Izv. **17** (1981), 355–390

[Sa2]   V. G. Sarkisov, *On the structure of conic bundles*, Math. USSR Izv. **20**, 1982, 355–390

[Sh]    V. V. Shokurov, *Prym varieties: theory and applications*, Math. USSR Izv. **23** (1984), 83–147

Alessio Corti
DPMMS, University of Cambridge,
Centre for Mathematical Sciences,
Wilberforce Road, Cambridge CB3 0WB, U.K.
e-mail: a.corti@dpmms.cam.ac.uk

# Twenty five years of 3-folds
# – an old person's view

Miles Reid

大器晩成　*

# 1 Introduction

This paper is an expanded version of a historical and autobiographical talk given at the closing conference of the Warwick 3-folds activity in December 1995 (at the time still entitled "Twenty years ... "). It discusses the development of 3-folds since the 1970s, particularly my work on canonical singularities and Mori theory since around 1975.

Castelnuovo and Enriques's work establishing the foundational results on algebraic surfaces took place over something like 12 years around 1900. The classification of surfaces has undoubtedly progressed since then, and remains a vital area of research, but the foundations of the theory were laid during this relatively short period. The comparable foundational work on 3-folds took place over the 20 years between 1970 and 1990, from the first papers of Iitaka and Ueno on Kodaira dimension through to Mori's solution of the flip problem and his Fields medal at the Kyoto international congress. While future generations will undoubtedly simplify and extend the theory of 3-folds, and find new applications, it seems to be a matter of historical record that the primary issues were settled during the 1970s and 1980s.

A survey based on autobiography may be a relatively painless way to gain insight into the subject, compared to the investment of effort involved in working through a more serious account of Mori theory. The narrative form allows me to stress that, while the subject is not without its technical aspects, the key issues are mostly very simple once grasped; like other revolutionary ideas, canonical singularities and Mori theory build on classical foundations, and, with the benefit of hindsight, stand out clearly as inevitable

---

*It takes a long time to make a big pot.

developments. Looking through my old letters and papers for a retrospective sketch of my early career in 3-folds has given me some new insights, a bit like preparing the index to Volume 1: Foundations. Talk of "inevitable lines of thought" prompts the hope that some bolshy youngsters will mount their own revolution on my generation's orthodoxy (but I don't see them coming).

I don't really need to apologise for the self-indulgence of my undertaking: a whole generation of algebraic geometers now work in the classification of varieties, and it is hard to read a paper or hear a lecture in the subject without recognising concepts and terminology that I had a hand in shaping.

**History**    It is widely appreciated that mathematicians usually treat history in a curiously dishonest way, rewriting the history of the subject as it should have been discovered, aiming to exploit past experience in building a modern worldview. The essential difficulty seems to be that the story in strictly chronological order will not make sense to anyone; the writer wants to give an explanation based on the logical layout of the subject, whatever violence it does to historical truth. Writing this paper has certainly confirmed this point for me: given the choice between saying things chronologically or according to the divisions of the subject matter, my sense of style always leads me to the latter. Another inevitable slant of autobiographical writing is to exaggerate the significance of the author's discoveries, and I also follow this tradition.

My first introduction to algebraic geometry took place in my first year as a graduate student at Cambridge in 1969–70 under the benevolent and liberal care of Peter Swinnerton-Dyer. I learned many things directly from him: the factorisation of birational maps between del Pezzo surfaces, Hodge theory and its application to Abelian varieties. A particular item that stands out is the derivation of the Weierstrass model of an elliptic curve from Riemann–Roch: the graded ring $R(E, P) = k[x, y, z]/(z^2 = y^3 + ax^4y + bx^6)$ and its Proj, the weighted hypersurface $E_6 \subset \mathbb{P}(1, 2, 3)$, has served as the model for all my graded ring calculations.

On Swinnerton-Dyer's advice I studied Mumford's "little red book" in a joint seminar with Jean-Louis Colliot-Thélène and Barry Tennison. My work on canonical 3-folds could reasonably be traced back to Mumford's question ([red], Chapter Three, §8; this follows a treatment of normalisation as a method of resolving surface singularities):

> *The existence of such a simple way of "making" every variety normal is one of the reasons why normal varieties are an important class. Life would be much simpler if there were an analogous way of canonically constructing a nonsingular variety birationally equivalent to any given variety.*

# 2 Canonical models are inevitable

The main discovery of my early career in 3-folds was the realisation that singularities arise as an inevitable part of higher dimensional life. They turn up in two different ways: by *equations* and by *cyclic quotient constructions*. Already the simplest cases are fun, and understanding them properly takes us a long way into the substantial issues. The ordinary quadric cone arises both as the simplest hypersurface singularity $X : (xy = z^2) \subset \mathbb{C}^3$ and as the simplest quotient singularity, the quotient of $\mathbb{C}^2$ by $(u, v) \mapsto (-u, -v)$. Its resolution $Y \to X$ is the ordinary blowup, having exceptional locus $E \cong \mathbb{P}^1$ with $E^2 = -2$ (a $-2$-curve), and it is well known that the canonical class of $Y$ is trivial. Because $K_Y = K_X = 0$ this is a crepant resolution of a strictly canonical singularity. The 3-fold ordinary double point $(x^2 + y^2 + z^2 + t^2 = 0)$ or $(xy = zt)$ discussed in Section 2.1 behaves in wholly different ways: it is not a finite quotient singularity, it is terminal, its blowup does not have trivial canonical class, it has the two small resolutions discussed in Section 3.2 that are related by the classic flop. It is completely unrelated to the simplest 3-fold quotient singularity $\frac{1}{2}(1, 1, 1)$, the quotient of $\mathbb{C}^3$ by $(u, v, w) \mapsto (-u, -v, -w)$ or the cone over the Veronese surface discussed in Section 2.2.

The next point concerns canonical rings. As I explain in Section 2.3, my interpretation of Enriques' work on surfaces of general type underlines the importance of working with pluricanonical graded rings, rather than just the 1-canonical or the individual $m$-canonical linear systems. Zariski and Mumford's proof of the finite generation of canonical rings of surfaces and Grothendieck's definition of Proj frees us from the obligation of working only with varieties with a fixed embedding in projective space and graded rings generated by elements of degree 1. The first canonical 3-folds generalise Enriques' models in a natural way. They make clear, for example, that the birational invariant $K^3$ that controls the asymptotic growth of the plurigenera is a rational number. Once this point is recognised, there is really no turning back to the nonsingular category.

## 2.1 Lefschetz theory and the ordinary double point

My first main point is the following:

> *The ordinary double point has entirely different behaviour and meaning in different dimensions.*

The ordinary double point (ODP)

$$X : \left( \sum_{i=0}^{n+1} x_i^2 = 0 \right) \subset \mathbb{C}^{n+1}.$$

is the simplest $n$-fold singularity. The function $\mathbb{C}^{n+1} \to \mathbb{C}$ given by $f(\underline{x}) = \sum x_i^2$ has a *nondegenerate critical point* at 0: critical because all $\partial f/\partial x_i = 0$, and nondegenerate because the Hessian matrix of second derivatives is a nondegenerate quadratic form. Over $\mathbb{R}$, these are just the properties of a Morse function. Lefschetz theory is a complex analog of Morse theory. However, in many contexts of classification, we may be more interested in the singular $n$-fold $X$ than in the function $f(\underline{x})$.

The basic idea of Lefschetz theory is to argue on the neighbouring fibre $X_t = f^{-1}(t)$ for $t$ close to 0. Let's work in a small neighbourhood $\sum |x_i|^2 \leq \delta$ of the singular point, and take $t = \varepsilon e^{i\theta}$ where $0 < \varepsilon \ll \delta \ll 1$. Then from the $C^\infty$ point of view, $X_t$ and $X_0$ practically coincide outside the $\delta$-ball, and $M = X_t \cap \delta$-ball (the *Milnor fibre* of the singularity) is given by quadratic equations. Scaling and separating real and imaginary parts puts the equations of $M$ in the form

$$\sum u_i^2 = 1 \text{ and } \sum u_i v_j = 0, \quad \text{where} \quad x_i = (\text{modulus}) \times e^{i\theta} \times (u_i + iv_i).$$

First, the things that are common to all dimensions: the Milnor fibre is the tangent bundle to $S^n$ (see Figure 2.1.1), the vanishing cycle is $\delta = S^n$, the monodromy as we follow the diffeomorphism of the fibres around $e^{i\theta}$ is given by the Picard–Lefschetz formula $x \mapsto x \pm (x \cdot \delta)\delta$.

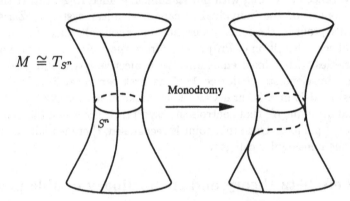

Figure 2.1.1: The ODP according to Lefschetz: the Milnor fibre is diffeomorphic to the tangent bundle $T_{S^n}$ of the vanishing cycle $\delta = S^n$, and the monodromy action is given by the Picard–Lefschetz formula $x \mapsto x \pm (x \cdot \delta)\delta$

For the purposes of birational geometry, we want to resolve the singularity; then each dimension is quite different. In the curve case, the ODP is the node, consisting of two branches meeting only at a point. The resolution of singularities or normalisation consists simply of separating the two branches.

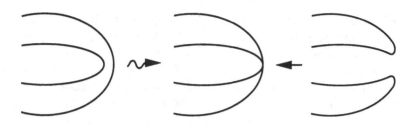

Figure 2.1.2: Picture of a curve losing 1 from its genus by acquiring an ODP, then disconnecting

This leads to the standard picture of genus change in a degeneration of Riemann surfaces (see Figure 2.1.2).

Something beautiful and completely unique happens in the surface case: the Milnor fibre $M$ is diffeomorphic to the resolution $Y \to X$. The ODP is the ordinary quadratic cone in $\mathbb{C}^3$. Its resolution is a neighbourhood of a copy of $\mathbb{P}^1_{\mathbb{C}} = S^2_{\mathbb{R}}$ with self-intersection $-2$. At the level of topology, the diffeomorphism is clear from Figure 2.1.1, because the Milnor fibre is the tangent bundle to the vanishing cycle $S^2_{\mathbb{R}}$, which also has self-intersection $-2$.

The 3-fold ODP demonstrates originality by having a nontrivial analytic class group: up to the obvious change of coordinates, it is the cone over the quadric surface $Q \cong \mathbb{P}^1 \times \mathbb{P}^1 : (xy = zt) \subset \mathbb{P}^3$, and the two linear systems of generators of $Q$ correspond to Weil divisors that are not factorial; for example, the principal divisor $x = 0$ breaks up as the union of two planes $x = z = 0$ and $x = t = 0$, neither of which is principal. The local class group can be viewed topologically as the second cohomology of the link of the singularity. As I discuss in Section 3.3, the two rulings of $Q$ give rise to the two small resolutions of the 3-fold ODP and the flop between them.

It is also interesting to run through a parallel discussion of the $A_2$ singularity $\sum_{i=1}^{n+1} x_i^2 + x_0^3 = 0$. In the curve case, this singularity is a cusp, which is locally irreducible. The Milnor fibre has two vanishing cycles, but in the 3-fold case, they have nondegenerate intersection pairing, and the resulting singularity is factorial; in fact, it is a homology manifold. It has no small modification. Its blowup is a nonsingular 3-fold containing an exceptional surface $E$ isomorphic to the quadric cone $Q \subset \mathbb{P}^3$, and with $\mathcal{O}_E(-E) \cong \mathcal{O}_Q(1)$. See Mirel Caibăr's thesis [Caibăr] for a much more detailed discussion of the relation between Milnor fibre, class group, and geometry of the resolution.

**Problem on higher dimensional ODPs**   I propose as a challenge to try to understand the new characteristic features of 4-fold and 5-fold ODPs. For example, the 5-fold ODP is the cone over the Plücker quadric $Q_6 =$ Grass$(2, 4) \subset \mathbb{P}^5$, and the analog of the class group is the two tautological

vector bundles of Grass(2, 4) or the spinor bundles of $Q_6$. My reason for saying this is that a number of people have looked for simple-minded extensions
to higher dimensions of features such as simultaneous resolution of Du Val
surface singularities and small resolutions of 3-fold cDV points, but I really
don't know of any reasonable class of higher dimensional examples where
this is likely to give useful results. For anyone wanting to study 4-fold or
5-fold analogs of the theory of surfaces and 3-folds, my problem seems a more
profitable direction.

**History**    The ODP $y_1y_2 = y_3y_4$ and its two rational functions $\frac{y_1}{y_3} = \frac{y_4}{y_2}$ and
$\frac{y_1}{y_4} = \frac{y_3}{y_2}$ occur in Mumford [red], Chap. I, §4, Remark IV after Proposition 1
in the discussion of rational functions. I was a graduate student at the IHES
between 1970 and 1972, and heard about the ideas of Picard–Lefschetz from
lectures by my great teacher Pierre Deligne, around the time he was preparing
Part II of SGA7 for publication [D].

The first project that Deligne offered me, global Torelli and the surjectivity
of the period map for K3 surfaces of degree 2 or 4, turned out to be beyond
my technical capabilities, but I learned a vast amount in my first year by
studying it, among other things: classification of surfaces, moduli spaces and
G.I.T., Hodge theory including variation and degeneration of Hodge structures, Satake compactifications. Deligne taught me about Du Val singularities
(including the name, in case you think I use it out of patriotism to the UK
or to Trinity College, Cambridge). An important point in the Torelli project
was the idea that K3s with only Du Val singularities should be viewed as
points "at finite distance" in the moduli space, since they have monodromy
of finite order, and their resolutions are still K3s. Equally well, you could
start learning to live with the singular variety.

## 2.2    The Veronese cone

I change tack, to discuss the first type of canonical singularity arising as a
quotient. Many classical construction of surfaces, such as those of Kummer
surfaces, Enriques surfaces or Godeaux surfaces, involve a group action having
no fixed points, or only involving ODPs.

*Simple constructions for surfaces involving quotients by a group
action often have 3-fold analogs having quotient singularities, for
example, the Veronese cone singularity.*

The first 3-fold quotient singularity to come to my attention was the
Veronese cone. As everyone knows, the Veronese surface $V_4 \subset \mathbb{P}^5$ is the image
of $\mathbb{P}^2$ under the embedding given by all quadratic monomials. To see it by

equations, we take coordinates $x_1, x_2, x_3$ on $\mathbb{P}^2$ and write down

$$M = \begin{pmatrix} u_{11} & u_{12} & u_{13} \\ u_{12} & u_{22} & u_{23} \\ u_{13} & u_{23} & u_{33} \end{pmatrix} \quad \text{where} \quad u_{ij} = x_i x_j.$$

Here $M$ is the generic symmetric $3 \times 3$ matrix. The Veronese surface is defined by rank $M \le 1$. On the other hand, it's obvious that the projective coordinate ring of $V_4$ is $k[V_4] = k[x_1^2, x_1 x_2, \ldots, x_3^2]$. This is, of course, the ring of even polynomials in $x_1, x_2, x_3$, or equivalently, the ring of invariants of $\{\pm 1\}$ acting by $(x_1, x_2, x_3) \mapsto (-x_1, -x_2, -x_3)$. Its Spec, the affine cone $X$ over $V_4$, is the quotient singularity $\frac{1}{2}(1, 1, 1)$ in modern terms. If you resolve it by a blowup $f \colon Y \to X$, you obtain a 3-fold $Y$ containing an exceptional surface $E \cong \mathbb{P}^2$ with normal bundle $\mathcal{O}_E(E) \cong \mathcal{O}_{\mathbb{P}^2}(-2)$; by the adjunction formula, $K_Y$ restricted to $E$ is $\mathcal{O}_{\mathbb{P}^2}(-1)$.

Since $K_Y$ has negative degree on $E$, all sections of $K_Y$ in a neighbourhood of $E$ vanish along $E$; it is an easy exercise in the same style to see that also all sections of $nK_Y$ vanish along $E$ with multiplicity $\ge n/2$. Thus

$$f_* \mathcal{O}_Y(nK_Y) = f_* \mathcal{O}_Y \left( nK_Y - \frac{n}{2} E \right) \quad \text{and} \quad 2K_Y = f^*(2K_X) + E.$$

If we start from an Abelian surface $A$ and pass to the quotient surface $\overline{S} = A/\{\pm 1\}$ by the action of $\{\pm 1\}$, then $\overline{S}$ has 16 ODPs at the fixed points of the group action, that is, the 2-torsion points of $A$. Resolving these leads to $-2$-curves on the minimal nonsingular model $S \to \overline{S}$, which is then a K3 surface, the *Kummer surface* of $A$. In contrast, the Kummer variety $\overline{X} = A/\{\pm 1\}$ of an Abelian 3-fold $A$ has 64 Veronese cone singularities at the 2-torsion points of $A$, and resolving these by the blowup $X \to \overline{X}$ is not necessarily such a good thing to do: you achieve nonsingularity at the expense of spoiling the property $2K_{\overline{X}} = 0$, replacing it by $2K_X = \sum E_i$ where $E_i$ are the exceptional $\mathbb{P}^2$s over the singularities.

The Veronese cone appears in many similar contexts: for example, a popular construction of the Enriques surface (possibly originally due to Burniat) is to divide a complete intersection of 3 quadrics $S = Q_1 \cap Q_2 \cap Q_3 \subset \mathbb{P}^5$ by the free action of $\mathbb{Z}/2$ acting by $\mathrm{diag}(1, 1, 1, -1, -1, -1)$ (think of diagonal quadrics for clarity). We can make $S$ avoid the two fixed planes $\mathbb{P}^2_{\pm}$ because we have three quadrics $Q_i$ to spare. If we try to generalise this construction to a 3-fold complete intersection of 3 quadrics in $\mathbb{P}^6$, one of the fixed components $\Pi$ must be $\mathbb{P}^3$ (at least), and we can't avoid having the 8 points $(Q_1 \cap Q_2 \cap Q_3) \cap \Pi$ as fixed points. This is one construction for Enriques–Fano 3-folds, that typically have 8 Veronese cone singularities.

**History** I shared an office for a few weeks with Bernard Saint-Donat on arriving at the IHES in 1970, and we both visited the Warwick Symposium run by David Mumford in 1970–1971; I gave a Cambridge Part III course around 1974 based on his paper on projective models of K3 surfaces [S-D]. This contains the example of the K3 surface $S_5 \subset \mathbb{P}(1^3, 2) \subset \mathbb{P}^6$ embedded in the Veronese cone; its general hyperplane section is a canonical curve of genus 6 with a $g_5^2$, that is, the Veronese embedding of a plane quintic, and so it is an exceptional case for which the projective image is not an intersection of quadrics (see [S-D], Theorem 7.2, (ii)). It has the $\mathbb{Q}$-Fano extension $X_5 \subset \mathbb{P}(1^4, 2)$ having the singularity $\frac{1}{2}(1, 1, 1)$, and when following Iskovskikh's work, I was at first surprised to find it excluded from the company of nonsingular Fanos ([I], Part II, Prop. 1.7 and §2).

As another example, in the context of Fano's study of anticanonical models of 3-folds $V_{2g-2} \subset \mathbb{P}^{g+1}$, Iskovskikh asked for an example of a nonsingular $V_{2g-2}$ containing a linearly embedded plane $\mathbb{P}^2$. If $E$ is such a plane then, by the adjunction formula, its normal bundle is given by $\mathcal{O}_E(-E) = \mathcal{O}_{\mathbb{P}^2}(2)$, so that $E$ should contract to a Veronese cone point. Examples such as $\mathbb{P}(1^3, 2)$, $X_3 \subset \mathbb{P}(1^4, 2)$ and $X_4 \subset \mathbb{P}(1^3, 2, 2)$ show that these really occur. (I wrote a letter to Iskovskikh in August 1978 attempting to prove that this cannot happen; but I found the examples soon after this.)

**The 1970–71 Warwick symposium** I followed Deligne to Warwick in the summer of 1971, and met Artin, Bombieri, Mumford, C.P. Ramanujam, Seshadri and many others; for example, I learnt more about G.I.T. from standing in a lunch queue with Seshadri than from several weeks' study of the book earlier that year. I took part in a seminar on etale cohomology organised by Saint-Donat and Anders Thorup, and had the embarrassing task of giving half-prepared lectures on the etale base change theorem with Mike Artin in the audience.

When Elaine Greaves threw me out of the old typewriter room (two clapped-out old mechanical typewriters, reserved for the use of graduate students), neither of us suspected that she would be my trusted henchman throughout the planning and organisation of the two subsequent Warwick algebraic geometry symposiums in 1982–83 and 1995–96. Several letters and rough drafts for [C3-f] were written on exactly the same typewriters when I arrived at Warwick in 1978. (In the meantime, our equipment budget has improved somewhat.)

## 2.3    Canonical rings and the first canonical 3-folds

There are many places in Enriques' work where he constructs a canonical surface by what seems to be an extraordinarily ingenious argument based on the geometry of the canonical or bicanonical map, but that becomes much more transparent when viewed in terms of graded rings.

*Graded rings such as the canonical ring of a variety of general type contain more information than individual linear systems (for example, m-canonical maps).*

The simplest illustration for my purposes is Enriques' construction of a surface $S$ with $p_g(S) = 4, q(S) = 0, K_S^2 = 6$ as the normalisation $S \to \overline{S}_6$ of a sextic surface $\overline{S}_6 \subset \mathbb{P}^3$ having an ordinary double curve along a plane cubic curve $C \subset \mathbb{P}^3$. My interpretation is as follows: the canonical ring of $S$ needs 4 generators $x_i \in H^0(K_S)$ in degree 1, then (at least) 1 more generator $y \in H^0(2K_S)$ in degree 2, simply because quadratic monomials $x_i x_j$ only provide 10 elements of $H^0(2K_S)$, whereas by Riemann–Roch

$$h^0(2K_S) = \chi(\mathcal{O}_S) + K^2 = 11.$$

Now we easily find that we need 2 relations in degree 3 and 4 between these generators, giving $S$ as the complete intersection $S_{3,4} \subset \mathbb{P}(1^4, 2)$. No ingenuity here, this is just the general complete intersection; we recover Enriques' model by assuming that the two relations are $x_0 y = a_3$ and $y^2 = b_4$, and eliminating $y$, giving the equation of the sextic as $\overline{S}_6 : (a_3^2 = x_0^2 b_4)$.

One of Enriques' most remarkable example of this kind is his construction of a surface of general type with $p_g(S) = 2, K_S^2 = 1$; in this case it is easy to see that the canonical system $|K_S|$ is a pencil of curves of genus 2 with a single transverse base point $P$, which is a Weierstrass point on each $C \in |K_S|$. Then $\mathcal{O}_C(2K_S) = K_C$, so that $|2K_S|$ maps each $C$ as a double cover of $\mathbb{P}^1$. Since the image $\mathbb{P}^1$s all contain the image of $P$, the image is the quadric cone $\varphi_{2K_S}(S) = Q \subset \mathbb{P}^3$, and $S$ is the double cover of $Q$ branched in the intersection of $C$ with a quintic and over the vertex. The modern construction is the weighted hypersurface $S_{10} \subset \mathbb{P}(1, 1, 2, 5)$.

Putting together the idea of canonical ring derived from Enriques' constructions and the experience of the Veronese cone described in Section 2.2, it was natural to look for examples such as $X_{14} \subset \mathbb{P}^4(1, 1, 2, 2, 7)$. This example can perfectly well be treated from the point of view of classical geometers: it has a canonical pencil $|K_X|$ consisting of surfaces $S_{14} \subset \mathbb{P}(1, 2, 2, 7)$. Each such surface has the equation $z^2 = f_7(x^2, y_1, y_2)$. If you substitute $X = x^2$ and $Z = xz$, and divide all the degrees by 2, this becomes the hypersurface $S_8 \subset \mathbb{P}(1, 1, 1, 4)$ (that is, $\mathbb{P}(2, 2, 2, 8)$) defined by $Z^2 = X f_7(X, y_1, y_2)$. In other words, my 3-fold has a pencil of surfaces with $p_g = 3$, $K^2 = 2$, each a double cover of $\mathbb{P}^2$ whose octic branch curve happens to split up as a septic plus a line, meeting of course in 7 nodes.

With what we know today, it is immediate that $X_{14}$ has 7 Veronese cone singularities along the $(y_1, y_2)$-axis. The striking thing here is that $K_X^3 = \frac{1}{2}$; in other words, the plurigenera $P_n(X)$ grow like $\frac{1}{6}n^3 \times \frac{1}{2}$. On the other hand, a 3-fold of general type that has a nonsingular model with $K_X$ nef has

$K_X^3$ a positive and even integer, so $K_X^3 \geq 2$. The plurigenera are manifestly birational invariants, and on a minimal model, we can use vanishing to deduce that $P_n$ grows like $\frac{1}{6}n^3 \times K_X^3$. Thus my example has no nonsingular model with $K_X$ nef. (Compare also Ueno [U].)

It might be a fun exercise to find a treatment in the old Italian style of examples such as $V_{6,10} \subset \mathbb{P}(1, 2^3, 3, 5)$, which has $p_g = 1$, $K^3 = 1/2$, and 15 Veronese cone points. The main point, however, is that canonical rings already give a simple construction of dozens of examples, with features that would be hard to study in other ways.

**History**    The inevitability of working with the canonical ring comes out clearly in the study of algebraic surfaces such as Godeaux surfaces, which I studied around 1974, following Bombieri's paper on canonical models and his lectures at the IHES and the Warwick symposium in 1970. For me, the pleasure of seeing how simple constructions with graded rings generate effortlessly many classes of algebraic surfaces was the main motivation for trying to do the same in dimension 3. It is clear that there is still a whole lot more mileage to be got out of these ideas.

After two years of my research fellowship at Christ's College, Cambridge, I applied to the British Council for a second year on their Soviet exchange, from September 1975. The Soviet ministry didn't come up with my placement by the due starting date for their own reasons, and when they did, it turned out to be at Minsk. It's not plausible that they excluded me from Moscow because of the bad company I kept during my first year, and much more likely that this was a tit-for-tat because a Cambridge department had refused to take some party stooge.

In October 1975, after finishing my paper on elliptic Gorenstein surface singularities [EllGor], and while hanging around Cambridge, waiting for my placement, I got interested in 3-fold examples of the above type, and even wrote a primitive computer program (in Fortran!) to search for examples of canonical 3-folds as hypersurfaces $X_d \subset \mathbb{P}^4(a_0, \dots, a_4)$ with all $a_i \mid d$ and $\sum a_i = d + 1$. Unfortunately, the only singularity I really believed in and was confident of being able to resolve was the Veronese cone, so that although my program found a few examples such as [Fl], 15.1, No. 14: $X_{18} \subset \mathbb{P}(1, 2, 2, 3, 9)$, which has $2 \times \frac{1}{3}(1, 1, 2)$ quotient singularities along the $z, t$ axis, I didn't recognise them as valid examples, and missed out on discovering terminal singularities of higher index. What a fool I was, Watson!

Notes written during my first month at Warwick in 1978 contain 21 cases of canonical complete intersections with Veronese cone points as their only singularities, including the example $V_{6,6,6} \subset \mathbb{P}(2^4, 3^3)$ with $p_g = 0$ discussed in [YPG], Example 2.9. The same notes and letters from the period conjectured falsely and repeatedly that the index of 3-fold canonical singularities is always $\leq 2$. In a letter around 1977, Ueno suggested I look at $\frac{1}{3}(1, 1, 1)$ for an index 3 singularity. Of course this has index 1, and I ignored the suggestion. But it is possible that what Ueno wrote was a misprint for $\frac{1}{3}(1,1,2)$; in any case, I would have stumbled on this somehow if I had interpreted his suggestion more sympathetically. Nick Shepherd-Barron discovered the singularities $\frac{1}{r}(1,1,r-1)$ of index $r$ in early 1979 in response to my false conjecture, and Dave Morrison provided the general case of terminal quotient singularities $\frac{1}{r}(1,a,r-a)$ in a letter of April 1980. I originally proposed to Nick that we should write [C3-f] as a joint paper, but he didn't want to, although he contributed several of the key ideas to it, in particular the first conjectural statement that the hyperplane section of a Gorenstein canonical singularity is either rational or elliptic. I realised at the Angers conference in July 1979 that we had a simple proof of this conjecture using nothing much more than the adjunction formula.

In the year after Angers I spent a couple of weeks in Bonn at the invitation of Van de Ven and Hirzebruch, where I discovered that I knew how to write out formally the "famous 95" list of K3 hypersurfaces.

# 3    Towards the definition of minimal model

The logic behind minimal models is much less clear-cut than for canonical models, and we arrived at the definition by interpolating between the canonical model and birational nonsingular models, with Mori and me working from opposite ends. (Compare [M3], Section 9 for a complementary historical discussion.) Whereas the preceding chapter argued the inevitability of canonical models, this chapter has the more modest task of putting together some considerations that make the definition of minimal model look at least reasonable. The material here does not use the Mori cone, and is mostly simpler than the eventual solution of the minimal model problem by Kawamata, Kollár, Mori, Shokurov and others (see Chapter 4).

Section 3.1 discusses Zariski decomposition of a divisor on a surface to explain the nef condition on a divisor. Section 3.2 goes back to the material of Section 2.1 to discuss small resolutions of ODPs and the classic flop, and Section 3.3 plays with examples of the construction of [C3-f] and [Pagoda] constructing partial resolutions of canonical singularities, in order to give some feeling for the condition that $K$ is nef. Section 3.4 discusses the special case of relative minimal models of surface fibrations.

## 3.1    Zariski's paper and nef divisors on surfaces

Zariski's paper [Z] on the asymptotic form of Riemann–Roch for a divisor on a surface forms a crucial bridge between the Italian tradition of surfaces and modern work on 3-folds. My attention was drawn to it by Mumford's appendix, referred to in Bombieri [B] for the proof of the finite generation of the canonical ring of a surface of general type. The paper stresses the notion of a divisor $D$ being *eventually free* (that is, the linear systems $|nD|$ is free for some $n > 0$), and its consequence *numerical eventually free* or *nef* (that is, $D\Gamma \geq 0$ for every curve $\Gamma \subset S$). Zariski introduces $\mathbb{Q}$-divisors, the idea of Zariski decomposition, and the related characterisation of quasieffective in terms of the existence of Zariski decomposition.

The degree of a divisor on a curve $C$ is an integer, so the notion of positive degree is not hard to find, and when $D$ has large degree, RR takes the simple form $h^0(C, D) = 1 - g + \deg D$. Zariski's aim is to study the Riemann–Roch space $H^0(S, nD)$ for a divisor on a surface, especially for $n \gg 0$; there is not much problem when $D$ is ample, so the whole point is to deal with negative features of $D$.

Assume that some multiple of $D$ is effective, say $mD = \sum d_i \Gamma_i$ for some $m > 0$ with $d_i > 0$. If $D\Gamma < 0$ for some curve on $S$ then $\Gamma$ must be one of the $\Gamma_i$, say $\Gamma_1$, and necessarily $\Gamma_1^2 < 0$. Then the fixed part of the linear system $|nD|$ contains $\Gamma_1$ with multiplicity $\geq na_1$, where $a_1 = (-D\Gamma_1)/(-\Gamma_1^2)$. In other words, decreasing $D$ down to $D_1 := D - a_1\Gamma_1$, where $a_1$ is chosen so that $D_1\Gamma_1 = 0$, leaves all the $H^0$ unchanged: $H^0(nD) = H^0(nD_1)$ for all $n > 0$. We thus find ourselves playing games with the quadratic intersection form on the components $\Gamma_i$, with the flavour of orthogonal complement or Gram–Schmidt orthogonalisation. The theory proceeds systematically via the *Zariski decomposition* of $D$, defined to be an expression $D = P + N$ where

(i) $P$ is a nef $\mathbb{Q}$-divisor;

(ii) $N$ is effective, negative definite and orthogonal to $P$.

The point is that the base part of the linear system $|nD|$ must be at least $nN$ for any $n > 0$, so that $h^0(nD) = h^0(nP)$. For any divisor $D$, the existence of a Zariski decomposition is an important dichotomy: it exists if and only if $D$ is quasieffective, and then $D\Gamma < 0$ happens for at most finitely many curves having a negative definite intersection pairing.

It seems to be a common misapprehension that just defining a Zariski decomposition in terms of orthogonal complement somehow makes it a one-off process in bilinear algebra, so I emphasise the following point:

*Calculating the Zariski decomposition of a divisor on a surface is an inductive process, and is a kind of minimal model program.*

Continuing the argument from above, if $D_1\Gamma_2 < 0$ we have to subtract off a multiple of $\Gamma_1$ and $\Gamma_2$ to make $D_2$ orthogonal to both $\Gamma_1$ and $\Gamma_2$, and so on by induction as long as $D_j$ is not nef. We don't know whether or not $\Gamma_2$ will fall in the negative part just by looking at $D$ and $\Gamma_2$ without running the inductive process. (I stress the point, at the risk of repetition: whether a rational curve $\Gamma \subset S$ on a surface with self-intersection $\Gamma^2 \leq -2$ is contracted on passing to the minimal model is not determined just by a neighbourhood of $\Gamma$, but by how many contracted $-1$-curves it meets in the course of $S \to S_1 \to \cdots$.)

If $S$ is a surface with $p_g > 0$ and $D = K_S$, then calculating the Zariski decomposition of $D$ is exactly the same thing as running a minimal model program on $S$. The first $\Gamma_1$ is a $-1$-curve because $K_S\Gamma_1 < 0$ and $\Gamma_1^2 < 0$. We subtract off a multiple of $\Gamma_1$, and from that point on always work in the orthogonal complement to $\Gamma_1$, which is equivalent to working only with the pullback of divisors from the contraction $S \to S_1$ of $\Gamma_1$. The negative part of the Zariski decomposition is the *discrepancy* $K_S - K_{S_{\min}}$ of the morphism to the minimal model $S \to S_{\min}$. Kawamata's work on log minimal models of log surfaces (or "noncomplete" surfaces) is a straightforward application of Zariski decomposition of log canonical divisors $K + D$.

**History**   I first visited Moscow for one year 1972–73 under the British Council exchange. By that time, Soviet citizens who talked to foreigners were no longer imprisoned or shot (or at least, not very much), but it was still not easy for a fairly ignorant and frivolous foreigner to get into serious conversation with people (except taxi drivers). Although I spent a lot of time with Moscow mathematicians, it was not until I got home that I realised that several of my colleagues had distinguished dissident connections. During my second visit in 1975–76 things were different: I was less ignorant, having read Solzhenitsyn and Robert Conquest on the Stalin terror and the even more terrible collectivisation holocaust. But more, I believe that most Russians were well informed and cynical about their political system, and keen to criticise it and to hear about the West. Even people who were party members would crack jokes such as "Is socialism scientific? No – they would have tried it out on animals first".

Minsk in 1975 was different; I was more or less the first capitalist seen in the town, and there were various confused reactions: some of my neighbours in the students' hostel were ordered by their komsomol organisers (communist youth league) not to consort with me. The foreign office in the university was splendidly friendly and incompetent, and cheerfully signed orders to the visa office to give me permission to travel to Moscow, where I stayed for several months, illegally as it turned out. The student hostel on Oktyabr'skaya Ulitsa, next to the vodka factory, was a real masterpiece of Mr Brezhnev's 5 year plans – the doors and windows didn't fit, and the elevator had to be turned off after 6 pm (in a 13 storey building) because of the inconvenience of calling out the repair man at night when it broke down. Over many months in Minsk, I searched everywhere to buy a teapot, but in vain; while making myself into a 3-folder by studying Zariski's paper, weighted projective spaces, the Riemann–Roch formula for 3-folds and simultaneous resolution of Du Val singularities, I made tea in a jam jar, and tried to pour it into the glass without scalding my hands. The jam jar was the big kind you got 5 kopecks back on; at the time vodka was 4 roubles and 12 kopecks a half-litre, with 12 kopecks back on the bottle; these prices were stable over several decades.

**1976–77 my first visit to Japan**    I was aware of young Japanese algebraic geometers at the AMS summer institute at Arcata in 1974, and I wrote to Kodaira to invite myself to Tokyo. He put me in touch with Miyaoka, who was also doing Godeaux surfaces. I was very lucky with my contacts, and made many friends at once. As my host at Tokyo University, Iitaka looked after me with great kindness; he introduced me to his young graduate student Kawamata, then working on log varieties. Miyaoka had worked on my notes on Bogomolov's inequality $c_1^2 \leq 4c_2$, and lectured on his proof of the famous inequality $c_1^2 \leq 3c_2$ during my first couple of weeks in Tokyo. Among many other kindnesses, Iitaka advised me to go to the March 1977 spring meeting of the Japan Math. Soc. at Kyoto "to meet new people"; this was when I first met my wife Nayo.

## 3.2    Simultaneous resolution and the classic flop

As mentioned in Section 2.1, an easy topological argument shows that the Milnor fibre $M$ and the minimal resolution $Y$ of the surface ordinary double point $xy + z^2 = 0$ are diffeomorphic. In the 1950s, Atiyah [A] discovered a much more convincing reason for the diffeomorphism $M \cong Y$, with an argument that was to have far-reaching consequences for 3-fold geometry.

For this, think of the family as $\mathcal{X} : (xy + z^2 = t) \subset \mathbb{C}^3 \times \mathbb{C}$, with $\mathcal{X} \to \mathbb{C}$ given by $t$. Replace the deformation parameter $t$ by $\sqrt{t} = \tau$, thus replacing $\mathcal{X}$ by $\mathcal{X}' : (xy + z^2 = \tau^2) \subset \mathbb{C}^3 \times T$. The fibres of $\mathcal{X}' \to T$ are the same fibres as before, but now each fibre $X_t$ with nonzero $t$ occurs twice as $X'_{\pm\tau}$. The new total space $\mathcal{X}'$ is the 3-fold ODP $xy + z^2 - \tau^2 = 0$, or $xy = zt$ after the trivial change of coordinates to $\tau \pm z$. Now the graph of the rational function $x/z = t/y$ provides a resolution of singularities $\mathcal{Y} \to \mathcal{X}'$, and moreover, one in which $X_\tau$ moves in a smooth 1-parameter family together with the minimal resolution $Y_0 \to X'_0$. (It is an easy exercise to see that the blowup of a line through a surface ODP has the same effect as the blowup of the point itself.) This is called a *simultaneous resolution* of the surface singularity.

Now consider the 3-fold ODP $X : (xy = zt)$, forgetting the family of surfaces for the moment. Let $Y \subset X \times \mathbb{P}^1$ be the graph of the rational function $x/z = t/y$. Then $f: Y \to X$ is a birational morphism with fibre $L \cong \mathbb{P}^1$ over 0. It is easy to check that $Y$ is nonsingular, and that the normal bundle to $L$ in $Y$ is $\mathcal{O}_{\mathbb{P}^1}(-1) \oplus \mathcal{O}_{\mathbb{P}^1}(-1)$. This is an example of a *small resolution*: the exceptional locus contains no divisors.[1]

We could have paired the factors $x, y$ and $z, t$ in the other order, and considered instead the rational function $x/t = z/y$. This gives a second resolution $f' : Y' \to X$ with exactly the same properties as $f$. It might seem at first sight that $Y = Y'$, but this is not so. Indeed, since $Y \subset X \times \mathbb{P}^1$ and the ratio $(x : z)$ is the coordinate on $Y$, the divisors of zeros and poles of $x/z = t/y$ are two disjoint surfaces cutting $L = \mathbb{P}^1$ transversally at 0 and $\infty$;

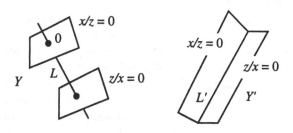

Figure 3.2.1: The flop $Y \dashrightarrow Y'$.

these are the birational transform on $Y$ of the planes $x = t = 0$ and $z = y = 0$ of $X$ (see Figure 3.2.1). On $Y'$ however, the same rational function $x/z = t/y$ has divisor of zeros and poles that meet transversally along $L'$, marking the two factors of the normal bundle $\mathcal{O}_{\mathbb{P}^1}(-1) \oplus \mathcal{O}_{\mathbb{P}^1}(-1)$. The "same surfaces" (more precisely, their birational transforms) have intersection number $+1$ with $L$ and $-1$ with $L'$.

The birational map $\varphi : Y \dashrightarrow Y'$ is the *classic flop*. To think of it as a correspondence, consider its closed graph $W \subset Y \times Y'$. Then $W$ fits into the traditional picture

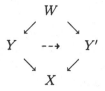

---

[1]The small morphism $f : Y \to X$ is closely related to the class group: $\mathcal{O}_X$ is not a UFD because $xy = zt$; the function $x/z = t/y$ can be written with two essentially different denominators, which was the point of the reference to [red] in Section 2.1.

An $n$-fold point $P \in X$ is *$\mathbb{Q}$-factorial* if every codimension 1 irreducible subvariety of $X$ through $P$ has a multiple that is a principal divisor. Although the $\mathbb{Q}$-factorial property may seems somewhat arcane at first sight, it is very simple, and plays a central role in Mori theory. To understand it, please do the following elementary exercise: if $P \in X$ is $\mathbb{Q}$-factorial, then for every partial resolution $f : Y \to X$ every component of the exceptional locus of $f$ has pure dimension $n - 1$. (This is called *van der Waerden purity*.)

with $W \to X$ the blowup of the ordinary double point having exceptional surface the quadric $Q = \mathbb{P}^1 \times \mathbb{P}^1$ with normal bundle $\mathcal{O}(-1,-1)$. The two sides $W \to Y$ and $W \to Y'$ are the respective blowups of $L \subset Y$ and $L' \subset Y'$ or the contractions of $Q$ along its two rulings.

Flops relate closely to the nef condition: let $\varphi \colon Y \dashrightarrow Y'$ be a classic flop as above. Because $\varphi$ is an isomorphism outside a set of codimension $\geq 2$, a divisor $D$ on $Y$ has a birational transform $D'$ on $Y'$. In other words, it is natural to identify the divisors on $Y$ and $Y'$. On the other hand, as we saw in Figure 3.2.1, if $DL > 0$ then $D'L' < 0$.

On a surface, Zariski had proved ([Z], Theorem 6.1) that if $|D|$ is a *mobile* linear system (that is, $|D|$ has no fixed part) then some multiple $|nD|$ is *free* (that is, $|nD|$ has no base points). I tried to prove the same thing for 3-folds until Iskovskikh and Bogomolov told me it was obvious nonsense, because of flops. The conclusion from this is that if you're hoping for Zariski decomposition for a divisor $D$ on a 3-fold $X$, you can't expect it to take place on the 3-fold itself. If it makes sense at all, you first subtract off some "negative part" $N$ so that $(D - N)\Gamma < 0$ holds for only finitely many curves $\Gamma_i$, and then hope to "flip" these by a birational modification $Y \dashrightarrow Y'$ so that the $\Gamma_i$ are replace by $\Gamma_i'$ with $(D - N)\Gamma_i' > 0$.

The classic flop is also central to the questions of the Weil divisor class group and projectivity of 3-folds. For example, let $\Pi$ be a plane in $\mathbb{P}^4$, and $X_d$ a general 3-fold hypersurface of degree $d \geq 3$ containing $\Pi$. The Weil divisor class group of $X_d$ is generated by $\Pi$ and the hyperplane class. Then $X_d$ can be written $xA + yB$ where $\deg A, B = d - 1$, and in general has $(d-1)^2$ ODPs at the points $x = y = A = B = 0$. Taking the graph of the rational map $x/y = -B/A$ amounts to blowing up $\mathbb{P}^4$ along $\Pi$, and is a projective small resolution of $X_d$ that introduces a flopping curve $L$ above each node. In terms of local coordinates, each curve $L$ can be flopped, but carrying out a flop of any strict subset of the $(d-1)^2$ lines $L$ results in a nonprojective variety.

It is known by work of Brieskorn [B] and Tyurina [Tyu] in the late 1960s that simultaneous resolution applies to any family of surfaces with at worst Du Val singularities (after passing to a suitable cover of the base, like taking the root $\tau = \sqrt{t}$ above). For example, a 1-parameter family of surfaces having nonsingular fibres when $t \neq 0$ and a Du Val singularity of type $A_{n-1}$ above $t = 0$ pulls back on taking the $n$th root $\tau = \sqrt[n]{t}$ to

$$xy = z^n - \tau^n = \prod_{i=1}^{n} z_i, \quad \text{where} \quad z_i = z - \varepsilon_i \tau$$

(and $\varepsilon_i$ runs through the roots of unity). The minimal resolution $Y_0 \to X_0$ of the central fibre has a chain on $n - 1$ copies of $\mathbb{P}^1$, and the family has

$n!$ different simultaneous resolutions corresponding to ordering the factors $z_i$. You pass between them by flops, corresponding to transpositions of the $z_i$.

The idea of simultaneous resolution means that you can allow ODPs in families of surfaces (K3 surfaces, surfaces of general type, del Pezzo surfaces, etc.), in the knowledge that they can be resolved in families. The moral for 3-folds, and an important reason underlying the success of my program in [C3-f] and [Pagoda] is that by taking cyclic covers, a canonical model is always only a finite distance away from a nonsingular minimal model.

**History**    Local Torelli says that moduli of nonsingular K3s is locally isomorphic to the period space, but the moduli problem is not locally separated near surfaces with extra $-2$-curves. This phenomenon is related to the need to make a cover to kill the monodromy before obtaining a simultaneous resolution, and to the ambiguity in the choice of the simultaneous resolution that gives flops: on the Hodge theory side of Torelli, you need the extra information of the effective cone (in other words, the lattice spanned by $-2$-curves has a distinguished Weyl chamber consisting of effective curves). If you're really careful with the definitions, you get a Torelli isomorphism between the moduli and period stacks; it is likely that Deligne had this aspect of the Torelli problem in mind for me in my first year. The problem was solved soon after by Burns and Rapoport [BR], and their solution was an important background item in my work on [Pagoda]. Meiki Rapoport also studied with Deligne, and was my office mate for 2 years at the IHES.

I always referred to the classic flop as *Atiyah's flop* during the 1970s and 1980s, until Gavin Brown pointed out to me that it occurred in Zariski's papers in the 1930s. As soon as you start looking back into the past, you see that it occurs in factoring the standard Cremona transformation

$$T_{\text{tet}} : \mathbb{P}^3 \to \mathbb{P}^3 \quad \text{given by} \quad (x_1, \ldots, x_4) \mapsto \left( \frac{1}{x_1}, \ldots, \frac{1}{x_4} \right),$$

which flops the 6 edges of the coordinate tetrahedron; it must have been well understood in this context by Cremona and Noether around 1870. (According to Hilda Hudson's bibliography, $T_{\text{tet}}$ was studied by Magnus in 1837 and by Beltrami in 1863.)

## 3.3    From canonical models to minimal models

As discussed above, there is no choice in [C3-f] about the definition of canonical models and canonical singularities. You must simply take whatever comes as the Proj of the canonical ring of a variety of general type (assumed finitely generated). The point I want to make here is this:

> *A little knowledge on canonical singularities shows that they admit a restricted class of blowup called* crepant partial resolution. *Applying this to a canonical model leads to a minimal model. This procedure is a close analog of the minimal resolution of Du Val surface singularities.*

The ordinary double point, the $A_2$ double point and the Veronese cone point described in Sections 2.1–2.2 are among the simpler kind of canonical singularities called *terminal singularities*. Any blowup of them leads to exceptional divisors that are *discrepant*.[2] For example, as we saw in Section 2.2, the blowup $Y \to X$ resolving the Veronese cone point has $K_Y = K_X + \frac{1}{2}E$; the discrepancy divisor $\frac{1}{2}E$ prevents $K_Y$ from being nef. See [YPG] for more details.

A simple example of a more typical canonical singularity would be a triple point such as $X : (x^3 + y^3 + z^3 + t^k = 0) \subset \mathbb{C}^4$ with $k \geq 3$. If $k = 3$, this is the cone over a nonsingular cubic surface $E$, a del Pezzo surface with $K_E = \mathcal{O}_E(-1)$. The blowup $X_1 \to X$ is nonsingular, and has $E$ as a divisor with normal bundle $\mathcal{O}_E(E) = \mathcal{O}_E(-1) = \mathcal{O}_E(K_E)$, so that $K_{X_1} = K_X$; we say that $X_1 \to X$ is a *crepant* resolution, and $E$ a crepant exceptional divisor. For higher $k$, the exceptional divisor $E$ is a cubic cone, and $X_1$ has in turn a singularity $X : (x^3 + y^3 + z^3 + t^{k-3} = 0)$, and we can repeat.

Consider again the crepant resolution $Y \to X$ of the cone over a nonsingular cubic surface $S \subset \mathbb{P}^3$, and write $E \cong S$ for the exceptional surface. Any of the 27 lines $L$ of $E$ is a $(-1, -1)$ curve on $Y$, and has a flop $Y \dashrightarrow Y_1$ that takes $E \to E_1$ by contracting the $-1$-curve $L$ and introducing a flopped curve $L_1$. Then $Y_1$ is an alternative resolution of $X$, and its exceptional locus consists of a del Pezzo surface $E_1$ of degree 4 together with a line $L_1$ that meets $E_1$ transversally. This illustrates the fact that flops occur all over the place, and that you almost never expect to get uniqueness of minimal models of 3-folds. In the present case, crepant resolutions $Y' \to X$ (that is, minimal models of $X$) correspond one-to-one with faces of the nef cone of the surface $E$, of which there are several hundred. For example, since there are 72 different contraction morphisms $E \to \mathbb{P}^2$ that contract 6 disjoint lines on $E$, there are 72 different crepant resolutions $Y \to X$ whose exceptional locus consists of the del Pezzo surface of degree 9 ($\mathbb{P}^2$ with normal bundle $\mathcal{O}(-3)$) and 6 flopping lines sticking out of it.

---

[2]A linguistic point to help the non-English reader: *discrepancy* means a difference or contradiction, as when the honest cop finds a discrepancy between the politician's small declared income and large detected expenditure. Here it means the difference $K_Y - f^*K_X$, that is, $1/r$ times the divisor of zeros on $Y$ of a local basis of $rK_X$. The word *crepant* is an English pun meaning not discrepant; it seems to have somewhat unpleasant reverberations in several European languages.

**Problem** We now know that flops account for the nonuniqueness of minimal models of 3-folds. Thus any two birational minimal models have the same Betti numbers, the same singularities, etc. All very tame stuff. But in dimension $\geq 4$, who knows?

**History** The general result highlighted above is the main content of my two papers [C3-f] and [Pagoda]. In [C3-f], I showed that a suitable chain of blowup can be used to extract all the crepant divisors living above a canonical but not terminal point. In [Pagoda], which was one of the main outcomes of my 1981 visit to Kyoto, I used the Brieskorn–Tyurina theory of simultaneous resolution to resolve nonisolated cDV points.

The flop $Y \dashrightarrow Y_1$ is clearly related to the projection of the del Pezzo surface $S_4$ of degree 4 to a cubic surface $S_3$, and should be viewed as a kind of affine cone over it: $X$ is the cone over $S_3$, and $Y_1$ contracts to a 3-fold $X_1$ having the cone over $S_4$ as its singularity, but with a projective line sticking through it. This picture is the original motivation for Francia's flip: consider the Veronese surface $V \subset \mathbb{P}^5$, and its projection from a point to the scroll $\mathbb{F}(1,2) \subset \mathbb{P}^4$. Write $X$ for the affine cone over $\mathbb{F}(1,2)$; then $X$ is the first flipping singularity. It has a ruling by planes that define a small resolution $X^+ \to X$ with exceptional locus a line $L^+$ having normal bundle $\mathcal{O}(-1) \oplus \mathcal{O}(-2)$, so that $K_{X^+} \cdot L^+ = 1$. On the other hand, it has a small partial resolution $X^-$ corresponding to the projection $V \dashrightarrow \mathbb{F}(1,2)$ where $X^-$ has a Veronese cone point with a line $L^-$ sticking through it, and $K_{X^-} \cdot L^- = -\frac{1}{2}$.

## 3.4 Minimal models via semistable degenerations of surfaces

Degeneration of surfaces is a special case of the problem of minimal models and classification of 3-folds that has attracted a lot of attention. Instead of just a 3-fold, we consider a 3-fold $X$ with a fibre space structure $X \to B$ over a base curve $B$. This has a number of advantages:

1. For most purposes we can treat the generic fibre $X_{\text{gen}}$ first, and assume that it is a minimal model, typically a K3 surface or a minimal surface of general type.

2. The minimal model problem then reduces to a neighbourhood of each degenerate fibre.

3. After making a base change by a ramified cover $B' \to B$ (replacing $X$ by a model birational to $X' = X \times_B B'$ as in Section 3.2), we can

assume that $X \to B$ satisfies the conclusions of Mumford's semistable degeneration theorem.

For reasons of space, I will not say very much about this big and technically difficult topic. Mumford raised the problem in the early 1970s of modifying a semistable degeneration of K3s to one of several normal forms corresponding to the possible unipotent filtrations of monodromy; in other words, if we believe that Torelli for K3s extends to infinity, we can use the compactification of the period space as a model for how to compactify the moduli problem.

Mumford's problem was solved around 1976 by Viktor Kulikov [Ku], who proved the existence of a minimal model for the 3-fold total space of a K3 fibre space with semistable degeneration; this was the first time that 3-fold minimal models were proved to exist in substantial cases. Kulikov works in the nonsingular category, without projective assumptions. His main aim is to get rid of any surface component of the degeneration that appears in the canonical class with positive multiplicity, first using 3-fold flops to reduce it to a minimal ruled surface and then contracting it out. The main difficulty is that using a flop to contract a $-1$-curve on a component of the degeneration has the effect of introducing another $-1$-curves somewhere else, possibly even on a different fibre of the same ruled surface, like a serpent that bites its own tail. (Think of Hironaka's nonprojective examples. This kind of thing would be excluded by projective assumptions.) Kulikov has to work hard with graphs and combinatorics and so on to exclude this kind of loop. His paper contained major innovations, together with some minor difficulties of exposition and inaccuracies, and figuring it out led to a flurry of activity in the USA around 1980 ([PP], [deg]).

I heard Kulikov's lectures in Moscow in 1976 while his program was in the unfinished state, and still involved formidable technicalities. At this time my ideas on canonical and minimal models were just forming. His systematic use of flops predate my published work by several years, and must have influenced me. However, since my own approach is always to run away from difficult things wherever possible, and I had my own agenda involving canonical and minimal models, I did not get involved in the details of his proof, and my main conclusion was that I really preferred to wait until the general minimal model program could be sorted out.

Several years later, after Kawamata's log surfaces, canonical models and Mori theory were all on-line, Tsunoda proposed a more general program to construct minimal models of semistable degenerations of surfaces with $K$ nef, by applying Mori theory to Kawamata's log surfaces. The experts seem to believe that Tsunoda's work over several years contains a more or less complete proof of semistable minimal models. However, in the intervening time Shokurov, Kawamata, Mori and most recently Corti have given more convincing versions of semistable flips and the semistable minimal model program. It remains a strong possibility that the future will see the easiest way to general Mori flips through semistable minimal models. (Compare Corti's proof given in [KM], Chapter 7.)

**The December 1981 Kinosaki meeting**  Japanese algebraic geometers run a regular workshop at Kinosaki, a famous spa town in the north of Kyoto prefecture. The tradition apparently started because of a family connection with the owners of one of the ryokan (traditional Japanese inn), for whom December is the off-season. At Kinosaki, you're supposed to visit all the other baths, walking around the streets in the evenings in yukata (evening kimono) and wooden clogs, and holding traditional greased paper umbrellas if it's snowing. Mathematicians gather in the hotel lobby in the evenings to discuss math and other matters, and to sample generous quantities of the finest sake and Japanese whisky. Up to the late 1980s, the lectures were still held in the intimate environment of the ryokan's dining room.

Tsunoda announced his solution to the minimal model problem for semi-stable degenerations of surfaces at the Kinosaki meeting in December 1981. After the workshop, Tsunoda, Maehara, Kurke and I spent an afternoon visiting the famous beauty spot Ama-no-hashidate. The Buddhist temple Nari-ai-ji at the top of the hill contained a Boddhisatva that is a Yuigon Bosatsu, that is, has the unusual power of granting any request, even if only prayed for a single time (the normal practice is to require 10,000 repetitions, a bit like trying to get your kids off the phone). On hearing of this, Tsunoda and I both pressed our palms together, bowed our heads, and solemnly prayed for 3-fold canonical rings to be finitely generated. We forgot to ask help in solving the problem for ourselves.

# 4   The Mori cone and Mori minimal models

Castelnuovo and Enriques, together with every geometer up to Zariski and Mumford, expressed the classification of surfaces in terms of birational invariants, especially the plurigenera and irregularity. In these terms, two of the main aims of the theory are Castelnuovo and Enriques' criteria

$$P_2(S) = q(S) = 0 \implies S \text{ is rational; and}$$
$$P_{12} = 0 \implies S \text{ is ruled or rational.}$$

Here the assumptions and the conclusions are birational, and for this reason, the classification of surfaces is often described as *birational classification*. This is one of several traditional preconceived notions you must unlearn if you wish to become a 3-folder.

The systematic approach to higher dimensional classification initiated by Iitaka and Ueno follows the same logic. This approach gives the right results (such as the Iitaka fibration) in cases when the plurigenera grow, or you have other information such as a nontrivial Albanese map. But if the plurigenera and irregularity are zero, it doesn't really give anything other than optimistic conjectures that the variety should be covered by rational curves, conjectures that are somewhat cheap because there is no plan for proving them. My approach described in the preceding chapter, although it set itself biregular aims, and was natural enough in view of my main interest in varieties of general type and their canonical rings, depended on even stronger assumptions than Iitaka's program: general type and finitely generated canonical ring.

Mori's interest seems to have started out from Fano varieties (varieties with $-K$ ample that classical geometers describe as "close to rational") and the existence of rational curves on them. The springboard for his first paper in minimal models [M1] was his earlier solution of the Hartshorne conjecture characterising $\mathbb{P}^n$ by ampleness of the tangent bundle. (I believe that this conjecture originated with a question of Deligne after Hartshorne's lecture on ample vector bundles at the 1970–71 Warwick symposium. I hope Mumford and Zeeman remembered to get it into the final SRC report.) A classical dimension count predicts that the space of rational curves is positive dimensional at any point, but we know that this type of argument cannot prove it is nonempty; Mori's key breakthrough is his extraordinary method of *bending-and-breaking* and *reduction to characteristic p* to prove the existence of rational curves.

The Kleiman–Mori cone described in Section 4.1 gives a strategy for minimal models based on contracting extremal rays (Section 4.2). In contrast to Iitaka's and my methods, these ideas work best of all for Fano varieties, whose Mori cone is a finite rational polyhedron with rays spanned by extremal rational curves. The contraction part can be done very generally by a circle of ideas that Kawamata's school call the *X method*, and modifications of its proof lead to the general cone theorem (Section 4.3). Flips were the last foundational problem to be solved by Mori in the late 1980s, and still lack a simple and wholly convincing argument.

## 4.1   Kleiman criterion and the Kleiman–Mori cone

The Riemann–Roch formula $\chi(\mathcal{O}_S(D)) = \frac{1}{2}D(D - K_S) + \chi(\mathcal{O}_S)$ for a divisor $D$ on a nonsingular surface $S$ is quadratic in $D$. It follows that for $D^2 > 0$, either $h^0(nD)$ or $h^0(-nD)$ grows quadratically with $n$. Grothendieck's proof of the algebraic Hodge index theorem follows easily from this: the index of the intersection form on the algebraic part of $H^2(S, \mathbb{R})$ is $(+1, -(\rho - 1))$, and effective divisors with $D^2 > 0$ form a half-cone in $H^2(S, \mathbb{R})$ that we draw as a future light cone.

The Kleiman criterion [Kl] starts with the question of the cone of nef divisor classes Nef $S \subset H^2(S, \mathbb{R})$. The class of an ample divisor is both in the interior of Nef $S$, and of the $D^2 > 0$ cone; thus ample divisors form a subcone of the future light cone. Conversely, it turns out that a nef divisor is in the *closure* of the ample cone in $H^2(S, \mathbb{R})$: if $D$ is nef and $H$ is ample then the index polynomial $p(\lambda) = (D + \lambda H)^2$ takes positive values for $\lambda \gg 0$, while $D + \lambda H$ is nef for any $\lambda \geq 0$. But $p(\lambda)$ cannot go from being positive to being negative while $D + \lambda H$ is nef, and it follows that $D^2 \geq 0$ and $D + \varepsilon H$ is an ample $\mathbb{Q}$-divisor for any $\varepsilon \in \mathbb{Q}$, $\varepsilon > 0$.

Thinking about effective and nef divisors on surfaces tends to obscure a key point that comes out naturally for a higher dimensional variety $V$:

> *Rather than* $H^2(S, \mathbb{R})$ *and its* quadratic *intersection form, it is better to think of the* bilinear *pairing between the two dual vector spaces* $H^2(V, \mathbb{R})$ *and* $H_2(V, \mathbb{R})$.

The first of these contains the classes of $\mathbb{Q}$-Cartier divisors, and the second the cone of effective curves NE $V$ and its closure $\overline{\text{NE}}\, V \subset H_2(V, \mathbb{R})$, the *Kleiman–Mori cone*.

The Kleiman criterion says that a Cartier divisor $L$ is ample on $V$ if and only if its divisor class in $H^2(V, \mathbb{R})$ is strictly positive on $\overline{\text{NE}}\, V \setminus \{0\}$. This condition is slightly stronger than saying $L\Gamma > 0$ for every curve $\Gamma$ in $V$; it requires equivalently that $L - \varepsilon H$ is still nef for any ample $H$ and some $\varepsilon > 0$, or that $Lz > 0$ for $z$ a *limit* of effective curves.

The Kleiman–Mori cone $\overline{\text{NE}}\, X \subset H_2(X, \mathbb{R})$ is a closed convex cone by definition. You don't expect to be able to calculate it in general, but enough cases are known to suggest that both $\overline{\text{NE}}\, X$ and the relation between NE $X$ and its closure $\overline{\text{NE}}\, X$ can be pretty well arbitrary in the half-space $K_X z \geq 0$. Mori's discovery concerns the half-space $K_X z < 0$: on a nonsingular $n$-fold $\overline{\text{NE}}\, X \cap ((K_X + \varepsilon H)z \leq 0)$ is finite rational polyhedral for any ample $H$ and any $\varepsilon > 0$. The next two sections sketch two approaches to proving this: Mori's argument by bending-and-breaking, and the rational threshold argument related to Kodaira vanishing.

**History** I learned the idea of the positive cone implicitly from Zariski [Z], from working with Bogomolov on asymptotic RR, and more especially from Mumford's comments on my notes on Bogomolov.

The use of cones such as the positive cone, Nef and NE can reasonably be traced back to Nagata's papers on rational surfaces [N]. If you look for Kleiman's paper [Kl] in the library of the Kyoto University math department, you will find that Vol. 84 of Ann. of Math. is particularly dirty and dog-eared, as if held over the photocopier dozens of times. I presume that Nagata has set

Kleiman's paper as background reading for successive generations of students. (Both Hironaka and Mori were Nagata's M.Sc. students.)

## 4.2   Mori cone theorem and Mori contractions

Mori first introduced bending-and-breaking under very strong assumptions in the context of the Hartshorne conjecture, where he must prove the existence of a rational curve in a variety $\Gamma \subset V$ (that is to be a straight line $\Gamma = \mathbb{P}^1 \subset \mathbb{P}^n$, so that $-K_V\Gamma = n + 1$). He extended the method in the proof of his original cone theorem, and later Kollár and Miyaoka generalised it further in several directions. I sketch Kollár's ideal form of the theorem.

> Let $C_0 \subset X$ be a curve with $-K_X C_0 > 0$ on a nonsingular $n$-fold $X$. Then there exists a rational curve $\Gamma \subset X$ passing through any point $Q \in C_0$ with $0 < -K_X\Gamma \le n + 1$.

STEP 1:   Replace the embedded curve $C_0 \subset X$ by the map $f \colon C \to C_0 \subset X$ from a nonsingular curve, with $P \in C$ such that $f(P) = Q$.

STEP 2:   The deformation of the map $f$ is controlled by the cohomology of the pullback $f^*T_X$ of the tangent bundle of $X$, a rank $n$ vector bundle on $C$ of degree $-K_X C_0 > 0$. Suppose that this degree is not just positive, but very large compared to $n$ and the genus of $C$. Then

**bending:** we can deform $f$ as a morphism $F_0 \colon C \times T_0 \to X$ over a parameter curve $T_0$ so that the image curve $f_t(C)$ sweeps out a surface, while fixing the value $f(P) = Q$.

**breaking:** let $T$ be the projective completion of $T_0$, so that $F \colon C \times T \dashrightarrow X$ is a rational map; then $F$ is *certainly not a morphism* in a neighbourhood of the section $\{P\} \times T$. Because this section has self-intersection 0, but $F$ contracts it to a point in $X$.

We get a rational curve on $X$ through $P$ by resolving the indeterminacy of $F$ by blowups and taking the image of a suitable exceptional curve, and we can arrange for it to have bounded degree and $0 < -K_X\Gamma \le n + 1$ by rerunning an appropriate version of the same argument if necessary.

STEP 3:   Now comes the really clever bit. If we are in characteristic $p$, we can replace $f$ by its composite $f^n$ with a suitable iteration of the Frobenius map. This allows us to pump up $\deg f^*T_X$ to be as large as we like, while fixing the genus. Thus the required rational curves exist in almost all characteristic $p$; an argument on finiteness of the Hilbert scheme implies that they also exist in characteristic 0.   Q.E.D.

It follows from this result that any cycle in $z \in \overline{\text{NE}}$ with $(K + \varepsilon H)z < 0$ necessarily splits off a rational curve of bounded degree. The Mori cone theorem (for a nonsingular variety) is a rather formal consequence of this. The kind of contraction morphisms $\varphi \colon X \to Y$ we are interested in are determined by certain faces of $\overline{\text{NE}}\, X$; namely if $Y$ is projective and $H$ ample on it, then $L = \varphi^* H$ is nef, and $L^\perp \cap \overline{\text{NE}}\, X$ is the cone of effective curves contracted by $\varphi$, that is, the relative cone $\overline{\text{NE}}(X/Y)$. A face of $\overline{\text{NE}}$ is a relative cone for a contraction if and only if it has a rational supporting hyperplane corresponding to an eventually free divisor $L$, and the morphism $\varphi_{nL} \colon X \to Y$ for $n \gg 0$ (or rather, its Stein factorisation) is uniquely determined up to isomorphism by the face.

In [M2], Mori only contracted extremal rays on nonsingular 3-folds in the $Kz < 0$ half of the cone, and obtained the famous classification that I'm sure you know about.

**History** I want to stress that the Mori cone came as a huge surprise to everyone, and that even now his leap of imagination seems quite incredible. I first heard that Mori was working on 3-fold contractions in a letter from Mumford around 1980 on practicalities of a visit to Warwick. Just in passing, he scribbled a note, that I paraphrase from memory: Mori had given a Harvard seminar, proving that if $K_X$ is not nef on a smooth 3-fold, there exists a contractible surface, either a scroll, or $\mathbb{P}^2$ with normal bundle $\mathcal{O}(-1)$ or $\mathcal{O}(-2)$, or a quadric surface. P.S. (clearly added as an afterthought) That is, unless we are in the uniruled case.

In other words, even after Mori had described his work at a seminar, none of us had really grasped the significance of his cone theorem, or the extra precision that his methods give us in the uniruled case. For about 3 years after this, I regularly wrote to Mori to bother him for more details, both for my own sake and for my Moscow correspondents. The research announcement [M1] only aroused our curiosity, and I clearly remember getting the preprint of [M2] and wondering what all this clever-clever stuff about cones had to do with contracting surfaces on 3-folds. My next reaction was to look among the varieties I knew for counterexamples to his conclusions for extremal contractions that were conic bundles or divisorial contractions to ODPs.

I learned Mori theory, and many of the details of Mori's paper [M2] from a series of lectures by Miyanishi at Ueno's seminar at Kyoto University in 1981, and its applications to Fanos from talking to Mukai who was at that time working with Mori [MM] on the classification of nonsingular Fano 3-folds with $B_2 \geq 2$. The 2-ray game plays an important role in their study (without the name). In 1982–83 I organised my first Warwick symposium. During the first half of this year, I got together with several long term symposium visitors (Beltrametti, Dolgachev, Palleschi and Tsunoda) to run a seminar on Mori's paper and subsequent developments.

## 4.3    The X method, contractions and thresholds

Two related issues presented themselves around 1981:

(i) If $X$ is a 3-fold minimal model of general type (that is, at worst terminal singularities, $K_X$ nef and $K_X^3 > 0$), can we prove that $K_X$ is eventually free? This is equivalent to the finite generation of the canonical ring of a minimal model $X$.

(ii) If $F \subset \overline{NE}\,X$ is a face of the Mori cone contained in the half-space $K_X z < 0$, can we prove that there exists a morphism $X \to Y$ contracting exactly the curves in $F$? Mori proved the contraction theorem for an extremal ray on a nonsingular 3-fold $X$, but his proof seems impossible to generalise, since it involves a detailed knowledge of $X$ in a neighbourhood of the contracted divisor.

Both these questions were answered at the same time by Kawamata's theorem [Ka]: if $D$ is a nef Cartier divisor and $D - \varepsilon K_X$ is ample for some $0 < \varepsilon \ll 1$ then $D$ is eventually free. This statement only has content if $D$ is a supporting hyperplane of $\overline{NE}\,X$ defining a nontrivial face $D^\perp \cap \overline{NE}\,X$ in the half-space $K_X z < 0$.

The method of proof is a complicated game that has subsequently been refined by many authors, notably Kawamata himself and Shokurov. Kawamata's graduate students, who have had to suffer this method more than most, have with characteristic Japanese linguistic inventiveness christened it the *X method*; the instruction manual is [KMM]. The method is powerful, we know how to use it pretty well in practice, and it's been around for 20 years without anyone thinking of a better one, but (in my opinion) it is still messy and not optimal, and we don't really understand it.

Kodaira vanishing is the theorem that if $X$ is a nonsingular projective $n$-fold over a field of characteristic 0 and $D$ and ample divisor, then

$$H^i(X, D + K_X) = 0 \quad \text{for all } i > 0.$$

Many generalisations are known, for example Kawamata–Viehweg vanishing: we can allow $D$ to be an ample $\mathbb{Q}$-divisor with fractional part supported on a normal crossing divisor, and get the conclusion $H^i(X, \ulcorner D \urcorner + K_X) = 0$. When using the X method, we blow up the $n$-fold $X$ together with any divisors we can see on it to a nonsingular variety $V$ with a normal crossings divisor $D$, then manoeuver into position to apply vanishing at the level of the different strata of $V, D$, thus obtaining surjective restriction maps.

*Kodaira vanishing and the X method give a reason of principle why life is simple around the K negative boundary of* $\overline{\text{NE}}$. *These ideas apply to give the Mori cone theorem and the contraction theorem for a very general class of singular varieties X.*

Very roughly, the argument is as follows: suppose $K_X$ is not nef, and let $D$ be an ample Cartier divisor. Consider $D + tK_X$ as $t$ increases; then there is some critical value or *nef threshold* at which $D + tK_X$ stops being ample:

$$t_0 = \sup\{t \mid D + tK_X \text{ is nef}\}.$$

Just before we reached $t_0$, the divisor $D + tK_X$ is ample. But if we add $K_X$, then $D + (t + 1)K_X$ is not nef; however, Kodaira vanishing applies to it, because $D + tK_X$ is ample. This implies that something very rational is going on, and, following [R], Kawamata and Kollár proved that $t_0$ is a rational number whose denominator can be bounded in terms of the dimension $n$ and the index of $X$.

**History**   During my 1981 stay at Kyoto University, I made extensive attempts to prove the contraction theorem for 3-folds of general type with $K_X$ nef. I was hoping to prove, for example, that $|10K|$ is free. I tried to use ideas of the following type: if $D$ is a divisor on $X$ passing through a point $P$ with multiplicity $\geq 3$, and $\sigma: X_1 \to X$ the blowup of $P$ with exceptional locus $E$, then $D^{(3)} = \sigma^*D - 3E$ is an effective divisor, and by exactly the same argument as in the Bombieri–Kodaira–Ramanujam method for surfaces,

$$P \in \text{Bs}\,|D + K_X| \iff H^1(\mathcal{O}_{D^{(2)}}) \to H^1(\mathcal{O}_{D^{(3)}}) \text{ is not surjective.}$$

For example, if $D$ is a reduced surface with an ordinary triple point at $P$, then $P$ is a base point of $D + K_X$ if and only if the blowup of the elliptic singularity $P$ increases the irregularity of $D$, rather than decreasing the geometric genus (so that $P$ "disconnects $D$", by analogy with a node disconnecting a curve in Bombieri's method). If $D$ is a little bit ample, then this should be impossible by some kind of Lefschetz hyperplane section argument in topology. I'm not sure why this idea still can't be made to work.

At the July 1981 conference in Tokyo, in connection with my [Pagoda] lecture, I discussed these kind of problems with Kawamata. This was just before he went to Berkeley to take up his Miller fellowship. I received several of his preprints (including [Ka]) during the Warwick symposium, and lectured on them and started developing my own version of them [R]. A particular aim was to massage the statement of Kawamata's theorem on eventually freedom to imply contractions of faces of the Mori cone.

Over many months during the Warwick symposium, Tsunoda came to discuss attempts to prove the Mori cone by deforming curves. I considered these to be doomed to failure because of the Francia flip: in that case, there is just a single curve on which $K$ is negative, and of course it doesn't deform. (I was wrong: a few years later, Kollár [Ko] found an ingenious argument involving Mori's idea of quotient stacks to extend the deformation argument to some singular $n$-folds, in particular those with hypersurface or quotient singularities.) I stumbled on the threshold proof of the Mori cone theorem in preparing lectures on [R], and in trying to find another way around the cone theorem to answer Tsunoda's questions. Shokurov, who discovered a version of the same argument independently of me [Sh], wrote in the summer of 1983 that he could do the nonvanishing trick that made these methods work for general $n$-folds. The cone and contraction theorems are now proved whenever it makes sense.

In the report to SRC on the 1982–83 year I wrote:

> ... *the framework for the study of 3-folds and higher dimensional varieties is just in the process of construction; we have just got to the stage when conventional wisdom was that 'there is no satisfactory extension to higher dimensions of the theory of minimal models of surfaces, or of the details of the classification' to the point when everyone now believes that 'there is a good theory provided that we work in the right class of singular varieties'. It is quite clear that the Warwick year 1982/83 will go down as a key period when substantial cases of the general theory of higher dimensional varieties were first proved.*

# References

[A] M. F. Atiyah, On analytic surfaces with double points, Proc. Roy. Soc. London. Ser. A **247** (1958) 237–244

[B] Egbert Brieskorn, Über die Auflösung gewisser Singularitäten von holomorphen Abbildungen, Math. Ann. **166** (1966) 76–102. Die Auflösung der rationalen Singularitäten holomorpher Abbildungen, Math. Ann. **178** (1968) 255–270

[BR] D. Burns, Jr. and M. Rapoport, On the Torelli problem for kählerian K3 surfaces, Ann. Sci. École Norm. Sup. (4) **8** (1975) 235–273

[Caibăr] Mirel Caibăr, Minimal models of canonical singularities and their cohomology, Warwick Ph.D. thesis, Feb. 1999, 78 + vii pp.

[D] P. Deligne, La formule de Picard–Lefschetz, and la théorie des pinceaux de Lefschetz, Exp. XV and XVIII of SGA7, Groupes de monodromie en géométrie algébrique, LNM**340** Springer (1973)

[DuV] Du Val, On the singularities which do not affect the condition of adjunction. I–III, Proc. Camb. Phil. Soc. **30** (1934) 453–491

[Fl] A. R. Iano-Fletcher, Working with weighted complete intersections, this volume, pp. 101–173

[deg] The birational geometry of degenerations (Harvard, 1981), Robert Friedman and David R. Morrison (eds.), Birkhäuser, Boston, 1983

[I] V.A. Iskovskikh, Fano threefolds I, II and III, Izv. Akad. Nauk SSSR Ser. Mat. **41** (1977) 516–562 = Math. USSR Izv. **11** (1977); **42** (1978) 506–549 = Math. USSR Izv. **12** (1978); Double projection from a line onto Fano threefolds of the first kind, Mat. Sb. **180** (1989) 260–278 = Math. USSR-Sb. **66** (1990) 265–284

[K] Y. Kawamata, On the classification of noncomplete algebraic surfaces, in Algebraic geometry (Copenhagen, 1978), Springer LNM **732** (1979) 215–232

[Ka] Y. Kawamata, On the finiteness of generators of a pluricanonical ring for a 3-fold of general type, Amer. J. Math. **106** (1984) 1503–1512

[KMM] Y. Kawamata, K. Matsuda and K. Matsuki, Introduction to the minimal model problem, in Algebraic geometry (Sendai, 1985), Adv. Stud. Pure Math., 10, North-Holland, Amsterdam-New York, 1987, pp. 283–360

[Kl] S. Kleiman, Towards a numerical theory of ampleness, Ann. of Math. **84** (1966) 293–344

[Ko] J. Kollár, Cone theorems and bug-eyed covers, J. Alg. Geom. **1** (1992) 293–323

[KM] J. Kollár and S. Mori, Birational geometry of algebraic varieties, C.U.P., 1998

[Ku] Viktor S. Kulikov, Degenerations of K3 surfaces and Enriques surfaces, Izv. Akad. Nauk SSSR Ser. Mat. **41** (1977) 1008–1042 = Math USSR Izv. **11** (1977) 957–989

[M1] S. Mori, Threefolds whose canonical bundles are not numerically effective, Proc. Nat. Acad. Sci. U.S.A. **77** (1980) 3125–3126

[M2] S. Mori, Threefolds whose canonical bundles are not numerically effective, Ann. of Math. **116** (1982) 133–176

[M3]  S. Mori, Classification of higher-dimensional varieties, in Algebraic Geometry, Bowdoin 1985, ed. S. Bloch, Proc. of Symposia in Pure Math. **46**, A.M.S. (1987), vol. 1, 269–331

[MM]  S. Mori and S. Mukai, On Fano 3-folds with $B_2 \geq 2$, in Algebraic varieties and analytic varieties (Tokyo, 1981), Adv. Stud. Pure Math., **1**, North-Holland, Amsterdam-New York, 1983, pp. 101–129

[red]  David Mumford, Introduction to algebraic geometry, preliminary version of first 3 Chapters, Notes distributed by Harvard from late 1960s. Reissued as The red book of varieties and schemes, LNM **1358** Springer, 1988

[N]  M. Nagata, On rational surfaces. I. Irreducible curves of arithmetic genus 0 or 1, Mem. Coll. Sci. Univ. Kyoto Ser. A Math. **32** (1960) 351–370. II, same J. **33** (1960/1961) 271–293

[PP]  H. Pinkham and U. Persson, Degeneration of surfaces with trivial canonical bundle, Ann. of Math. (2) **113** (1981) 45–66

[Q]  M. Reid, The intersection of two quadrics, Ph.D. Thesis, Cambridge 1972 (unpublished)

[EllGor]  M. Reid, Elliptic Gorenstein surface singularities, 1975 (unpublished)

[C3-f]  M. Reid, Canonical 3-folds, in Journées de géométrie algébrique d'Angers, ed. A. Beauville, Sijthoff and Noordhoff, Alphen 1980, 273–310

[Pagoda]  M. Reid, Minimal models of canonical 3-folds, in Algebraic varieties and analytic varieties (Tokyo, 1981), Adv. Stud. Pure Math., **1**, Kinokuniya and North-Holland, 1983, 131–180

[R]  M. Reid, Projective morphisms according to Kawamata, Warwick preprint 1983 (unpublished)

[YPG]  M. Reid, Young person's guide to canonical singularities, in Algebraic Geometry, Bowdoin 1985, ed. S. Bloch, Proc. of Symposia in Pure Math. **46**, A.M.S. (1987), vol. 1, 345–414

[S-D]  B. Saint-Donat, Projective models of K3 surfaces, Amer. J. Math. **96** (1974) 602–639

[Sh]  V. V. Shokurov, The closed cone of curves of algebraic 3-folds, Izv. Akad. Nauk SSSR Ser. Mat. **48** (1984) 203–208 = Math. USSR Izv. **24** (1985) 193–198

[Tyu] G. N. Tjurina, Resolution of singularities of flat deformations of rational double points, Funkcional. Anal. i Priložen. **4** (1970) 77–83

[U] K. Ueno, On the pluricanonical systems of algebraic manifolds, Math. Ann. **216** (1975) 173–179

[Z] Oscar Zariski, The theorem of Riemann–Roch for high multiples of an effective divisor on an algebraic surface, Ann. of Math. (2) **76** (1962) 560–615

Miles Reid,
Math Inst., Univ. of Warwick,
Coventry CV4 7AL, England
e-mail: miles@maths.warwick.ac.uk
web: www.maths.warwick.ac.uk/~miles

# Index

2-ray game, 11, 217–219, 249–253, 269–271, 274, 300, 304, 309, 337

amazing example, 229–232, 246, 255, 256
amplitude, 111–115, 134, 136
anticanonical
  divisor, 8
  map, 197, 253, 299, 301
  model, 103, 131, 186, 193, 218, 244, 248, 251, 301, 320
  ring, 177, 182–183, 193, 197–199, 202, 214, 244, 248, 250, 251, 265, 266, 310
  system, 80, 182–183, 252

bad line, 305
bad link, 11, 186, 203, 212–215, 218, 232, 244, 248–256, 271
bending-and-breaking, 334–336
Big Table, 178, 189, 194–198, 201–204, 214–215, 221, 223, 225, 232–243
birational rigidity, 9, 10, 13–15, 82, 177–178, 248, 254, 260–267, 275, 277, 286, 292–293, 295–305, 310
  over a base $S$, 262–263, 295
birationally superrigid, 82, 94
bizarre quartet, 221, 255, 256

canonical, 180, 185, 187, 208, 265, 269, 289
  3-fold, 103, 130, 133, 151, 163, 314, 320, 322

model, 322, 323, 329, 332
multiplicity (discrepancy), 79
ring, 102, 315, 321, 322, 324, 329, 333, 334, 338
sheaf, 109, 110
singularities, 102, 110, 123, 128, 132, 137, 140, 143, 155, 187, 217, 219, 252, 268, 313, 318, 322, 323, 329, 330
singularity, 315
surface, 320
threshold, 187, 269, 273, 274
centre of a valuation, 78, 187, 208, 210, 265, 266, 279
conic bundle, 7–9, 14, 17–18, 51–53, 78, 82, 94, 95, 177, 178, 262, 263, 266, 267, 286–289, 293–296, 300, 337
contraction, 325, 337, 339
  theorem, 338–340
contraction, see divisorial, extremal, flipping, small
crepant, 315, 330, 331
critical point, 55–57, 316
  degenerate, 55
  in positive characteristic, 55
cyclic cover, 51–55, 57–59

deformation theory, 8
del Pezzo
  fibration, 14, 51, 78, 82, 177, 254, 262, 263, 265–267, 283, 289, 293, 295–300, 309
  surface over a nonclosed field, 54, 63–69

346

discrepancy, 79, 96, 204, 210, 217,
  265, 268, 276, 283, 285, 325,
  330
discrepancy, *see* canonical multiplicity
discrete valuation, *see* valuation
divisorial contraction, 2, 6, 11, 15–
  17, 188, 190, 191, 194, 204,
  205, 212, 214, 217, 218, 253,
  254, 261, 264, 266, 270, 271,
  273, 281, 283, 301, 302, 309,
  337

eccentric trio, 219, 245, 256
effective
  cone, 329
  curve, 335, 337
  divisor, 324, 334, 339
  threshold, 287, 294
excluding, 75, 76, 81–83, 87, 90,
  186, 189, 194, 196, 197, 202–
  232, 243, 244, 248, 249, 251–
  256, 266
extremal
  contraction, 6, 11, 16, 188, 190,
  191, 204–206, 212, 214, 218,
  253, 254, 261, 263, 266, 270,
  274, 281, 294, 334, 337
  extraction, 10, 11, 16, 187, 188,
  190, 207, 217, 249, 251, 252,
  273
  ray, 1, 4–7, 10, 11, 184, 186,
  213, 215, 217, 261, 270, 274,
  294, 334, 337, 338
  RC fibration, 292

Q-factorial, 11, 17, 77, 177, 190,
  192, 261, 270, 271, 310, 327
"famous 95", 9, 14, 17, 103, 138–
  140, 155–158, 175–178, 189,
  232–243, 250, 261, 263, 267,
  281, 300, 310, 323
Fano

3-fold, 1, 7, 9, 10, 12, 14, 15, 17,
  76, 131, 133, 154, 155, 158,
  163, 175, 178, 188, 189, 219,
  243, 248, 250, 253, 260, 263,
  264, 266, 300, 304, 310, 320,
  337
  hypersurface, 198, 219, 249,
  263, 310
  fibration, 76, 78, 82
  variety, 5, 13, 76, 78, 177, 179,
  180, 334
Fano fibration, *see* Mori fibre space
fibration, *see* conic bundle, del Pezzo,
  Fano, Mori fibre space
flip, 2, 6–8, 15–17, 22, 188, 218,
  261, 271, 328, 331, 332, 340
  conjecture, 7
  theorem, 6, 313
flip, *see* inverse flip
flipping contraction, 16, 218
fundamental
  generator, 23, 25, 35, 36, 38, 41
  simplex, 108, 137, 141, 146, 149

geometric valuation, *see* valuation
geometrically
  irreducible, 61, 66
  rational, 55
  ruled, 60
  ruled, uniruled, etc., 55
graph method, *see* resolution graph

Hilbert
  basis, 23, 37, 38, 47, 48
  function, 103, 133, 162, 163, 245
Hodge numbers, 115

index, 78
index, *see* quasieffective threshold
infinitely near maximal singularity,
  73, 76, 80, 81, 87, 90
inverse flip, 11, 188, 218, 219, 245,
  248–250, 252, 253, 256, 271

involution, 67, 86, 250, 255, 264,
    300–302, 305, 307
  Cremona, 13
  Geiser and Bertini, 14, 178, 179,
    181–186, 294
  of Types I and II, 178
  quadratic and elliptic, 178, 193–
    203, 232, 245, 255

$K^2$ condition, 263, 265, 288, 293,
    295, 297, 300, 310
Kawamata blowup, 16, 190–197, 199,
    202, 217, 232, 244–246, 248,
    281, 304

lattice polytope, 21, 30, 34, 36–39,
    41
linear system without fixed part, see
    mobile linear system
link, 11–15, 17, 18, 183, 185–187,
    189, 190, 217, 248–250, 252,
    253, 260, 261, 266, 267, 269–
    271, 273–275, 288, 294, 301,
    304, 305, 309, 310
  of Type I–IV, 11, 188, 189, 193,
    197, 250, 269–271, 274, 288
locus of log canonical singularities,
    278
log
  canonical, 208–211, 275, 276, 278–
    280, 303, 308
  category, 10, 251, 261, 269, 292,
    325, 326, 332
  surface method, 10, 244, 278

maverick, 212–216, 218, 253, 256
maximal
  centre, 183, 186, 189, 190, 194,
    196, 202–204, 206
  cycle, 80–83, 85–87, 94
  multiplicity, 74
  point, 83, 87

singularity, 12, 76, 77, 80, 87,
    90, 93, 180, 186–188, 190,
    194, 206
  triple, 74
  method of maximal singularities, see
    maximal singularity
mini-definition of rigidity, 82, 177,
    248
minimal model, 323–333
Minkowski
  decomposition, 30, 32, 44
  mixed volume, 123, 125, 127
  sum, 8, 25, 29, 30, 39, 41, 48
    generalised, 26
  summand, 21, 25–28, 35
mobile linear system, 12, 13, 73, 77,
    79, 82–84, 87, 90, 96, 180,
    181, 183–185, 187–189, 209–
    211, 265–267, 273, 275, 278–
    280, 283, 284, 289, 293, 296,
    303, 328
Mori
  category, 2, 3, 6, 7, 10, 11, 186–
    188, 190, 191, 193, 204, 217,
    219, 252, 261, 269–271, 294
  cone, 3–5, 184, 188, 204, 212,
    213, 263, 270, 272, 295, 323,
    333–340
  fibre space, 2, 5, 7, 8, 11–15, 17,
    18, 175, 177, 180, 188, 189,
    217, 251, 260–263, 265–267,
    269–274, 286, 287, 289, 293,
    296, 302, 304, 310, 337
  flip, 333
  theory, 1–8, 10, 11, 15, 177, 179,
    183, 186, 251, 260, 261, 265,
    313, 327, 332, 337
Morse lemma in positive character-
    istic, 56

nef, 1, 4, 5, 7, 8, 184, 193, 204, 205,
    215, 250, 252, 253, 272, 273,

278, 290, 294, 295, 301, 302,
321, 323–324, 328, 330, 332,
335, 337–339
threshold, 185, 294, 339
Newton polyhedron (polygon), 125,
127, 128, 198, 220
Noether–Fano–Iskovskikh inequalities, 12, 74, 179, 180, 265,
267, 272, 289–291, 296, 304
Noether's inequality, 74
nonrational variety, 9, 51, 52, 54,
295
nonruled variety, 51–54

pliability, 14, 15
plurigenus formula, 162, 245, 321
Poincaré series, see Hilbert function
projection, 10, 12, 17, 18, 32, 51,
52, 54, 60, 67, 84, 86, 96,
181, 193, 201, 211, 219, 246,
288, 301, 309, 331

quasieffective, 272, 287, 289–291, 294,
324
threshold, 78, 80, 185
quasireflection, 107, 133
quasismooth, 105, 109, 111–123, 129,
130, 133, 136, 137, 141–143,
145–150, 154, 155, 158, 168,
169, 177, 192, 195–197, 199,
203, 211, 220, 232, 245, 253,
254, 264, 302
quotient singularity, 2–4, 24, 26, 27,
29, 36, 102, 105, 108, 110,
137, 141, 145, 146, 149, 167,
281, 288, 315, 318–323, 340

rainbow, 251–253
rational
over $F$, 54
variety, 54, 293, 294
rationality problem, 9, 15, 51, 52,
54, 64, 65, 78, 262, 293

rationally connected, 9, 15, 51, 59,
292
realisation of a valuation, 79
regular versus smooth, 54, 55, 64–
66, 68
renegade, 208, 210, 219, 256
resolution
graph, 88–90, 92, 97, 246, 260,
266
of a valuation, 88
of singularities, 80, 315, 316, 318,
323, 326, 330
rigid, see birational rigidity
(infinitesimally), 46
boundary, 14–15, 180, 250–251,
309–310
Russian matrix, 182, 185, 194

Sarkisov
category, 193, 250, 262, 275
degree, 12, 13, 189, 267–269
decreases, 274
isomorphism, 262, 272, 273
link, see link
program, 9–17, 76, 179, 186, 187,
189, 217, 248, 251, 252, 260,
261, 265–275, 289, 290, 302,
305
theorem on conic bundles, 14,
52, 94, 178
semistable
degeneration, 331–332
del Pezzo fibration, 295, 298–
300
flip, 332
Shokurov connectedness, 9, 208–210,
260, 261, 264, 266, 275, 277–
279, 283
simultaneous resolution, 318, 326,
328, 329, 331
small
contraction, 6, 7, 270, 271

resolution, 315, 317, 318, 323, 327, 328, 331
special surface, 214–217, 223–225, 254, 256
square birational map, 14, 262, 272, 287–289, 293–296, 299, 300
standard conic bundle, 94, 287–288, 293–295
starred monomial, 196, 197, 256
strict Mori fibre space, 2, 7, 8, 10, 11, 78, 178, 250, 267, 310
strong maximal
  centre, 193
  singularity, 187–189
superrigid, see birationally super-rigid

table method, 162, 163, 165
terminal
  quotient sing. $\frac{1}{r}(1, a, r - a)$, 2, 3, 124, 145, 147, 151, 167, 177, 189, 191–193, 202, 281, 315, 318, 319, 322, 323
  singularities, 2–3, 6, 7, 11, 15, 17, 77, 123, 130, 131, 147, 148, 150, 154, 155, 168, 177, 178, 186, 190, 191, 193, 250, 260, 261, 270, 271, 281, 295, 309, 310, 315, 330, 338
test
  class, 203, 204, 206, 207, 212, 213, 219, 225, 227–229, 244, 251, 256
  method, 76, 203–213, 216, 252–254
  pair, 77–82
  surface, 211, 303
threshold, see canonical, nef, quasi-effective
toric
  blowup, 192
  deformation, 22, 29, 30, 32

Gorenstein singularities, 8, 9
modification, 218
stratum, 108, 117, 120, 128
variety, 8, 21–24, 29, 30, 32–34, 247

uniformisation of a valuation, 268
unprojection, 33, 246–248
untwisting, 12–13, 74, 76, 81, 82, 86, 87, 179, 183, 185, 186, 189, 190, 193–202, 232, 243, 249, 251, 252, 254–256, 260, 266, 267, 272–274, 307

valuation, 78–80, 87, 97, 187, 188, 190, 191, 210, 265, 266, 268, 269, 273, 276, 279, 281–283, 285
very general, 51

weak maximal singularity, 187, 191
weighted
  blowup, 16, 191, 197, 199, 281, 283, 300, 309
  hypersurfaces and c.i., 102–171, 177, 192, 261, 264, 267, 314, 321
    their cohomology, 114–116
  projective space, 22, 38, 325
well formed, 105, 107, 110–113, 129, 130, 133, 134, 136, 137, 141–143, 146–148, 150

Zariski decomposition, 323–325

Printed in the United States
By Bookmasters